Dynamic Analysis of Offshore Wind Turbine Foundations in Soft Clays

Xinglei Cheng · Dechun Lu · Piguang Wang

Dynamic Analysis of Offshore Wind Turbine Foundations in Soft Clays

Springer

Xinglei Cheng
Tianjin, China

Dechun Lu
Beijing, China

Piguang Wang
Beijing, China

ISBN 978-981-97-9453-9 ISBN 978-981-97-9454-6 (eBook)
https://doi.org/10.1007/978-981-97-9454-6

This work was supported by National Natural Science Foundation of China.

This Springer imprint is published by the registered company Springer Nature Singapore Pte Ltd.
The registered company address is: 152 Beach Road, #21-01/04 Gateway East, Singapore 189721, Singapore

Preface

Offshore wind energy has experienced rapid growth in recent years, driven by its abundant, clean, and renewable advantages. In China's nearshore waters, deep soft clay layers are widespread, characterized by high sensitivity, strong compressibility, and low shear strength. Furthermore, China's coastal regions lie within the circum-Pacific seismic belt, making most nearshore waters prone to earthquakes. In such demanding offshore environments, offshore wind turbines (OWTs) installed in soft clays face prolonged environmental loads including wind, waves, currents, and occasional seismic events over their typical 25-year design lifespan. The combination of weak soil conditions and complex loading scenarios poses a considerable risk to the safe operation of OWTs. Current guidelines for OWTs place significant emphasis on the Service Life Limit State (SLS) requirements in foundation design. Stringent displacement control criteria present substantial hurdles for designers, prompting extensive research into the dynamic responses of offshore foundations under these diverse conditions.

The authors and their research team systematically investigated dynamic analysis methods and response characteristics of various offshore wind turbine (OWT) foundations, including single pile, multi-pile, single bucket, and multi-bucket configurations, under marine environmental loads. Nonlinear constitutive simulation of soil was crucial for analyzing the dynamic behavior of these foundations. The authors developed a simple bounding surface constitutive model of clays and proposed a refined time-domain global analysis approach that integrates calculations of the soft clay-foundation-wind turbine superstructure system. Furthermore, they devised an elastic-plastic p-y model specifically for analyzing lateral dynamic responses of monopile foundations is soft clays, simplifying calculations for monopile-based wind turbines. This book systematically introduces dynamic analysis methods and load-bearing characteristics of offshore wind turbine foundations in soft clay environments, offering valuable insights and guidance for technicians and researchers in marine geotechnical engineering.

This book is structured as follows: Chap. 1 provides an introduction to the research background and motivation. Chapter 2 details a simple single bounding surface model for the undrained cyclic behavior of saturated clays. Chapter 3 introduces a novel cyclic p-y elastoplastic model applied to laterally loaded piles in soft clays. Chapter 4 discusses the lateral cyclic responses of large-diameter monopiles for OWTs in soft clays. Chapter 5 explores the seismic response of monopile-supported OWTs in clays. Chapter 6 presents the dynamic analysis of tripod pile-supported OWTs in clays. Chapter 7 discusses the seismic response analysis of single suction bucket foundations for OWTs in clays. Chapter 8 introduces the lateral cyclic response of tripod suction bucket foundations for OWTs in clays. Chapter 9 details the response of tripod suction bucket foundations for OWTs to seismic and environmental loading. Finally, Chap. 10 summarizes the main conclusions and outlines future work.

The authors gratefully acknowledge the financial support from the following research projects: the National Natural Science Foundation of China (52378361 and 52108334), the Natural Science Foundation of Tianjin (20JCYBJC00570 and 22JCYBJC01360), the Chinese Postdoctoral Science Foundation (2019M650411), and the scholarship from China Scholarship Council. The authors express gratitude to Prof. Xiuli Du from the Beijing University of Technology and Prof. M. Hesham El Naggar from the Western University in Canada for their valuable guidance and insightful suggestions during the research. The authors also appreciate the contributions of the postgraduate students in the research group, namely Wenlong Cheng, Tianju Wang, Jiaqing Lu, Maolin Li, Ziwen Fang, and Mengmeng Liu.

Due to the authors' limited knowledge, this book may contain a few minor inconsistencies or errors. The authors welcome valuable suggestions from experts, scholars, and other readers to enhance the book's quality.

Tianjin, China — Xinglei Cheng
Beijing, China — Dechun Lu
Beijing, China — Piguang Wang
July 2024

Contents

Chapter 1
Introduction

1.1 Background and Motivation

The ocean serves as a crucial foundation for human survival and sustainable social development. Ocean resources, energy, and space development are integral components of oceanic strategy. Offshore wind energy, with its vast developmental potential, has emerged as a pivotal player. In light of goals such as 'Carbon peak and carbon neutrality', strategies like 'Ocean power', and initiatives such as the 'Maritime Silk Route', offshore wind power stands poised to play a pivotal role in supplying energy and facilitating China's transition to green energy. China's offshore wind power industry has witnessed rapid growth in recent years. Between 2015 and 2022, cumulative installed capacity surged from 1.62 to 32.5 million kW, with projections suggesting it will surpass 60 GW by 2025. Presently, China leads the world in cumulative offshore wind power capacity. This rapid industry expansion has heightened demands for the safe operation of offshore wind turbines (Koh and Ng 2016). Foundations constitute a critical component of wind turbine systems, accounting for approximately 30% of total investment (Oh et al. 2018; Zhang and Wang 2022). Ensuring the safe and cost-effective design of these foundations is essential to guaranteeing the reliable operation of offshore wind turbines.

Foundations for offshore wind turbines are primarily categorized into two types: fixed and floating. Among the currently operational turbines, the majority are supported on fixed foundations, including gravity base, monopile, suction caisson, tripod, and jacket foundations (refer to Fig. 1.1). Fixed foundations are typically installed in water depths less than 50 m (Bhattacharya 2014). The monopile (Fig. 1.1b), a single large-diameter hollow pipe, is the most widely adopted choice due to its ease of installation, cost-effectiveness, and logistical advantages, accounting for approximately 75% of all installed offshore wind turbines (Malhotra 2010; Wu et al. 2019). Monopiles are commonly used in shallow water depths under 30 m. For depths exceeding 30 m, monopiles can become excessively flexible, prompting

X. Cheng et al., *Dynamic Analysis of Offshore Wind Turbine Foundations in Soft Clays*,
https://doi.org/10.1007/978-981-97-9454-6_1

consideration of alternatives such as guyed monopiles, or tripod and jacket structures for their economic viability (O'Kelly and Arshad 2016). Tripods (Fig. 1.1d) consist of a large-diameter central steel tubular section supported by three braces, while jacket structures (Fig. 1.1e) are lattice frames composed of small-diameter steel struts. Both tripod and jacket foundations are typically anchored to the seabed using suction buckets or piles, enabling axial transfer of loads from the offshore wind turbine and its supporting structure to the seabed foundation. Looking ahead, floating structures are anticipated to gain commercial traction, particularly for water depths exceeding 50 m, although they are currently in the research and development phase.

In many instances, offshore wind turbine foundations are constructed on sites characterized by deep soft clay. The high sensitivity, strong compressibility, and low shear strength of soft clay present significant challenges for these foundations. China's coastal regions lie within the circum-Pacific seismic belt, with many nearshore areas prone to earthquakes. Wind turbine foundations installed in deep soft seabeds are therefore exposed to long-term environmental loads such as wind, waves, and currents, alongside the potential seismic risks throughout their typical 25-year design life. These dynamic loads can induce considerable overturning moments on the wind turbine foundation, leading to excessive rotational displacement. Such displacements can impair the normal operation of the upper wind turbine structure

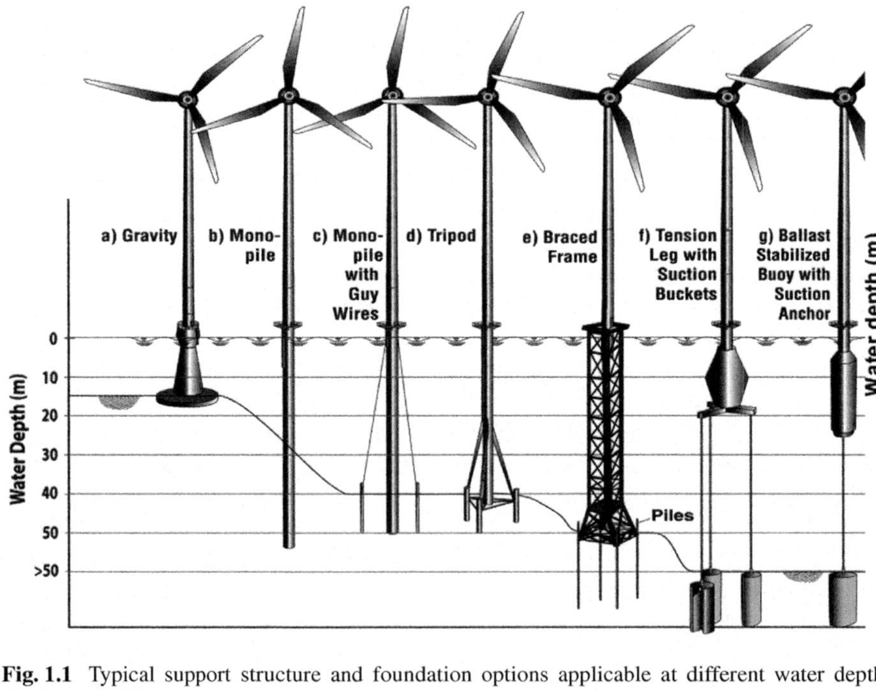

Fig. 1.1 Typical support structure and foundation options applicable at different water depths (Malhotra 2010)

and may even result in the system overturning (Watson 2000). In the current guidelines for OWTs, the requirements for the serviceability limit state (SLS) typically dominate the whole design of foundations (Arany et al. 2015). For instance, specifications in the guidelines mandate a maximum permanent accumulated rotation of 0.5° at the mudline for monopiles under SLS conditions (Det Norske Veritas (DNV) 2013). Stringent requirements for displacement control present substantial challenges for designers, prompting extensive research into the dynamic responses of offshore foundations under these loads.

Currently, there is a notable deficiency in precise and efficient dynamic analysis theories and methodologies tailored for wind turbine foundations situated in deep soft soil areas. Additionally, research on their dynamic load-bearing characteristics remains inadequate. In response to this gap, the authors of this book have undertaken comprehensive and systematic investigations into the dynamic characteristics of offshore wind turbine foundations within soft clay environments subjected to both seismic and environmental loads. Their efforts have led to the development of pertinent analysis theories and computational methods, elucidating the dynamic response patterns and failure mechanisms of these foundations. The outcomes of this research are poised to provide critical technological support for the engineering design and safety assessment of offshore wind turbine foundations.

1.2 Numerical Analysis Method for Dynamic Response of Offshore Wind Turbine Foundation

1.2.1 Numerical Method Based on Advanced Soil Constitutive Models

The foundation-soil dynamic interaction is crucial in the dynamic analysis of offshore wind turbine systems, as it has a significant impact on the natural frequency and dynamic response of the upper structure of the wind turbine. The time-domain numerical analysis of foundation-soil interaction based on soil constitutive model provides an effective approach for simulating the dynamic response of wind turbine systems. For this method, the time-domain global numerical models are commonly established using commercial finite element software such as ABAQUS, ANSYS, and OPENSEES, and the soil domain is discretized using three-dimensional solid elements (see Fig. 1.2). In these numerical models, the dynamic stress–strain response of soil is simulated employing advanced constitutive models, while the foundation-soil interface behavior is simulated employing the contact elements. The lateral dynamic responses of the wind turbine foundations have been analyzed using this method (Hong et al. 2017; Chen et al. 2020; Cao et al. 2020). In parallel, the authors of this book have developed a simple single bounding surface constitutive model within the framework of bounding surface plasticity theory (Cheng et al. 2020a, b). This model can consider the stiffness degradation and the plastic strain accumulation of soft clay and has been applied to analyze the dynamic response of offshore wind turbine foundations under lateral loads such as earthquake, wind, and waves (Cheng et al. 2023). Furthermore, researchers have analyzed the axial dynamic responses of monopile foundations for offshore wind turbine through finite element simulations employing advanced soil constitutive models (D'Aguiar et al. 2011; Huang and Liu 2015). This numerical method based on advanced soil constitutive models can track the stress–strain history of soil elements and effectively capture the dynamic failure process of wind turbine foundations in soft clays.

In the time-domain finite element analysis of foundation-soil interaction, a significant challenge lies in accurately simulating the complex nonlinear dynamic response of soil. Currently, many researchers rely on commercial finite element software models such as the Mohr–Coulomb, Drucker-Prager, and modified Cam-clay models. However, these models, originally designed for static conditions, fail to capture the nonlinear hysteresis response of soil under cyclic loading. The widely used equivalent linear model, although it accounts for soil nonlinearity and hysteresis in seismic response calculations, does not adequately describe the accumulation of soil plastic deformation. Other advanced constitutive models found in literature, like the multi-yield surface and two-yield surface models, often involve intricate kinematic hardening rules and numerous model parameters, which limit their application in dynamic

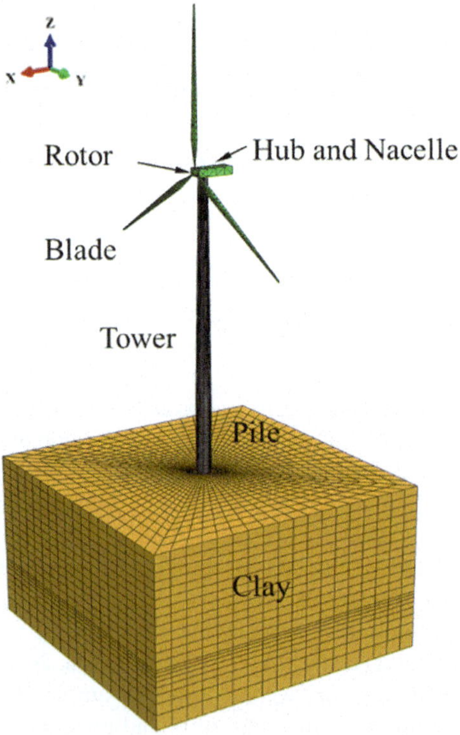

Fig. 1.2 Three-dimensional numerical model of monopile offshore wind turbine

analyses featuring strong material and contact nonlinearities. Numerical methods for foundation-soil interaction based on soil constitutive models offer a more comprehensive approach, addressing continuity, nonlinearity, special boundary conditions, and foundation-soil interface behaviors. However, this approach demands substantial computational resources, leading to low computational efficiency and hindering its practical engineering application. For dynamic analyses of wind turbine foundations in soft clays, there is a need to develop relatively simple and practical constitutive models that capture the main dynamic characteristics of soft clays. Moreover, an emphasis should be placed on developing high-precision and efficient time-domain global finite element analysis methods tailored to these specific conditions.

1.2.2 Numerical Method Based on Empirical Soil Stiffness Degradation Model

Numerous studies have demonstrated that soft clay exhibits stiffness degradation and reduced strength when subjected to cyclic loading (Idriss et al. 1978; Anderson et al. 1980; Vucetic and Dobry 1988; Huang and Li 2010). The extent of this degradation depends on factors such as the number of cycles, initial static stress level, cyclic stress magnitude, and several influencing variables including overconsolidation ratio, principal stress orientation, strain rate, and vibration frequency (Yasuhara et al. 1992; Hyodo et al. 1994; Zhou and Gong 2001; Li et al. 2011; Guo et al. 2013). Dynamic soil tests reveal that under cyclic loading, soft clays experience gradual reduction in shear modulus along the stress–strain hysteresis curve, coupled with increasing cyclic deformation with successive cycles. Accumulation of plastic strain occurs particularly under initial static deviatoric stress conditions.

Researchers have developed soil stiffness degradation models to depict how soil modulus around foundations decreases progressively with increasing cyclic loading cycles. Integrating these models with static constitutive relationships, three-dimensional finite element models of pile-soil interaction have been established to simulate bearing capacity degradation and calculate cumulative foundation displacement. For instance, these models have been applied to assess the long-term performance of offshore wind turbine monopiles under cyclic lateral loading, investigating factors such as pile geometry, loading conditions, embedment length, and cycle numbers (Achmus et al. 2009; Kuo et al. 2011; Depina et al. 2015; Ma et al. 2017; Yang et al. 2018). This approach utilizes soil modulus degradation to reflect the impacts of cyclic loading on soil strength and stiffness (see Fig. 1.3). While these methods cannot capture the time history of cyclic deformation in monopiles, thus limiting their ability to reveal the detailed mechanisms of pile-soil interaction during cyclic loading, they offer higher computational efficiency compared to numerical approaches based on advanced soil constitutive models. This advantage is particularly evident in simulations of long-term cyclic loading scenarios.

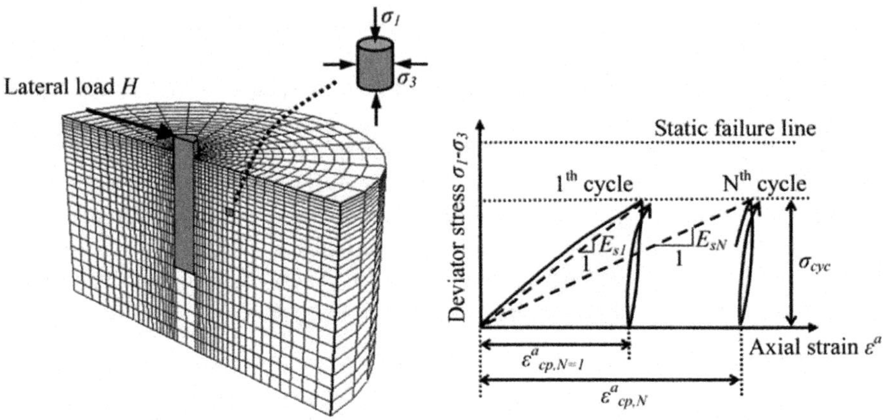

Fig. 1.3 Degradation of secant modulus under cyclic loading in the pile-soil model (Achmus et al. 2009)

1.2.3 Numerical Method Based on Soil p-y Curve Model

The analysis of pile-soil interaction holds significant importance in dynamic response studies of offshore wind turbines supported by pile foundations. In numerous numerical simulations focusing on these foundations' dynamic response, the beam on nonlinear Winkler foundation (BNWF) method is commonly employed (Zhao and Maisser 2006; Zuo et al. 2018; Yan et al. 2021). In this method, the pile is modeled as an elastic beam, while the soil is represented by discrete nonlinear springs distributed along the pile (see Fig. 1.4). These springs include p-y springs (representing lateral soil resistance-deflection relationships), t-z springs (accounting for mobilized axial soil-pile adhesion-local pile deflection relationships), and Q-z springs (reflecting mobilized end-bearing resistance-axial tip deflection relationships). Although the BNWF method does not precisely simulate soil continuity due to its discrete spring representation, it remains widely used for its conceptual simplicity and high computational efficiency.

P-y springs are a prevalent tool for simulating the lateral response of pile foundations. Matlock (1970) introduced a p-y model specifically for piles embedded in soft clay, which was later incorporated into the API RP 2GEO (American Petroleum Institute 2014) and has since gained widespread industry use. However, criticisms of this model's limitations have been extensively discussed in the literature. Studies by Jeanjean (2009) and Zhu et al. (2017) have shown that the model tends to be overly conservative, resulting in underestimated initial stiffness and ultimate soil reaction. Furthermore, in cases involving cyclic loading, the API typically incorporates an empirical reduction factor to adjust the ultimate lateral soil resistance, aiming to account for cyclic degradation effects. However, this reduction factor is often considered too simplistic and fails to adequately capture factors such as the level of cyclic loading, loading patterns, or loading history. To address the varying number

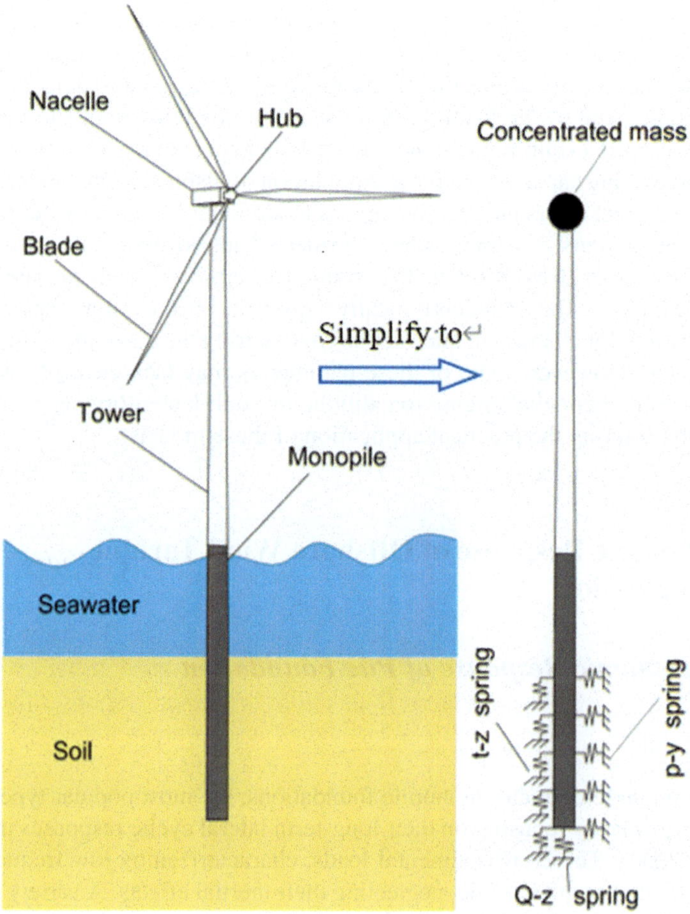

Fig. 1.4 Numerical model of monopile offshore wind turbine based on BNWF method

of load cycles and their amplitudes, researchers have proposed cyclic degradation factors aimed at refining traditional monotonous p-y models (Fan and Long 2005; Dewaikar and Patil 2006; Zhu et al. 2017). However, these models remain empirical, developed based on limited cyclic load tests, and do not fully capture the lateral cyclic hysteretic responses of piles. An alternative approach involves analyzing the cyclic lateral response of piles using cyclic p-y hysteresis curves within the framework of the beam on nonlinear Winkler foundation (BNWF) method. These models incorporate specific components—such as springs, dashpots, drag cells, and gap cells—to simulate key aspects of pile-soil interaction and capture dynamic hysteretic behavior at the soil-pile interface (El Naggar and Novak 1996; Gerolymos and Gazetas 2005b; Allotey and El Naggar 2008; Heidari et al. 2014; Liang et al. 2018). However, the inclusion of numerous components can complicate model implementation.

The t-z spring is widely employed to model the vertical response of pile foundations. The API clay t-z model, originally proposed by Coyle and Reese (1966), is a prominent example but is specifically designed for static axial loading and does not accommodate axial cyclic loading. In response to this limitation, Chin and Poulos (1991) developed a simplified t-z model capable of accounting for both the degradation of pile bearing capacity and the accumulation of displacement under axial cyclic loads. This model offers an effective means to simulate the axial cyclic response of piles. In recent years, researchers have further advanced pile-soil interface models (Dias and Bezuijen 2018; Xu et al. 2019) aimed at capturing the axial cyclic behavior of pile foundations. These models typically require the introduction of specific parameters to account for changes in interface strength during the axial interaction between piles and soil. However, some of these parameters may lack clear physical significance, making them challenging to calibrate through laboratory geotechnical tests and thereby limiting the practical application of these models.

1.3 Dynamic Response of Offshore Wind Turbine Foundation

1.3.1 Dynamic Response of Pile Foundation

(1) Monopile Foundation

Research on large-diameter monopile foundations, the most popular type, has been extensive, primarily focusing on their long-term lateral cyclic responses under wind and wave loads. These environmental loads, characterized by low frequencies, are typically treated as quasi-static, neglecting their inertial effects. A variety of studies, including field tests (Zhu et al. 2017; Xu et al. 2020), centrifuge model tests (Jeanjean 2009; Zhang et al. 2011; Hong et al. 2017; Lai et al. 2020), and scaled model tests (Liao et al. 2018), have investigated the lateral cyclic responses of monopiles in clay. Additionally, three-dimensional (3D) finite element simulations incorporating stiffness degradation models (Achmus et al. 2009; Kuo et al. 2011) and advanced soil constitutive models (Bourgeois et al. 2010; Giannakos et al. 2012; Ma et al. 2017; Corciulo et al. 2017; Cheng et al. 2021; Ding et al. 2023) have been employed to analyze the lateral cyclic responses of monopiles. Significant progress has been achieved in understanding the load-bearing capacity and deformation characteristics of large-diameter monopile foundations under horizontal cyclic loads. Insights have been gained into factors affecting cumulative rotation angle, lateral deflection, and bending moment, including cycle number, amplitude, frequency, and direction of cyclic loading.

For offshore wind turbines (OWTs) situated in seismically active regions, seismic loads are crucial considerations in their design (Kaynia 2019). Unlike wind and wave loads, seismic loads necessitate accounting for inertial effects due to their high-frequency nature. Seismic centrifuge tests on monopile-supported OWTs (Yu et al. 2015; Wang et al. 2019; Li et al. 2021) have shown significant tilting during earthquakes, with amplified responses observed at the tower head. Similarly, findings from 3D numerical simulations (Vacareanu et al. 2019; Cheng et al. 2023) indicate that seismic loading leads to rotational accumulation and settlement of monopiles. Several studies have assessed the seismic vulnerability of monopile-supported OWTs using nonlinear Winkler foundation models to simulate pile-soil interaction (Kim et al. 2014a; De Risi et al. 2018). Mo et al. (2021) examined the influence of seismic motion direction on monopile seismic responses, while Kaynia (2021) studied the impact of kinematic interaction on these responses. The dynamic responses of monopile-supported OWTs under combined wind, wave, and earthquake actions have been investigated. These studies have analyzed various responses such as displacement, bending moment, acceleration, and other dynamic characteristics of the OWT system (Zheng et al. 2015; Wang et al. 2018a; Yang et al. 2019; Zhang et al. 2021; Xi et al. 2022). Overall, these studies have provided comprehensive insights into the seismic dynamic response characteristics of monopiles, encompassing their load-bearing capacity, deformation, and the influence of pile-soil interaction on seismic responses of OWT structures.

(2) Multi-pile Foundation

Different from the load resistance mechanism of monopile foundations, multi-pile foundations counteract large overturning moments through axial resistance—specifically, the axial push–pull forces exerted by individual piles—alongside horizontal resistance (Lu et al. 2017). Studies have investigated the axial cyclic behavior of multi-pile foundations through both model tests (Jardine and Standing 2012; Tsuha et al. 2012) and numerical simulations (D'Aguiar et al. 2011; Huang and Liu 2015). These inquiries have underscored the significant influence of pile shaft friction on the axial dynamic response of the piles. Furthermore, research has delved into the behavior of the pile-soil interface through interface shear tests, evaluating the evolution mechanism of soil-pile interface strength (Liu et al. 2019; Zhou et al. 2020). Given that each pile within a multi-pile foundation experiences simultaneous axial and horizontal loads, it is essential to thoroughly examine the overall dynamic response of such foundations.

Research into the overall dynamic response of four-pile foundations for jacket offshore wind turbines (OWTs) under horizontal cyclic loads has been explored through centrifugal model tests (Kong et al. 2019) and numerical simulations (Wei et al. 2013). Wang et al. (2020a) further investigated the evolution characteristics of natural frequency and cumulative tilt of these foundations under long-term horizontal cyclic loads. Abhinav and Saha (2018) investigated the response of a jacket-supported 5-MW OWT under the hydrodynamic and aerodynamic loading by incorporating soil-structure interaction using a Winkler-spring model. Additionally, Ma et al. (2018) examined the impact of scouring on the dynamic response of tripod pile foundations using 3D finite element simulations. Asumadu et al. (2022) investigated wave-induced oscillatory seabed response around tripod pile foundations of OWTs using an integrated 3D numerical model. However, relatively limited research exists on the overall dynamic response of multi-pile foundations in soft clays, especially concerning their seismic dynamic behavior.

1.3.2 Dynamic Response of Suction Bucket Foundation

(1) Monopod Bucket Foundation

The lateral cyclic behavior of monopod suction buckets has been extensively investigated through various approaches: (a) prototype and reduced-scale filed tests (Houlsby et al. 2005b, 2006; Barari and Ibsen 2012; Zhang et al. 2015); (b) scaled 1-g laboratory tests (Wang et al. 2006; Zhu et al. 2013, 2018; Foglia and Ibsen 2016; Hung et al. 2018); (c) centrifugal model tests (Zhang et al. 2007; Cox et al. 2014; Wang et al. 2017a, b); and (d) numerical simulations (Kourkoulis et al. 2014; Gelagoti et al. 2018; Cheng et al. 2016, 2018, 2020; Yi et al. 2021; Zhang et al. 2021). These studies have significantly enhanced our understanding of the lateral cyclic behavior of monopod suction buckets, including the evolution of cumulative rotation and unloading stiffness with cyclic load amplitude, frequency, direction, and number of cycles. In recent years, research has also focused on new types of bucket foundations, such as hybrid foundations combining a traditional monopile with a wide-shallow bucket. Studies have demonstrated their advantages over traditional bucket foundations (Chen et al. 2020; Li et al. 2021; Ma and Yang 2020; Wang et al. 2020b; Trojnar et al. 2021).

Most studies on the seismic response of suction bucket-supported OWTs have primarily focused on the performance of monopod suction bucket foundations installed in sandy soils. A significant emphasis has been placed on understanding the pore pressure accumulation in sand, as well as assessing the impact of parameters such as skirt length, bucket diameter, and dead weight on the seismic rotation and settlement of monopod suction buckets (Yamazaki et al. 2003; Yu et al. 2014; Olalo et al. 2016; Wang et al. 2017a; Zayed et al. 2019). These investigations have consistently shown that increasing the diameter, skirt length, or dead weight of the monopod suction bucket can enhance its resistance to seismic liquefaction, thereby effectively reducing both its rotation and settlement under seismic loading conditions.

(2) Multi-bucket Foundation

The multi-bucket foundation enhances the overturning resistance of OWTs by converting the system's overturning moment into axial push–pull forces distributed across individual buckets (Kim et al. 2014b; Zhang et al. 2016). Research has focused on investigating the vertical pullout and compression behaviors of each bucket element to understand the cyclic responses of multi-suction buckets under lateral loading (Kelly et al. 2006; Lu et al. 2007; Jeong et al. 2019). Centrifugal model studies have compared the horizontal cyclic behaviors of monopod suction bucket and tripod suction bucket foundations installed in sands (Kim et al. 2014c; Wang et al. 2018b). Jeong et al. (2021) further explored the moment-rotation responses, cyclic stiffness, and permanent displacements of tripod foundations in sands through centrifuge model tests. These studies revealed that the cyclic behavior of tripod foundations is influenced significantly by loading amplitude and direction. Initially, the foundation's rotation angle increases during cyclic loading, followed by a gradual decrease, indicative of favorable 'self-healing' characteristics. In contrast to sandy soil foundations, cohesive soil foundations are less susceptible to liquefaction but may experience significant total settlement and uneven settlement under earthquake conditions, potentially affecting OWT operation. Therefore, investigating the dynamic behavior of multi-bucket foundations in cohesive soils under seismic and environmental loading is crucial for ensuring their stability and performance in offshore applications.

1.4 Objectives and Organization of This Book

This book introduces in detail the bounding surface constitutive model of soft clay, and the cyclic p-y elastoplastic model applied to lateral loaded pile in soft clays. This book also elaborates in detail on the refined time-domain global finite element method for analyzing the dynamic response of offshore wind turbine foundations in soft clays under wind, wave, current and earthquake loads. This book systematically reveals the dynamic response laws and failure mechanisms of typical offshore wind turbine foundations, including monopile, multi-pile, monopod bucket, and multi-bucket, through large-scale numerical simulation calculations. This book aims to demonstrate the dynamic response laws and corresponding analysis methods of typical offshore wind turbine foundations in soft clays. The research findings in the book can provide guidance for the design and disaster prevention of offshore wind turbine foundations in soft clays. The book can be divided into ten chapters.

This book provides a comprehensive introduction to several key topics in OWT foundation engineering. It details the bounding surface constitutive model for clays and explores the cyclic p-y elastoplastic model used for analyzing lateral loaded piles in soft clays. Additionally, the book presents a refined time-domain global finite element method designed to analyze the dynamic response of OWT foundations in soft clays under varying environmental loads including wind, waves, currents, and earthquakes. Moreover, the book systematically uncovers the dynamic response behaviors and failure mechanisms of various OWT foundation types such as monopiles, multi-piles, monopod buckets, and multi-buckets through extensive numerical simulations. Its primary objective is to elucidate the dynamic response laws and associated analytical approaches pertinent to typical OWT foundations situated in soft clays. The research findings presented herein are intended to offer valuable insights for the design and disaster prevention strategies concerning OWT foundations in soft clay environments. Structured into ten chapters, this book serves as a comprehensive guide to understanding and addressing the challenges posed by OWT foundations under complex marine dynamic loads.

This chapter provides an introduction to the research background and motivations, offering a concise review of literature concerning dynamic response and numerical analysis methods applicable to offshore wind turbine foundations.

Chapter 2 outlines the theoretical framework of a simplified single bounding surface model designed to capture undrained cyclic behaviors in saturated clays. It details the model's performance and its numerical implementation method.

Chapter 3 introduces a novel cyclic p-y elastoplastic model tailored for analyzing the lateral response of piles in soft clays. The chapter evaluates the model's efficacy in predicting cyclic lateral behaviors of installed piles.

Chapter 4 presents a three-dimensional finite element method for assessing the lateral cyclic responses of large-diameter monopiles. It explores how different loading patterns impact the lateral behavior of monopiles.

Chapter 5 introduces a numerical approach for simulating the dynamic interaction of the soil-pile-structure system. This chapter specifically examines the seismic dynamic responses of monopile offshore wind turbines situated in clay environments.

Chapter 6 discusses a numerical method focused on analyzing the dynamic responses of tripod pile foundations installed in clay. It covers their behaviors under lateral monotonic, cyclic, and seismic loading conditions.

Chapter 7 delves into a numerical method for studying the seismic responses of single suction bucket foundations used in offshore wind turbines. It explores the seismic bearing mechanisms of bucket foundations in clays and associated influencing factors.

Chapter 8 presents a numerical method aimed at analyzing the lateral cyclic responses of tripod suction bucket foundations for offshore wind turbines in clays. It discusses the rotational mechanisms of tripod suction buckets under lateral monotonic and cyclic loading scenarios.

Chapter 9 addresses a numerical method for evaluating the seismic responses of tripod suction bucket foundations for offshore wind turbines within clay environments. It examines the dynamic behaviors of tripod foundations under seismic and environmental loading conditions.

Chapter 10 concludes the book by summarizing the main findings and limitations of the methodologies employed throughout. It outlines future directions for advancing dynamic analysis in the realm of offshore wind turbine foundations.

References

Abhinav, K.A., and N. Saha. 2018. Nonlinear dynamical behaviour of jacket supported offshore wind turbines in loose sand. *Marine Structures* 57: 133–151.

Achmus, M., Y.S. Kuo, and K. Abdel-Rahman. 2009. Behavior of monopile foundations under cyclic lateral load. *Computers and Geotechnics* 36 (5): 725–735.

Allotey, N., and M.H. El Naggar. 2008. Generalized dynamic Winkler model for nonlinear soil-structure interaction analysis. *Canadian Geotechnical Journal* 45 (4): 560–573.

American Petroleum Institute (API). 2014. Recommended Practice 2GEO, Geotechnical and Foundation Design Considerations, 1st Edition Addendum 1, Published by the American Petroleum Institute.

Andersen, K.H., W.F. Rosenbrand, S.F. Brown, et al. 1980. Cyclic and static laboratory tests on Drammen clay. *Journal of Geotechnical and Geoenvironmental Engineering* 106 (5): 499–529.

Arany, L., S. Bhattacharya, J.H.G. Macdonald, et al. 2015. A critical review of serviceability limit state requirements for monopile foundations of offshore wind turbines. In *Offshore Technology Conference Houston.*

Asumadu, R., J. Zhang, and H. Osei-Wusuansa. 2022. 3-D numerical study of offshore tripod wind turbine pile foundation on wave-induced seabed response. *Ocean Engineering* 255: 111421.

Barari, A., and L.B. Ibsen. 2012. Undrained response of bucket foundations to moment loading. *Applied Ocean Research* 36: 12–21.

Bhattacharya, S. 2014. Challenges in design of foundations for offshore wind turbines. *Engineering and Technology Journals* 1 (1): 922.

Bourgeois, E., M.H.J. Rakotonindriana, A. Le Kouby, et al. 2010. Three-dimensional numerical modelling of the behaviour of a pile subjected to cyclic lateral loading. *Computers and Geotechnics* 37 (7–8): 999–1007.

Cao, G., Z. Chen, C. Wang, et al. 2020. Dynamic responses of offshore wind turbine considering soil nonlinearity and wind-wave load combinations. *Ocean Engineering* 217 (2): 108155.

Chen, L., X. Yang, L. Li, W. Wu, et al. 2020. Numerical analysis of the deformation performance of monopile under wave and current load. *Energies* 13 (23): 6431.

Cheng, X.L., J.H. Wang, and Z.X. Wang. 2016. Incremental elastoplastic FEM for simulating the deformation process of suction caissons subjected to cyclic loads in soft clays. *Applied Ocean Research* 59: 274–285.

Cheng, X.L., A.W. Yang, and G.N. Li. 2018. Model tests and finite element analysis for the cyclic deformation process of suction anchors in soft clays. *Ocean Engineering* 151: 329–341.

Cheng, X.L., X.L. Du, D.C. Lu, et al. 2020a. A simple single bounding surface model for undrained cyclic behaviours of saturated clays and its numerical implementation. *Soil Dynamics and Earthquake Engineering* 139: 106389.

Cheng, X.L., P.G. Wang, N. Li, et al. 2020b. Predicting the cyclic behaviour of suction anchors based on a stiffness degradation model for soft clays. *Computers and Geotechnics* 122: 103552.

Cheng, X.L., T.J. Wang, J.X. Zhang, et al. 2021. Finite element analysis of cyclic lateral responses for large diameter monopiles in clays under different loading patterns. *Computers and Geotechnics* 134: 104104.

Cheng, X.L., T.J. Wang, J.X. Zhang, et al. 2023. Dynamic response analysis of monopile offshore wind turbines to seismic and environmental loading considering the stiffness degradation of clay. *Computers and Geotechnics* 155: 105210.

Chin, J.T., and H.G. Poulus. 1991. A "T-Z" approach for cyclic axial loading analysis of single piles. *Computers and Geotechnics* 12 (4): 289–320.

Corciulo, S., O. Zanoli, and F. Pisanò. 2017. Transient response of offshore wind turbines on monopiles in sand: Role of cyclic hydro–mechanical soil behaviour. *Computers and Geotechnics* 83: 221–238.

Cox, J.A., C.D. O'Loughlin, M. Cassidy, et al. 2014. Centrifuge study on the cyclic performance of caissons in sand. *International Journal Physical Modelling Geotechnics* 14: 99–115.

Coyle, H.M., and L.C. Reese. 1966. Load transfer for axially loaded piles in clay. *Journal of the Soil Mechanics and Foundations Division* 92 (2): 1–26.

D'Aguiar, S.C., A. Modaressi, A. Jaime, et al. 2011. Piles under cyclic axial loading: Study of the friction fatigue and its importance in pile behavior. *Canadian Geotechnical Journal* 48 (10): 1537–1550.

De Risi, R., S. Bhattacharya, and K. Goda. 2018. Seismic performance assessment of monopile-supported offshore wind turbines using unscaled natural earthquake records. *Soil Dynamics and Earthquake Engineering* 109: 154–172.

Depina, I., T.M.H. Le, G. Eiksund, et al. 2015. Behavior of cyclically loaded monopile foundations for offshore wind turbines in heterogeneous sands. *Computers and Geotechnics* 65: 266–277.
Det Norske Veritas (DNV). 2013. *DNV-OS-J101 Design of Offshore Wind Turbine Structures*. Det Norske, Veritas AS.
Dewaikar, D.M., and P.A. Patil. 2006. Analysis of a laterally loaded pile in cohesion-less soil under static and cyclic loading. *Indian Geotechnical Journal* 36: 181.
Dias, T.G.S., and A. Bezuijen. 2018. Load-transfer method for piles under axial loading and unloading. *Journal of Geotechnical and Geoenvironmental Engineering* 144 (1): 04017096.
Ding, X., S.C. Chian, J. Lian, et al. 2023. Wind-wave combined effect on dynamic response of soil-monopile-OWT system considering cyclic hydro-mechanical clay behavior. *Computers and Geotechnics* 154: 105124.
El Naggar, M.H., and M. Novak. 1996. Nonlinear analysis for dynamic lateral pile response. *Soil Dynamics and Earthquake Engineering* 15 (4): 233–244.
Fan, C.C., and J.H. Long. 2005. Assessment of existing method for predicting soil response of laterally loaded piles in sand. *Computers and Geotechnics* 32 (4): 274–289.
Foglia, A., and L.B. Ibsen. 2016. Monopod bucket foundations under cyclic lateral loading. *International Journal Offshore and Polar Engineering* 26 (2): 109–115.
Gelagoti, F., I. Georgiou, R. Kourkoulis, et al. 2018. Nonlinear lateral stiffness and bearing capacity of suction caissons for offshore wind-turbines. *Ocean Engineering* 170: 445–465.
Gerolymos, N., and G. Gazetas. 2005. Phenomenological model applied to inelastic response of soil-pile interaction systems. *Soils and Foundations* 45 (4): 119–132.
Giannakos, S., N. Gerolymos, and G. Gazeta. 2012. Cyclic lateral response of piles in dry sand: Finite element modeling and validation. *Computers and Geotechnics* 44: 116–131.
Guo, L., J. Wang, Y. Cai, et al. 2013. Undrained deformation behavior of saturated soft clay under long-term cyclic loading. *Soil Dynamics and Earthquake Engineering* 50: 28–37.
Heidari, M., M.H. El Naggar, M. Jahanandish, et al. 2014. Generalized cyclic p-y curve modeling for analysis of laterally loaded piles. *Soil Dynamics and Earthquake Engineering* 63: 138–149.
Hong, Y., B. He, L.Z. Wang, et al. 2017. Cyclic lateral response and failure mechanisms of semi-rigid pile in soft clay: Centrifuge tests and numerical modelling. *Canadian Geotechnical Journal* 54 (6): 806–824.
Houlsby, G.T., R.B. Kelly, J. Huxtable, et al. 2005. Field trials of suction caissons in clay for offshore wind turbine foundations. *Géotechnique* 55 (4): 287–296.
Houlsby, G.T., R.B. Kelly, J. Huxtable, et al. 2006. Field trials of suction caissons in sand for offshore wind turbine foundations. *Géotechnique* 56 (1): 3–10.
Huang, M.S., and S. Li. 2010. Degradation of stiffness and strength of offshore saturated soft clay under long-term cyclic loading. *Chinese Journal of Geotechnical Engineering* 32 (10): 1491–1498.
Huang, M.S., and Y. Liu. 2015. Axial capacity degradation of single piles in soft clay under cyclic loading. *Soils and Foundations* 55 (2): 315–328.
Hung, L.C., S.H. Lee, S. Vicent, et al. 2018. An experimental investigation of the cyclic response of bucket foundations in soft clay under one-way cyclic horizontal loads. *Applied Ocean Research* 71: 59–68.
Hyodo, M., Y. Yamamoto, and M. Sugiyama. 1994. Undrained cyclic shear behaviour of normally consolidated clay subjected to initial static shear stress. *Soils and Foundations* 34 (4): 1–11.
Idriss, I.M., R. Dobry, and R.D. Singh. 1978. Nonlinear behavior of soft clays during cyclic loading. *Journal of the Geotechnical and Geoenvironmental Engineering Division* 104 (12): 1427–1447.
Jardine, R.J., and J.R. Standing. 2012. Field axial cyclic loading experiments on piles driven in sand. *Soils and Foundations* 52 (4): 723–736.
Jeanjean, P. 2009. Re-assessment of P-Y curves for soft clays from centrifuge testing and finite element modelling. In *Proceedings of the Offshore Technology Conference*, Paper number: OTC-20158-MS.
Jeong, Y.H., J.H. Kim, H.J. Park, et al. 2019. Cyclic behavior of unit bucket for tripod foundation system supporting offshore wind turbine via model tests. *Wind Energy* 22 (2): 257–268.

Jeong, Y.H., K.W. Ko, D.S. Kim, et al. 2021. Studies on cyclic behavior of tripod suction bucket foundation system supporting offshore wind turbine using centrifuge model test. *Wind Energy* 24: 515–529.

Kaynia, A.M. 2019. Seismic considerations in design of offshore wind turbines. *Soil Dynamics and Earthquake Engineering* 124: 399–407.

Kaynia, A.M. 2021. Effect of kinematic interaction on seismic response of offshore wind turbines on monopiles. *Earthquake Engineering & Structural Dynamics* 50 (3): 777–790.

Kelly, R.B., G.T. Houlsby, and B.W. Byrne. 2006. Transient vertical loading of model suction caissons in a pressure chamber. *Geotechnique*: 665–675.

Kim, D.H., S.G. Lee, and I.K. Lee. 2014a. Seismic fragility analysis of 5 Mw offshore wind turbine. *Renewable Energy* 65: 250–256.

Kim, S.R., L.C. Hung, and M. Oh. 2014b. Group effect on bearing capacities of tripod bucket foundations in undrained clay. *Ocean Engineering* 79: 1–9.

Kim, D.J., Y.W. Choo, J.H. Kim, et al. 2014c. Investigation of monotonic and cyclic behavior of tripod suction bucket foundations for offshore wind towers using centrifuge modeling. *Journal of Geotechnical and Geoenvironmental Engineering* 140 (5): 04014008.

Koh, J.H., and E.Y.K. Ng. 2016. Downwind offshore wind turbines: Opportunities, trends and technical challenges. *Renewable & Sustainable Energy Reviews* 54: 797–808.

Kong, D., K. Wen, and B. Zhu. 2019. Centrifuge modeling of cyclic lateral behaviors of a tetrapod piled jacket foundation for offshore wind turbines in sand. *Journal of Geotechnical and Geoenvironmental Engineering* 145 (11): 04019099.

Kourkoulis, R.S., P.C. Lekkakis, et al. 2014. Suction caisson foundations for offshore wind turbines subjected to wave and earthquake loading: Effect of soil–foundation interface. *Géotechnique* 64 (3): 171–185.

Kuo, Y.S., M. Achmus, and K. Abdel-Rahman. 2011. Minimum embedded length of cyclic horizontally loaded monopiles. *Journal of Geotechnical and Geoenvironmental Engineering* 138 (3): 357–363.

Lai, Y., L. Wang, Y. Hong, et al. 2020. Centrifuge modeling of the cyclic lateral behavior of large-diameter monopiles in soft clay: Effects of episodic cycling and reconsolidation. *Ocean Engineering* 200: 107048.

Li, L.L., H.B. Dan, and L.Z. Wang. 2011. Undrained behavior of natural marine clay under cyclic loading. *Ocean Engineering* 38 (16): 1792–1805.

Li, X., X. Zeng, X. Yu, et al. 2021. Seismic response of a novel hybrid foundation for offshore wind turbine by geotechnical centrifuge modeling. *Renewable Energy* 172: 1404–1416.

Liang, F., H. Chen, and Y. Jia. 2018. Quasi-static p-y hysteresis loop for cyclic lateral response of pile foundations in offshore platforms. *Ocean Engineering* 148: 62–74.

Liao, W.M., J.J. Zhang, J.B. Wu, et al. 2018. Response of flexible monopile in marine clay under cyclic lateral load. *Ocean Engineering* 147: 89–106.

Liu, J.W., L. Cui, and N. Zhu. 2019. Investigation of cyclic pile-sand interface weakening mechanism based on large-scale CNS cyclic direct shear tests. *Ocean Engineering* 194 (6): 106650.

Lu, X., Y. Wu, B. Jiao, et al. 2007. Centrifugal experimental study of suction bucket foundations under dynamic loading. *Acta Mechanica Sinica* 23 (6): 689–698.

Lu, W., G. Zhang, and A. Wang. 2017. Bearing behavior of multiple piles for offshore wind driven generator. *Ocean Engineering* 129: 538–548.

Ma, H., and J. Yang. 2020. A novel hybrid monopile foundation for offshore wind turbines. *Ocean Engineering* 198: 106963.

Ma, H., J. Yang, and L. Chen. 2017. Numerical analysis of the long-term performance of offshore wind turbines supported by monopiles. *Ocean Engineering* 136: 94–105.

Ma, H., J. Yang, and L. Chen. 2018. Effect of scour on the structural response of an offshore wind turbine supported on tripod foundation. *Applied Ocean Research* 73: 179–189.

Malhotra, S. 2010. Design and construction considerations for offshore wind turbine foundations in North America. In *Proceedings of GeoFlorida 2010: Advances in Analysis, Modeling & Design*, eds. by D.O. Fratta, A.J. Puppala, and B. Muhunthan. Orlando, Florida, GSP 199, 1533–1542.

Matlock, H. 1970. Correlations for design of laterally loaded piles in soft clay. *Proceedings of the Offshore Technology Conference, Houston, Paper OTC* 1204: 577–594.

Mo, R., R. Cao, and M. Liu. 2021. Effect of ground motion directionality on seismic dynamic responses of monopile offshore wind turbines. *Renewable Energy* 175: 179–199.

Oh, K.Y., W. Nam, M.S. Ryu, et al. 2018. A review of foundations of offshore wind energy convertors: Current status and future perspectives. *Renewable and Sustainable Energy Reviews* 88: 16–36.

O'Kelly, B., and M. Arshad. 2016. Offshore wind turbine foundations: analysis and design. In *Offshore Wind Farms*, 589–610. Woodhead Publishing.

Olalo, L.T., Y.W. Choo, and K.T. Bae. 2016. Influence of the skirt on the seismic response of bucket foundations for offshore wind tower using dynamic centrifuge model tests. In *Proceedings of the ASME 2016 35th International Conference on Ocean, Offshore, and Arctic Engineering (OMAE 2016)*, June 19–24. Busan, South Korea.

Trojnar, K., 2021. Simplified design of new hybrid monopile foundations for offshore wind turbines. *Ocean Engineering* 219: 108046.

Tsuha, C., P.Y. Foray, and R.J. Jardine. 2012. Behaviour of displacement piles in sand under cyclic axial loading. *Soils and Foundations* 52 (3): 393–410.

Vacareanu, V., E. Kementzetzidis, and F. Pisano. 2019. 3D FE seismic analysis of a monopile-supported offshore wind turbine in a non-liquefiable soil deposit. In *Proceedings of the 2nd International Conference on Natural Hazards & Infrastructure (ICONHIC2019)*, 23–26.

Vucetic, M., and R. Dobry. 1988. Degradation of marine clays under cyclic loading. *Journal of Geotechnical Engineering* 114 (2): 133–149.

Wang, Y.H., X.B. Lu, S.Y. Wang, et al. 2006. The response of bucket foundation under horizontal dynamic loading. *Ocean Engineering* 33: 964–973.

Wang, X., X. Yang, and X.W. Zeng. 2017a. Seismic centrifuge modelling of suction bucket foundation for offshore wind turbine. *Renewable Energy* 114: 1013–1022.

Wang, X., X. Yang, and X.W. Zeng. 2017b. Lateral response of improved suction bucket foundation for offshore wind turbine in centrifuge modelling. *Ocean Engineering* 141: 295–307.

Wang, P., M. Zhao, and X. Du. 2018a. Wind, wave and earthquake responses of offshore wind turbine on monopile foundation in clay. *Soil Dynamics and Earthquake Engineering* 113: 47–57.

Wang, L.Z., H. Wang, B. Zhu, et al. 2018b. Comparison of monotonic and cyclic lateral response between monopod and tripod bucket foundations in medium dense sand. *Ocean Engineering* 155: 88–105.

Wang, X., X. Zeng, X. Yang, et al. 2019. Seismic response of offshore wind turbine with hybrid monopile foundation based on centrifuge modelling. *Applied Energy* 235: 1335–1350.

Wang, L., W. Zhou, and Z. Guo. 2020a. Frequency change and accumulated inclination of offshore wind turbine jacket structure with piles in sand under cyclic loadings. *Ocean Engineering* 217: 108045.

Wang, X., X. Zeng, X. Li, et al. 2020b. Liquefaction characteristics of offshore wind turbine with hybrid monopile foundation via centrifuge modelling. *Renewable Energy* 145: 2358–2372.

Watson, G. 2000. *Structure and Foundations Design of Offshore Wind Installations*. CLRC Rutherford Appleton Laboratory.

Wei, S., H. Park, and J. Han. 2013. A study on the effect of different modeling parameters on the dynamic response of a jacket-type offshore wind turbine in the Korean Southwest Sea. *Renewable Energy* 58: 50–59.

Wu, X., Y. Hu, Y. Li, et al. 2019. Foundations of offshore wind turbines: A review. *Renewable & Sustainable Energy Reviews* 104: 379–393.

Xi, R., C. Xu, X. Du, et al. 2022. Framework for dynamic response analysis of monopile supported offshore wind turbine excited by combined wind-wave-earthquake loading. *Ocean Engineering* 247: 110743.

Xu, L.Y., W.D. Shao, Y.Y. Xue, et al. 2019. A simplified piecewise-hyperbolic softening model of skin friction for axially loaded piles. *Computers and Geotechnics* 108: 7–16.

Xu, D.S., X.Y. Xu, W. Li, et al. 2020. Field experiments on laterally loaded piles for an offshore wind farm. *Marine Structures* 69: 102684.

Yamazaki, H., Y. Morikawa, and F. Koike. 2003. Study on design method of suction foundation using model tests. In *Soft Ground Engineering in Coastal Areas*, 419–422.

Yan, Y., C. Li, and Z. Li. 2021. Buckling analysis of a 10 MW offshore wind turbine subjected to wind-wave-earthquake loadings. *Ocean Engineering* 236: 109452.

Yang, M., R. Luo, and W. Li. 2018. Numerical study on accumulated deformation of laterally loaded monopiles used by offshore wind turbine. *Bulletin of Engineering Geology and the Environment* 77: 911–921.

Yang, Y., M. Bashir, C. Li, et al. 2019. Analysis of seismic behaviour of an offshore wind turbine with a flexible foundation. *Ocean Engineering* 178: 215–228.

Yasuhara, K., K. Hirao, and A.F. Hyde. 1992. Effects of cyclic loading on undrained strength and compressibility of clay. *Soils and Foundations* 32 (1): 100–116.

Yi, J.T., F. Liu, and T.B. Zhang. 2021. Determination of the ultimate consolidation settlement of jack-up spudcan footings embedded in clays. *Ocean Engineering* 236: 109509.

Yu, H., X. Zeng, B. Li, et al. 2015. Centrifuge modeling of offshore wind foundations under earthquake loading. *Soil Dynamics and Earthquake Engineering* 77: 402–415.

Yu, H., X.W. Zeng, and J.J. Lian. 2014. Seismic behavior of offshore wind turbine with suction caisson foundation. In *Geo-Congress 2014: Geo-characterization and Modeling for Sustainability*, 234, 1206–1214.

Zayed, M., K. Kim, and A. Elgamal. 2019. Seismic response of suction caisson in large-scale shake table test. In *VII International Conference on Earthquake Geotechnical Engineering*, 17–20.

Zhang, J., and H. Wang. 2022. Development of offshore wind power and foundation technology for offshore wind turbines in China. *Ocean Engineering* 266: 113256.

Zhang, J.H., L.M. Zhang, and X.B. Lu. 2007. Centrifuge modeling of suction bucket foundations for platforms under ice-sheet-induced cyclic lateral loadings. *Ocean Engineering* 34 (8–9): 1069–1079.

Zhang, C., D. White, and M. Randolph. 2011. Centrifuge modeling of the cyclic lateral response of a rigid pile in soft clay. *Journal of Geotechnical and Geoenvironmental Engineering* 137 (7): 717–729.

Zhang, P., Y. Guo, Y. Liu, et al. 2015. Model tests on sinking technique of composite bucket foundations for offshore wind turbines in silty clay. *Journal of Renewable and Sustainable Energy* 7 (3): 033113.

Zhang, P.Y., J.C. Shi, H.Y. Ding, et al. 2016. Design of offshore wind power foundation with multi-bucket. *Transactions of Tianjin University* 22 (6): 502–507.

Zhao, X., and P. Maisser. 2006. Seismic response analysis of wind turbine towers including soil-structure interaction. *Proceedings of the Institution of Mechanical Engineers, Part k: Journal of Multi-Body Dynamics* 220 (1): 53–61.

Zhang, Q. Y., Zhang, Y., Lin, H. D., et al., 2021. Numerical investigation on bearing capacity of owt foundation with large diameter monopile under seismic load. *Applied Ocean Research* 108(7): 102518.

Zheng, X.Y., H. Li, W. Rong, et al. 2015. Joint earthquake and wave action on the monopole wind turbine foundation: An experimental study. *Marine Structures* 44: 125–141.

Zhou, J., and X. Gong. 2001. Strain degradation of saturated clay under cyclic loading. *Canadian Geotechnical Journal* 38 (1): 208–212.

Zhou, W., Z. Guo, and L. Wang. 2020. Sand-steel interface behaviour under large-displacement and cyclic shear. *Soil Dynamics and Earthquake Engineering* 138: 106352.

Zhu, B., B.W. Byrne, and G.T. Houlsby. 2013. Long-term lateral cyclic response of suction caisson foundations in sand. *Ournal of Geotechnical and Geoenvironmental Engineering* 139 (1): 73–83.

Zhu, B., Z. Zhu, T. Li, et al. 2017. Field tests of offshore driven piles subjected to lateral monotonic and cyclic loads in soft clay. *Journal of Waterway, Port, Coastal, and Ocean Engineering* 143 (5): 05017003.

Zhu, F.Y., C.D. O'Loughlin, B. Bienen, et al. 2018. The response of suction caissons to long-term lateral cyclic loading in single-layer and layered seabeds. *Géotechnique* 68 (8): 729–741.
Zuo, H., K. Bi, and H. Hao. 2018. Dynamic analyses of operating offshore wind turbines including soil-structure interaction. *Engineering Structures* 157: 42–62.

Chapter 2
A Simple Single Bounding Surface Model of Undrained Saturated Clays

A simple single bounding surface constitutive model is developed to predict the undrained behaviors of saturated clays under cyclic loads. The new model does not involve complex kinematic hardening rules, and it is only required to memorize important stress reverse events; therefore, the simplicity should be the largest advantage of the model. A new interpolation function of an elastoplastic shear modulus is proposed based on bounding surface theories. The evolution of a hardening modulus is described in the deviatoric stress space by the movement and updating of a mapping center based on the new interpolation function, which enables the model to describe the stress–strain hysteretic responses of clays under cyclic loading. The new model can be regarded as an improvement of some classical one-dimensional soil dynamic models and a generalization in three-dimensional stress space. The model parameters can usually be determined by performing triaxial tests. The model performance has been verified by a comparative analysis on clays subjected to one-way and two-way cyclic loading at different stress levels. The developed model can capture the essential features of behaviors in saturated clay, including reverse plastic flow, evolution of hysteretic loops, accumulation of plastic deformations, and soil stiffness degradation. The newly developed constitutive model has been successfully encoded into the ABAQUS software package by the secondary development interface of UMAT. The ability of the model to calculate boundary value problems, such as clay foundations subjected to seismic loads, has been verified to some extent by simulating the seismic responses of homogeneous horizontal sites.

2.1 Introduction

It is of great importance to capture the nonlinear behaviors of soils in site response analysis and in soil-structure interaction problems. Thus, substantial effort has been devoted to developing analytical techniques and numerical methods for evaluating

X. Cheng et al., *Dynamic Analysis of Offshore Wind Turbine Foundations in Soft Clays*,
https://doi.org/10.1007/978-981-97-9454-6_2

the response of soil deposits to strong earthquake motions. The majority of practical methods used for describing nonlinear soil amplification are classified into two categories: frequency domain equivalent linear and time-domain nonlinear methods. Although equivalent linear types of analyses are the most popular, they have certain well-known limitations under strong seismic shaking (Gerolymos and Gazetas 2005). Considering more desirable approaches, the key to time-domain nonlinear analysis lies in the selection of soil constitutive models. Unfortunately, many of the commercially available nonlinear models, such as the Mohr–Coulomb model and Duncan-Chang model, are incapable of capturing complex nonlinear stress–strain behavior under seismic loads. Therefore, it is essential to develop a relatively simple and practical constitutive model that is suitable for geotechnical earthquake simulation.

It is observed in soil experiments that the soils show obvious nonlinear, hysteresis properties and strain accumulation characteristics under cyclic loading. The classical isotropic elastoplastic theory assumed that no plastic deformation occurs during the unloading process, and thus, it cannot describe the nonlinear hysteretic characteristics of stress–strain curves under cyclic loading. Mroz (1967) introduced the concept of 'field of workhardening modulus' by generalizing the known rules of isotropic and kinematic workhardening. Based on this concept, multi-surface models were proposed by Mroz et al. (1978, 1981) and Prevost (1977, 1978), in which the stress–strain behavior of clays was determined by calculating the translation and contraction (or expansion) of each yield surface in the modulus field. Bounding surface plasticity theory was proposed by Dafalias (1986), Dafalias and Herrmann (1986). The key of the theory is to describe the evolution of the hardening modulus by the interpolation function of a plastic modulus between the outer bounding surface and inner loading surface, which significantly simplified the multi-surface models. The two-surface models include a bounding surface and a loading surface and were developed by Dafalias and Popov (1975), Krieg (1975), and Bardet (1986), based on bounding surface theory. The single-surface model was developed by Dafalias and Popov (1977), in which the elastic domain is reduced to a point within the bounding surface, which is a further simplification of the two-surface models.

Thus far, more research has focused on effective stress-based constitutive models to predict the cyclic behavior of clays based on bounding surface plasticity theory (Tabbaa and Wood 1989; Liang and Ma 1992; Crouch and Wolf 1994; Li and Meissner 2002; Yu et al. 2007; Huang et al. 2011; Kimoto et al. 2015; Yin et al. 2013; Seidalinov and Taiebat 2014; Hong et al. 2014; Ni et al. 2015; Yao and Wang 2018; Yang et al. 2019; Cheng et al. 2020b). Such models describe the deformation and failure of soft clays from the perspective of effective stress based on critical state theory. The models often involve complex interpolation functions of plastic modulus and hardening rules. Too many model parameters are required, and the volume change constraints or the consolidation equation must be introduced to conduct finite element analysis on the undrained condition, which makes the models difficult for numerical implementation. Compared to the effective stress-based constitutive models, research about the total stress-based constitutive models is less frequent (Prevost 1977; Borja and Amies 1994; Anastasopoulos et al. 2011; Huang and Liu 2014; Wang and Yao 1996; Zhuang et al. 2006; Cheng and Wang 2016). For the latter, the hardening modulus field is usually constructed in the deviatoric stress space, and then, the relationship between the deviatoric stress increment and the deviatoric strain increment is built. Since the total stress-based models involve only the hardening mechanism of plastic deviatoric strain, the expression of the hardening modulus is relatively simple and there are fewer model parameters, and as a result, they are much easier to implement in 3D finite element codes. However, phenomena such as pore-pressure buildup and dissipation cannot possibly be captured.

It is assumed that the behaviors of saturated clay foundations occur with no volume change under undrained conditions because of low permeability when they are subjected to rapid seismic loads (Prevost 1977). For this case, from the perspective of total stress, the yielding of the clays is independent of normal stress and depends only on deviatoric stress. The deviatoric deformation of undrained clays can be calculated by using a total stress-based model during seismic loading. The most important thing is the simplicity of the numerical implementation and the high computing efficiency for the total stress-based model. The main objective of this chapter is to develop a simple total stress-based single-surface model to predict the undrained behavior of clays under cyclic loads (Cheng et al. 2020a). A new interpolation function of elastoplastic shear modulus was developed based on the bounding surface theories. The evolution of a hardening modulus is described in a deviatoric stress space by the movement and updating of the mapping center to describe the cyclic hysteretic stress–strain responses. The proposed model can be regarded as an improvement of some classical one-dimensional soil dynamic models (Hardin and Drnevich 1972; Pyke 1979) and their generalization in three-dimensional stress space. Determining the methods for the model parameters is discussed in detail. The model performance for predicting the cyclic behaviors of clays is verified. The numerical implementation and applications of the model are introduced in detail in the present work.

2.2 New Ideas on Elastoplastic Constitutive Theory and Stress Update Algorithm

Developing a rational constitutive model is crucial for describing the nonlinear mechanical behavior of soils. Constructing a plastic constitutive model entails critical factors such as determining the magnitude and direction of plastic strain increments, along with criteria for loading and unloading. The numerical implementation of constitutive models forms the basis for conducting numerical simulations in marine geotechnical engineering. The stress update algorithm is critical to the numerical implementation of constitutive models, and it typically faces challenges such as non-smoothness and nonlinearity. The authors of this book determine the plastic flow direction using the fractional gradient of the yield function and propose the 3D non-orthogonal flow rule suitable for geomaterials. Based on the idea of numerical optimization and numerical differentiation method, the unconstrained stress update algorithms with strong convergence are developed. The author's innovative work in plastic constitutive modeling and stress update algorithms has provided new insights for numerical analysis of dynamic responses in offshore wind turbine foundations in clay, enhancing both computational efficiency and accuracy. A brief introduction to the relevant research work is as follows.

2.2.1 *Multi-physical and Non-orthogonal Geotechnical Plasticity*

The authors of this book propose a multi-physical and non-orthogonal plastic modeling method to establish the incremental stress–strain-environment relationship in unsteady environments (Lu et al. 2023a). When determining the magnitude of plastic strain increment, the environmental factors are used as the constitutive variables to describe the plastic strain behavior in unsteady environments. When determining the direction of plastic strain increment, a non-orthogonal flow rule without plastic potential function is proposed to obtain the non-orthogonal gradient of yield function. In addition, a multi physics loading and unloading criterion is presented to determine the loading and unloading states of materials under the combined action of mechanical and environmental loads. Many non-orthogonal elastoplastic models (Li et al. 2023; Lu et al. 2019a) of soil and concrete (Lu et al. 2019b, 2020; Zhou et al. 2020) have been established based on this new modeling method, which will provide strong support for the dynamic analysis of offshore wind turbine foundation in soft clays.

Let's start with the classical generalized Hooke's law. The incremental stress–strain relationship of materials can be expressed as

$$\mathrm{d}\boldsymbol{\sigma} = \mathbb{D}:\mathrm{d}\boldsymbol{\varepsilon}^{e}, \tag{2.1}$$

where $\mathrm{d}\boldsymbol{\sigma}$ is the stress increment. $\mathrm{d}\boldsymbol{\varepsilon}^{p}$ represents the elastic strain increment. $\mathbb{D}$ is the elastic stiffness matrix. When the material undergoes plastic strain, the elastic strain increment can be obtained by deducting the plastic part from the total strain increment.

$$\mathrm{d}\boldsymbol{\sigma} = \mathbb{D}:\left(\mathrm{d}\boldsymbol{\varepsilon} - \mathrm{d}\boldsymbol{\varepsilon}^{p}\right), \tag{2.2}$$

where $\mathrm{d}\boldsymbol{\varepsilon}$ and $\mathrm{d}\boldsymbol{\varepsilon}^{p}$ represent the total strain increment and irreversible plastic strain increment. The plastic strain increment $\mathrm{d}\boldsymbol{\varepsilon}^{\mathrm{p}}$ can be expressed as the product of magnitude and direction.

$$\mathrm{d}\boldsymbol{\varepsilon}^{p} = \mathrm{d}\phi\mathbf{r}, \tag{2.3}$$

where $\mathrm{d}\phi$ is a nonnegative plastic multiplier, which can be determined by applying consistency conditions to the yield function. **r** denotes the direction of plastic strain increment, which represents the proportional relationship among the components of plastic strain increment and can be determined by the plastic flow rule.

2.2.1.1 Magnitude of Plastic Strain Increment

In a broad sense, the yield function is the function of the stress and environmental variables (e.g., temperature, saturation, time, strain rate, ionic concentration, etc.) that can affect the plastic deformation process of materials.

$$f\left[\boldsymbol{\sigma}, H\left(\boldsymbol{\varepsilon}^{p}\right), \mathbf{M}(\psi)\right] = 0, \tag{2.4}$$

where H indicates the hardening function that is generally a function of plastic strain. ψ denotes one or more environmental factors and is also constitutive variables similar to stress or strain. The vector **M** represents the material parameters that depend on the environmental variables.

$$\mathrm{d}f = \frac{\partial f}{\partial \boldsymbol{\sigma}}:\mathrm{d}\boldsymbol{\sigma} + \frac{\partial f}{\partial H}\frac{\partial H}{\partial \boldsymbol{\varepsilon}^{\mathrm{p}}}:\mathrm{d}\boldsymbol{\varepsilon}^{p} + \frac{\partial f}{\partial \mathbf{M}}\frac{\partial \mathbf{M}}{\partial \psi}\mathrm{d}\psi = 0. \tag{2.5}$$

Substituting Eqs. (2.2) and (2.3) into Eq. (2.5), we can obtain the expression of plastic multiplier in unsteady environments

$$\mathrm{d}\phi = \frac{\frac{\partial f}{\partial \boldsymbol{\sigma}}:\mathbb{D}:\mathrm{d}\boldsymbol{\varepsilon} + \frac{\partial f}{\partial \psi}\mathrm{d}\psi}{\frac{\partial f}{\partial \boldsymbol{\sigma}}:\mathbb{D}:\mathbf{r} - \frac{\partial f}{\partial H}\frac{\partial H}{\partial \boldsymbol{\varepsilon}^{\mathrm{p}}}:\mathbf{r}}, \tag{2.6}$$

where $\partial f / \partial \psi \mathrm{d}\psi$ indicates that an increase in environmental variables can also produce plastic flow.

2.2.1.2 Direction of the Plastic Strain Increment

In the non-orthogonal plastic flow rule, the plastic flow direction is determined by the non-orthogonal gradient of existing yield function with the aid of fractional derivative. The non-orthogonal flow rule (Lu et al. 2019a, b) can be expressed as follows:

$$\mathbf{r} = \left(\frac{\partial^{\mu_1} f}{\partial S_1^{\mu_1}}, \frac{\partial^{\mu_2} f}{\partial S_2^{\mu_2}}, \frac{\partial^{\mu_3} f}{\partial S_3^{\mu_3}}\right)\frac{\partial \mathbf{S}}{\partial \boldsymbol{\sigma}}, \tag{2.7}$$

where μ_1, μ_2, and μ_3 are the fractional orders. $\mathbf{S} = (S_1, S_2, S_3)$ is a set of coordinate bases consisted of stress invariants for constructing the yield function

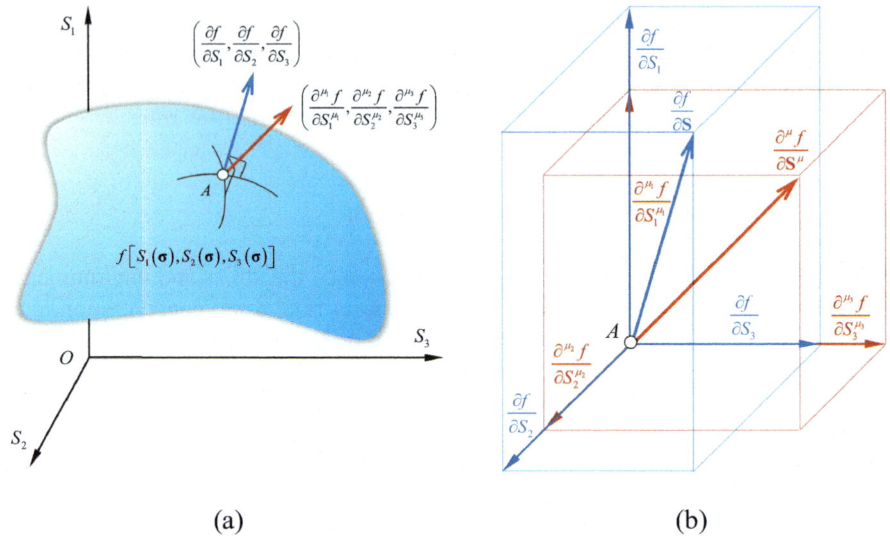

Fig. 2.1 Non-orthogonal flow rule: **a** global view; **b** local view

f. For example, the mean stress p, the generalized shear stress q, and the lode angle θ. $\partial^{\mu_1} f / \partial S_1^{\mu_1}$, $\partial^{\mu_2} f / \partial S_2^{\mu_2}$, and $\partial^{\mu_3} f / \partial S_3^{\mu_3}$ are the fractional derivatives of yield function f with respect to stress invariants S_1, S_2, and S_3, respectively. $\left(\partial^{\mu_1} f / \partial S_1^{\mu_1}, \partial^{\mu_2} f / \partial S_2^{\mu_2}, \partial^{\mu_3} f / \partial S_3^{\mu_3}\right)$ is the fractional gradient of yield surface in $\mathbf{S}$ coordinate space, which is generally non-orthogonal to the yield surface. The geometric interpretation of non-orthogonal flow rule is shown in Fig. 2.1a, b from the local and global views.

In the case of $\mu_1 = \mu_2 = \mu_3 = \mu$, Eq. (2.7) will degenerate into the following form:

$$\mathbf{r} = \frac{\partial^{\mu} f}{\partial S_1^{\mu}} \frac{\partial S_1}{\partial \boldsymbol{\sigma}} + \frac{\partial^{\mu} f}{\partial S_2^{\mu}} \frac{\partial S_2}{\partial \boldsymbol{\sigma}} + \frac{\partial^{\mu} f}{\partial S_3^{\mu}} \frac{\partial S_3}{\partial \boldsymbol{\sigma}}. \tag{2.8}$$

The fractional order μ is the model parameter, and its physical meaning depends on the form of the yield function and the properties of the material. When μ equals to 1, $\partial^{\mu} f / \partial \mathbf{S}^{\mu}$ is orthogonal to the yield surface and the non-orthogonal flow rule degenerates into the orthogonal flow rule.

2.2.1.3 Multi-physical Loading/Unloading Criterion

The loading–unloading criterion evaluates whether the plastic flow occurs in the material according to the geometric relationship between the load vector and the yield surface at the current state point. For the elastoplastic calculation of strain softening materials, the load vector can be composed of strain increment and environmental variable increment. The stress–environmental variable coordinate space is used to fully represent the geometric relationship between the load vector and the yield surface, as shown in Fig. 2.2.

Here, F represents the yield function expressed by the strain and environmental variable, i.e., $F(\boldsymbol{\varepsilon}, \psi, H) = f(\boldsymbol{\sigma}, \psi, H)$. The changes in both the strain and the environmental variable determine the loading–unloading states under the current increment step. When the included angle between the load vector $(\mathrm{d}\boldsymbol{\varepsilon}, \mathrm{d}\psi)$ and the yield surface's external normal direction $(\partial F / \partial \boldsymbol{\varepsilon}, \partial f / \partial \psi)$ is acute, the material is in a loaded state. When the included angle is a right angle or obtuse angle, the material is in neutral loading or unloaded state respectively. The multi-physical loading/unloading criterion can be expressed by

$$\left(\frac{\partial F}{\partial \boldsymbol{\sigma}}, \frac{\partial F}{\partial \psi}\right)(\mathrm{d}\boldsymbol{\sigma}, \mathrm{d}\psi) = \begin{cases} \dfrac{\partial F}{\partial \boldsymbol{\sigma}}:\mathrm{d}\boldsymbol{\sigma} + \dfrac{\partial F}{\partial \psi}\mathrm{d}\psi > 0 & \text{Loading} \\ \dfrac{\partial F}{\partial \boldsymbol{\sigma}}:\mathrm{d}\boldsymbol{\sigma} + \dfrac{\partial F}{\partial \psi}\mathrm{d}\psi = 0 & \text{Neutral loading} \\ \dfrac{\partial F}{\partial \boldsymbol{\sigma}}:\mathrm{d}\boldsymbol{\sigma} + \dfrac{\partial F}{\partial \psi}\mathrm{d}\psi < 0 & \text{Unloading} \end{cases}. \tag{2.9}$$

Generally, the yield function is more of a function of stress rather than strain. The multi-physical loading/unloading criterion based on stress increment and environmental variable increment can be obtained by using the chain rule for Eq. (2.9).

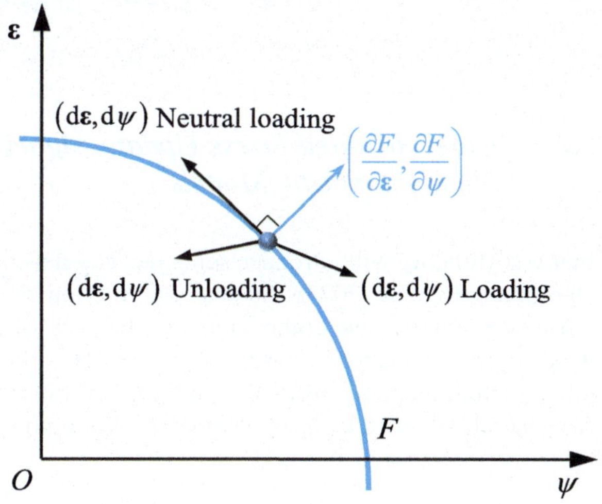

Fig. 2.2 Loading–unloading criterion under the stress–environmental variable coordinate space

$$\frac{\partial F}{\partial \boldsymbol{\varepsilon}}:\mathrm{d}\boldsymbol{\varepsilon} + \frac{\partial F}{\partial \psi}\mathrm{d}\psi = \begin{cases} \frac{\partial f}{\partial \boldsymbol{\sigma}}:\mathbb{D}:\mathrm{d}\boldsymbol{\varepsilon} + \frac{\partial f}{\partial \psi}\mathrm{d}\psi > 0 & \text{Loading} \\ \frac{\partial f}{\partial \boldsymbol{\sigma}}:\mathbb{D}:\mathrm{d}\boldsymbol{\varepsilon} + \frac{\partial f}{\partial \psi}\mathrm{d}\psi = 0 & \text{Neutral loading} \\ \frac{\partial f}{\partial \boldsymbol{\sigma}}:\mathbb{D}:\mathrm{d}\boldsymbol{\varepsilon} + \frac{\partial f}{\partial \psi}\mathrm{d}\psi < 0 & \text{Unloading} \end{cases}. \tag{2.10}$$

2.2.1.4 Constitutive Equations

Combining Eqs. (2.2), (2.3), and (2.6) can obtain the incremental stress–strain–environment constitutive relation during the plastic loading as follows:

$$\mathrm{d}\boldsymbol{\sigma} = \left[\mathbb{D} - \frac{(\mathbb{D}:\mathbf{r}) \otimes \frac{\partial f}{\partial \boldsymbol{\sigma}} : \mathbb{D}^e}{\frac{\partial f}{\partial \boldsymbol{\sigma}}:\mathbb{D}:\mathbf{r} - \frac{\partial f}{\partial H}\frac{\partial H}{\partial \boldsymbol{\varepsilon}^{\mathrm{p}}}:\mathbf{r}}\right]:\mathrm{d}\boldsymbol{\varepsilon} - \frac{(\mathbb{D}:\mathbf{r}) \otimes \frac{\partial f}{\partial \psi}}{\frac{\partial f}{\partial \boldsymbol{\sigma}}:\mathbb{D}:\mathbf{r} - \frac{\partial f}{\partial H}\frac{\partial H}{\partial \boldsymbol{\varepsilon}^{\mathrm{p}}}:\mathbf{r}}\mathrm{d}\psi. \tag{2.11}$$

Equation (2.11) can describe the mechanical response caused by environmental changes. In particular, the incremental stress–strain–environment relationship degenerates into the incremental stress–strain relationship when the environmental variable is constant, i.e., $\mathrm{d}\psi = 0$.

$$\mathrm{d}\boldsymbol{\sigma} = \left[\mathbb{D} - \frac{(\mathbb{D}:\mathbf{r}) \otimes \frac{\partial f}{\partial \boldsymbol{\sigma}}:\mathbb{D}^e}{\frac{\partial f}{\partial \boldsymbol{\sigma}}:\mathbb{D}:\mathbf{r} - \frac{\partial f}{\partial H}\frac{\partial H}{\partial \boldsymbol{\varepsilon}^{\mathrm{p}}}:\mathbf{r}}\right]:\mathrm{d}\boldsymbol{\varepsilon}. \tag{2.12}$$

2.2.2 *Unconstrained Stress Update Algorithm for Elastoplastic Models*

In this section, we will introduce some recently developed advanced stress updating algorithms (Lu et al. 2023b; Su et al. 2023; Zhou et al. 2022a, b, 2023), which can effectively improve the computational efficiency of dynamic analysis of offshore wind turbine foundation in clays. For a classical rate-independent elastoplastic constitutive model, the mathematical equations describing the stress–strain relationship are generally defined by a set of ordinary differential equations with constraints as follows:

$$\begin{cases} \mathrm{d}\boldsymbol{\sigma} = \mathbb{D}:\mathrm{d}\boldsymbol{\varepsilon}^e = \mathbb{D}:\left(\mathrm{d}\boldsymbol{\varepsilon} - \mathrm{d}\boldsymbol{\varepsilon}^p\right) \\ \mathrm{d}\boldsymbol{\varepsilon}^p = \mathrm{d}\phi\mathbf{r} \\ \mathrm{d}H = \mathrm{d}\phi h^p \\ \mathrm{d}\phi \geq 0, f \leq 0, \mathrm{d}\phi f = 0 \end{cases}, \tag{2.13}$$

where the four parts of Eq. (2.13) are known as Hooke's Law, flow rule, hardening law, and the KKT conditions, respectively. h^p denotes the gradient of plastic internal variable, respectively. Note that the KKT conditions constrain the allowable state variables, namely, $\mathrm{d}\phi \geq 0$ and $f(\boldsymbol{\sigma}, H) = 0$ for loading, and $\mathrm{d}\phi = 0$ and $f \leq 0$ for unloading.

2.2.2.1 Unconstrained Stress Updating Strategy

Based on the backward Euler integration formula, Eq. (2.13) can be written as follows:

$$\begin{cases} \boldsymbol{\sigma}_{n+1} = \boldsymbol{\sigma}_n + \mathbb{D}:\left(\Delta\boldsymbol{\varepsilon}_{n+1} - \Delta\boldsymbol{\varepsilon}^p_{n+1}\right) \\ \Delta\boldsymbol{\varepsilon}^p_{n+1} = \Delta\phi_{n+1}\mathbf{r}_{n+1} \\ \Delta H_{n+1} = \Delta\phi_{n+1}h^p_{n+1} \\ \Delta\phi_{n+1} \geq 0,\ f_{n+1} \leq 0, \Delta\phi_{n+1}f_{n+1} = 0 \end{cases}, \tag{2.14}$$

where Δ represents the increment with the finite size. The aim of the stress updating algorithm is to solve Eq. (2.14) based on a given set of state variables $(\boldsymbol{\sigma}_n, H_n)$ at step n and the strain increment $\Delta\boldsymbol{\varepsilon}_{n+1}$ to obtain the state variables $(\boldsymbol{\sigma}_{n+1}, H_{n+1})$ at step $n+1$.

In classical return mapping stress update strategy (Simo and Hughes 2006), the constitutive equations constrained by inequality constraints are solved by the elastic prediction and plastic correction. First, the trial stress is computed by $\boldsymbol{\sigma}^{\mathrm{trial}}_{n+1} = \boldsymbol{\sigma}_n + \mathbb{D}:\Delta\boldsymbol{\varepsilon}_{n+1}$. Then, the trial stress inside the yield surface (i.e., $f\left(\boldsymbol{\sigma}^{\mathrm{trial}}_{n+1}, H_n\right) \leq 0$) is accepted as the true stress at step $n+1$, whereas the trial stress on the outside of the yield surface (i.e., $f\left(\boldsymbol{\sigma}^{\mathrm{trial}}_{n+1}, H_n\right) > 0$) is pulled back to the yield surface by the plastic correction. The above judgment process makes elastoplastic calculations cumbersome. In this book, we introduced an unconstrained stress updating strategy without the need for loading/unloading judgments (Zhou et al. 2022a, b). The *KKT* conditions are replaced equivalently by the smoothing function. Then, Eq. (2.14) is transformed into the smooth version shown below:

$$\begin{cases} \boldsymbol{\sigma}_{n+1} - \boldsymbol{\sigma}_n - \mathbb{D}:\left(\Delta\boldsymbol{\varepsilon}_{n+1} - \Delta\boldsymbol{\varepsilon}^p_{n+1}\right) \\ \boldsymbol{\varepsilon}^p_{n+1} - \boldsymbol{\varepsilon}^p_n - \Delta\phi_{n+1}\mathbf{r}_{n+1} = 0 \\ H_{n+1} - H_n - \Delta\phi_{n+1}h_{n+1} = 0 \\ \sqrt{(c_{\mathrm{d}}\Delta\phi_{n+1})^2 + f^2_{n+1} + 2\beta} - c_{\mathrm{d}}\Delta\phi_{n+1} + f_{n+1} = 0 \end{cases}, \tag{2.15}$$

where Eq. (2.15) is the *Fischer–Burmeister* (FB) function. c_{d} is a dimensional parameter. The parameter β controls the approximation degree of smooth function to KKT conditions. Figure 2.3 shows the effect of β on the FB smoothing curve. The smoothing function converges to the KKT conditions when β trends to 0.

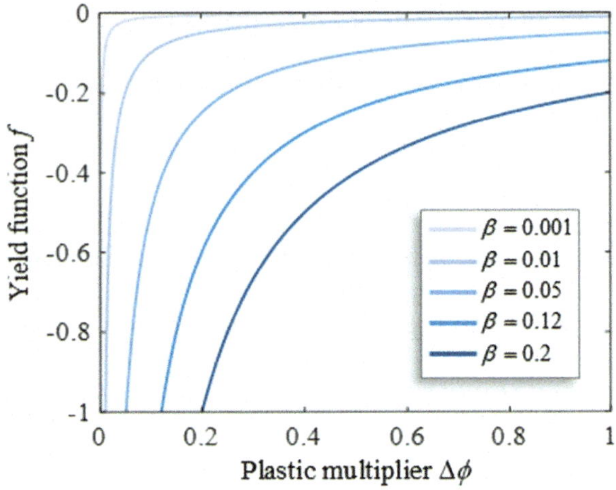

Fig. 2.3 Smoothing curves with the different values of β

2.2.2.2 Line Search Method

The iteration procedure is likely to diverge for the large $\Delta\boldsymbol{\varepsilon}_{n+1}$ if the standard Newton method is directly used to solve Eq. (2.15). In the field of numerical optimization, the convergence of solutions to nonlinear equation systems can be improved through the line search method or the trust region method (Zhou et al. 2022a). In some sense, the search strategies for the two methods are dual to each other. The former optimizes the step size with the given search direction, while the latter first choose the step size (i.e., trust region radius) and then determines the search direction. The details of the former method will be introduced below.

The iterative procedure of line search method is defined as follows:

$$\{x\}_{n+1}^{k+1} = \{x\}_{n+1}^{k} + \alpha^k \{d\}_{n+1}^{k}, \tag{2.16}$$

where the subscript '$n+1$' will be omitted hereafter for the convenience of writing. $\{d\}^k$ and α^k represent the size and direction of the search step at the kth iteration, respectively. The default value for α^k is set to 1 in the Newton method.

In order to ensure the quadratic convergence speed, the Newton search direction is used as follows:

$$\{d\}^k = -[\mathbf{J}]_k^{-1} \{f(x)\}^k, \tag{2.17}$$

where $[\mathbf{J}]$ is the Jacobian matrix, i.e., the derivatives of the nonlinear equations $\{f(x)\}$ with respect to independent variables $\{x\}$. α^k is calculated by the following iterative formula and acceptance criterion.

$$\begin{cases} \text{Accept}\,\alpha_j^k \text{ and exit} & \text{IF } \psi\left(\alpha_j^k\right) < \left(1 - 2\rho\alpha_j^k\right)\psi(0) \\ \alpha_{j+1}^k = \max\left\{\varsigma\alpha_j^k, \frac{\psi(0)}{\psi(0)+\psi\left(\alpha_j^k\right)}\right\} & \text{ELSE} \end{cases}, \tag{2.18}$$

where α_0^k is set to 1. $\rho = 10^{-4}$ and $\varsigma = 0.1$ are recommended (Zhou et al. 2022b). $\psi\left(\alpha_j^k\right)$ is the merit function and is defined by

$$\psi\left(\alpha_j^k\right) = \frac{1}{2}\left\{f\left(\mathbf{x}^k + \alpha_j^k\mathbf{d}^k\right)\right\}^T\left\{f\left(\mathbf{x}^k + \alpha_j^k\mathbf{d}^k\right)\right\}. \tag{2.19}$$

It is worth noting that since Eq. (2.19) only involves the operation of the scalar function, the computational cost of optimizing step size is relatively small for the line search method.

2.2.2.3 Numerical Differentiation Methods

In the implicit stress update algorithm, the determination of Jacobian matrix in Eq. (2.17) requires a substantial number of 1st and 2nd derivatives. For simple constitutive models, the derivative terms can be obtained by the manual differentiation method (MAM) that is accurate and efficient. However, the MAM can quickly become difficult, time-consuming, and error-prone as the constitutive models increase in complexity. The numerical differentiation methods are increasingly favored by researchers since such methods are easy to implement and not sensitive to the complexity of the functions. Compared with other methods, the forward difference method (FDM) and central difference method (CDM) were used earlier by researchers for the derivative evaluation in stress update problems than other ones. These two methods, however, suffer from truncation and subtractive cancelation errors. Even worse, as the perturbation value decreases, the former decreases while the latter increases. The complex step derivative approximation (CSDA) only suffers from the truncation error due to the absence of subtraction operation in the 1st differentiation formula. It is a pity that this advantage is lost when computing the 2nd derivative using the CSDA. The hyperdual step derivative approximation (HDSDA) almost achieves error-free differentiation (Fike 2013), without truncation or subtractive cancelation errors, so its calculation results are comparable to the MAM. Zhou et al. (2023) proposed an implicit stress updating algorithm based on the HDSDA for elastoplastic models, in which the numerical consistency tangent stiffness matrix exhibits the same quadratic convergence speed as the analytical one, and the accuracy of the numerical derivative is independent of the magnitude of the perturbation value.

Table 2.1 provides a summary of the differentiation formulas for the 1st and 2nd derivatives in the four numerical differentiation methods. As an example, the 1st and 2nd derivatives of $f(x) = xe^x \Big/ \sqrt{\sin^3 x + \cos^3 x}$ at $x = \pi / 12$ are calculated by using the above four numerical differentiation methods. Figure 2.4 shows the relative numerical error under different perturbation values for the 1st and 2nd derivatives, respectively. As the h begins to decrease, the numerical errors of FDM, CDM, and CSDA decrease, and their decreasing rates are consistent with the orders of the truncation error presented in Table 2.1. At this stage, the truncation error dominates the overall numerical error. Further decreasing the perturbation value leads to an increase in the numerical errors of CDM and FDM since the subtractive cancelation error begins to dominate the overall numerical error. In the case of the 1st derivative, the numerical error of CSDA approaches the machine accuracy when $h < 10^{-8}$. However, for the 2nd derivative, the numerical error of CSDA increases again as the perturbation value decreases. This is due to the reintroduction of the subtraction operation in the differentiation formula for the 2nd derivative in the CSDA, resulting in the subtractive cancelation error. On the other hand, the HDSDA always maintains the numerical errors of the 1st and 2nd derivatives near the machine accuracy. The numerical error of the HDSDA is independent of the perturbation value h, as the HDSDA does not suffer from truncation error or subtractive cancelation error.

Table 2.1 Differentiation formulas for different methods

Differentiation methods	$\frac{\mathrm{d}f}{\mathrm{d}x} = (\cdot)$	Truncation error	$\frac{\mathrm{d}^2 f}{\mathrm{d}x^2} = (\cdot)$	Truncation error
FDM	$\frac{f(x+h)-f(x)}{h}$	$O(h)$	$\frac{f(x+2h)-2f(x+h)+f(x)}{h^2}$	$O(h)$
CDM	$\frac{f(x+h)-f(x-h)}{2h}$	$O(h^2)$	$\frac{f(x+h)-2f(x)+f(x-h)}{h^2}$	$O(h^2)$
CSDA	$\frac{\Re_{\vec{i}}\left[f\left(x+\vec{i}h\right)\right]}{h}$	$O(h^2)$	$\frac{2}{h^2}\left[f(x) - \Re_{\vec{i}}\left[f\left(x+\vec{i}h\right)\right]\right]$	$O(h^2)$
HDSDA	$\frac{\Re_{\epsilon_1}[f(x+h_1\epsilon_1+h_2\epsilon_2)]}{h_1}$	None	$\frac{\Re_{\epsilon_1\epsilon_2}[f(x+h_1\epsilon_1+h_2\epsilon_2)]}{h_1 h_2}$	None

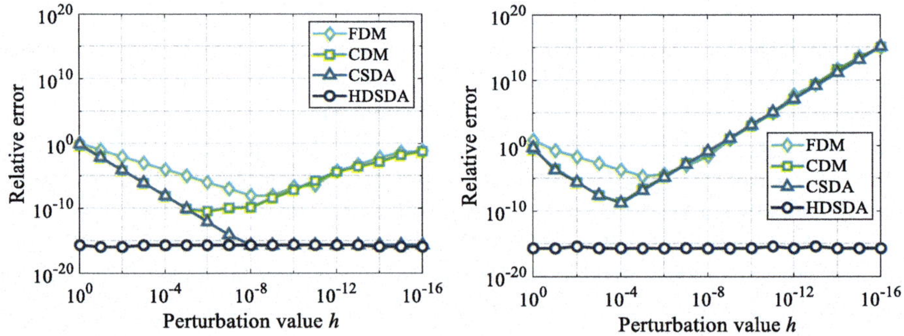

Fig. 2.4 Errors change with the perturbation value h: **a** 1st derivative; **b** 2nd derivative

2.3 Single Bounding Surface Model

It has been observed in experiments that soil in general undergoes both elastic and plastic deformations simultaneously upon shearing regardless of whether there is loading or unloading. Therefore, there is no pure elastic domain in soil. However, classical isotropic hardening bounding surface models regard the unloading process as an elastic process, and thus, they cannot capture the hysteresis property of the stress–strain curve. Hence, kinematic plastic hardening rules are introduced to capture the hysteresis behaviors of soils, in which the yield surface is allowed to change in size and translate in stress space. However, this introduction also makes the constitutive model too complex to be used in numerical analysis. The present single bounding surface model includes only one bounding surface without any loading surface or yield surface. The model does not involve complex kinematic hardening rules, but only memorizes important stress reverse events, which makes the model simple and easy to implement numerically.

2.3.1 Bounding Surface Equation

The present model is based on assumptions that the elastic domain enclosed by the bounding surface is reduced to a point and that plastic flow occurs immediately for any stress increment within the bounding surface, which is inspired by the concept of 'vanishing elastic domain' proposed by Dafalias and Popov (1977). The bounding surface represents the ultimate strength of the soil element, which means that the failure occurs when the stress point reaches the bounding surface. The bounding surface equation is represented by the Drucker-Prager function:

$$F = \sqrt{\overline{J}_2} - \alpha \overline{I}_1 - k = 0 \tag{2.20}$$

where $\overline{I}_1$ and $\overline{J}_2$ are the first stress invariant and the second deviatoric stress invariant on the bounding surface, and it should be noted that the superposed bar indicates that the variables are related to the bounding surface in the whole article; α and k are two material constants, which can be determined by the following formula:

$$\alpha = \frac{2\sin\varphi}{\sqrt{3}(3 - \sin\varphi)}, \quad k = \frac{6c\cos\varphi}{\sqrt{3}(3 - \sin\varphi)}, \tag{2.21}$$

where φ and c is the internal friction angle and the cohesion of clays, respectively. The bounding surface is a conical surface in the principal stress space, as shown in Fig. 2.5. The cross-section of the bounding surface by any deviatoric plane ($\overline{I}_1$ = constant) is circular. It should be acknowledged that better predictions can be obtained by using those surfaces that can reflect the influence of intermediate principal stress, such as Matsuoka-Nakai or Lade criterion, but it will also make the

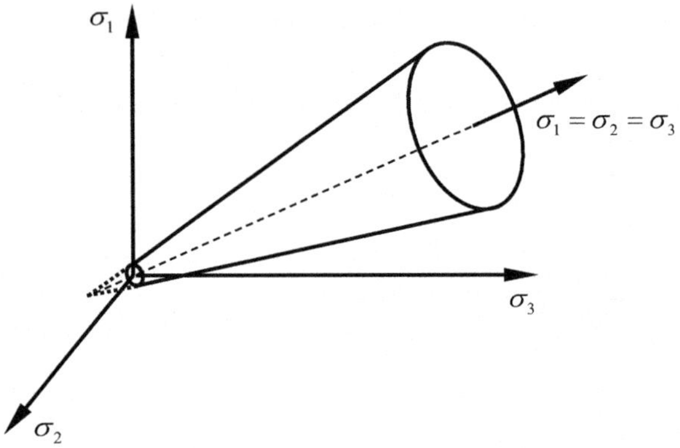

Fig. 2.5 Bounding surface in principal stress space

model more complicated. Here, Drucker-Prager surface is adopted mainly because of its simplicity, that is, it has the simplest circular shape in the π plane, although ignoring the influence of intermediate principal stress. The model based on Drucker-Prager surface is relatively simple and practical to describe the undrained responses of clay foundations subjected to seismic loads despite its limitations.

If the soil element is in the simple shear stress state with a vertical stress, it satisfies $\overline{\sigma}_z \neq \overline{\sigma}_x = \overline{\sigma}_y$, $\overline{\tau}_{zy} = \overline{\tau}_{xy} = 0$, and Eq. (2.20) can be simplified to

$$F = (\overline{\sigma}_z - \overline{\sigma}_x)^2 + 3\overline{\tau}_{zx}^2 - [\alpha(\overline{\sigma}_z + 2\overline{\sigma}_x) + k]^2 = 0, \tag{2.22}$$

where $\overline{\sigma}_z$, $\overline{\sigma}_x$ and $\overline{\tau}_{zx}$ denote the vertical normal stress, lateral normal stress and shear stress in the z-x plane, respectively.

If the soil element is in the simple shear stress state without vertical stress (pure shear), it satisfies $\overline{\sigma}_z = \overline{\sigma}_x = \overline{\sigma}_y = 0$, $\overline{\tau}_{zy} = \overline{\tau}_{xy} = 0$. Equation (2.20) can be simplified to

$$F = 3\overline{\tau}_{zx}^2 - k^2 = 0. \tag{2.23}$$

If the soil element is in the triaxial stress state, $\overline{\sigma}_z \neq \overline{\sigma}_x = \overline{\sigma}_y$, and $\overline{\tau}_{zx} = \overline{\tau}_{zy} = \overline{\tau}_{xy} = 0$, the bounding surface in Eq. (2.20) can be further simplified to

$$F = (\overline{\sigma}_z - \overline{\sigma}_x)^2 - [\alpha(\overline{\sigma}_z + 2\overline{\sigma}_x) + k]^2 = 0. \tag{2.24}$$

2.3.2 *Hardening Modulus and Rules*

A typical stress–strain curve obtained in cyclic triaxial tests is shown in Fig. 2.6a, in which $\sigma_z - \sigma_x$ is axial deviatoric stress and ε_z is axial strain. The bounding surface represented by Eq. (2.22) is a circle of radius $\alpha(\overline{\sigma}_z + 2\overline{\sigma}_x) + k$ in $\sigma_z - \sigma_x$ versus the

$\sqrt{3}\tau_{zx}$ stress plane, as shown in Fig. 2.6b, c. In the triaxial stress state, the shear stress components $\tau_{zx} = 0$, and hence, the stress point continues moving consistently along the $\sigma_z - \sigma_x$-axis during the cyclic loading. In the present work, the radial mapping rule proposed by Dafalias (1986) is adopted, as is schematically shown in Fig. 2.6b, where s_{R}^0, s_{R}' and s_{R} denote the initial loading point, stress reversal point, and current stress point, respectively. Additionally, s_{R}^0 is considered to be the mapping center, and $\bar{s}_{\mathrm{R}}$ denotes the image stress point, which is the intersection point of the bounding surface and a line that passes through the mapping center and the current stress point. In the unloading path after the stress reversal, the stress reversal point s_{R}' is considered to be the mapping center, $\bar{s}_k$ denotes the image stress point for unloading, and s_k denotes any point on the unloading stress path, as schematically shown in Fig. 2.6c. The positions of each point in the stress–strain curve are shown in Fig. 2.6a. An illustration of the bounding surface and mapping rule in multidimensional stress space is shown in Fig. 2.7.

The maximum elastoplastic shear modulus of the stress–strain curve is obtained at the beginning of the loading (the mapping center) and can be represented as $H_{\max}$. Failure occurs for the soil element when the stress point reaches the bounding surface

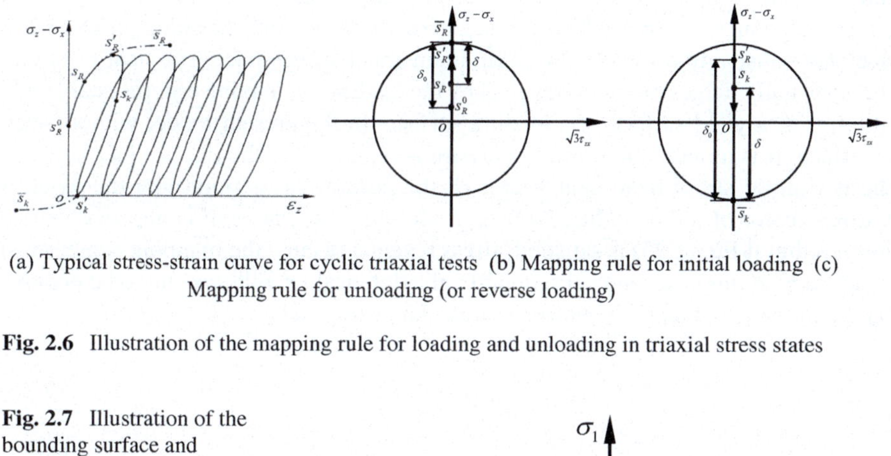

(a) Typical stress-strain curve for cyclic triaxial tests (b) Mapping rule for initial loading (c) Mapping rule for unloading (or reverse loading)

Fig. 2.6 Illustration of the mapping rule for loading and unloading in triaxial stress states

Fig. 2.7 Illustration of the bounding surface and mapping rule in multidimensional stress space

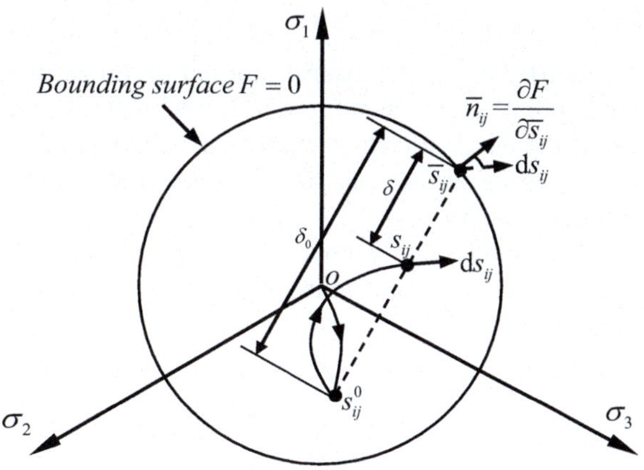

(the image stress point). At this exact moment, the modulus should approach zero, which is denoted as $H = 0$. The elastoplastic shear modulus gradually decreases to a certain value from the maximum value when the stress point moves from the initial loading point to the stress reversal point. It is assumed that the elastoplastic shear modulus gets the maximum $H_{\max}$ upon stress reversal. In the unloading path after stress reversal, the elastoplastic shear modulus gradually again decreases to a certain value at the next stress reversal point from the $H_{\max}$. Furthermore, for the reloading and subsequent re-unloading stress paths, the stress reversal points where the stress paths abruptly change direction are defined as the mapping center, which means that the mapping center moves and updates constantly as the stress reverses. The elastoplastic shear modulus reaches its maximum upon stress reversal during cyclic loading.

There are two key problems: one problem is how to judge the stress reversal that is the key to updating the mapping center, and the other problem is how to determine the elastoplastic shear modulus of any stress point in the loading and unloading stress path, although we know that the maximum and minimum value of the elastoplastic shear modulus is $H_{\max}$ and 0, respectively. The present model includes only one outer bounding surface without any inner yield surface or loading surface. In the one-dimensional stress space, the stress reversal is easily defined by the direction of the stress rate, which is either positive or negative; therefore, the vanishing yield surface does not cause any difficulty. In multidimensional stress space, classical plasticity theory usually judges the stress reversal by the loading–unloading criteria established in terms of a yield surface or a loading surface. In the present model without such a surface, it is assumed that the stress path reverses when the dot product between the deviatoric stress increment vector of the current stress point and the exterior normal vector of the bounding surface at the image stress point is negative, which implies that if Eq. (2.25) is satisfied, stress reversal occurs; the mapping center must be updated at this case, and subsequently, the dot product will turn out to be positive based on the new mapping center after the stress reversal.

$$\frac{\partial F}{\partial \bar{s}_{ij}} \mathrm{d}s_{ij} < 0, \tag{2.25}$$

where $\partial F / \partial \bar{s}_{ij}$ is the exterior normal vector of the bounding surface at the image stress point, as shown in Fig. 2.7. It is actually the direction of the plastic flow. Here, $\mathrm{d}s_{ij}$ is the deviatoric stress increment vector of the current stress point. In this way, the bounding surface can play the same role that the yield surface plays in classical plasticity theory.

A new interpolation function represented as Eq. (2.26) was developed to determine the elastoplastic shear modulus H of the current stress point in the loading and unloading stress path.

$$H = \left(\frac{\delta}{\delta_0}\right)^{\mu} H_{\max} e^{-\xi \int \left|\mathrm{d}\bar{\varepsilon}_d^p\right|}. \tag{2.26}$$

In Eq. (2.26), δ denotes the distance between the current stress point and the image stress point, and δ_0 denotes the distance between the mapping center and the image stress point, as shown in Fig. 2.7. H_{max} represents the maximal elastoplastic shear modulus; μ is a parameter that reflects the shape of the stress–strain curve of the initial loading; e is natural constant; $d\overline{\varepsilon}_d^p$ is the incremental equivalent plastic shear strain, which is represented as

$$d\overline{\varepsilon}_d^p = \sqrt{\frac{2}{3} de_{ij}^p de_{ij}^p} = \sqrt{\frac{2}{9}\left[\left(d\varepsilon_1^p - d\varepsilon_2^p\right)^2 + \left(d\varepsilon_2^p - d\varepsilon_3^p\right)^2 + \left(d\varepsilon_3^p - d\varepsilon_1^p\right)^2\right]}. \quad (2.27)$$

$\int \left| d\overline{\varepsilon}_d^p \right|$ is the cumulative equivalent plastic shear strain during cyclic loading. It is an internal variable that is always a positive value and monotonically increases with time, which reflects the constant degradation of the soil stiffness with the accumulation of plastic shear strain. ξ is a coefficient that controls the cumulative rate of equivalent plastic strain. Here, $e_{ij}^p = \varepsilon_{ij}^p - \frac{1}{3}\delta_{ij}\varepsilon_{kk}^p$, where e_{ij}^p is the plastic deviatoric strain tensor component, and ε_{ij}^p is the plastic strain tensor component ($i, j = 1, 2, 3$).

2.3.3 *Model Simplification and Interpretation*

It is worthwhile to note that the current model can be simplified into certain classical one-dimensional dynamic constitutive models, especially the interpolation function shown in Eq. (2.26), which is closely related to the establishment of a one-dimensional model. For example, Hardin and Drnevich (1972) proposed a hyperbolic dynamic stress–strain backbone curve of soils, which is represented as

$$\tau = \frac{G\gamma}{1 + \frac{\gamma}{\gamma_f}}, \quad (2.28)$$

where G is the elastic shear modulus. Here, γ_f is the reference shear strain, τ is the shear stress. γ is the shear strain. Additionally, δ_0 and δ in Eq. (2.26) equal τ_f(maximum shear stress) and $\tau_f - \tau$, respectively, as shown in the initial loading backbone curve in Fig. 2.8. The tangent shear modulus of the hyperbolic backbone curve can be determined based on Eq. (2.28), which is represented as

$$G_t = \frac{\partial \tau}{\partial \gamma} = \frac{G}{\left(1 + \frac{\tau}{G\gamma_f - \tau}\right)^2} = \frac{G}{\left(1 + \frac{\tau}{\tau_f - \tau}\right)^2} = \frac{G}{\left(1 + \frac{\delta_0 - \delta}{\delta}\right)^2} = \left(\frac{\delta}{\delta_0}\right)^2 G. \quad (2.29)$$

It is easy to construct the stress–strain relationship for unloading and reloading by using the massing criterion for constant amplitude periodic cyclic loads. However, the numerical implementation of the massing criterion is more complex for variable amplitude loads such as seismic loads. Thus, Pyke (1979) used the following

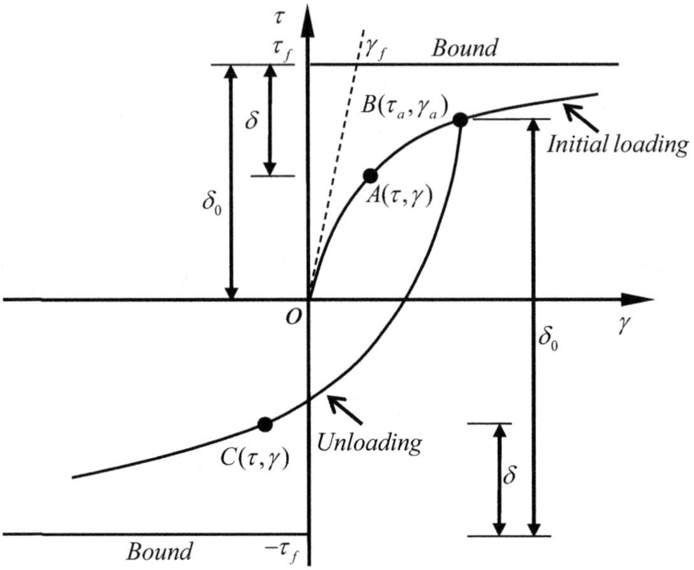

Fig. 2.8 Hardin curve and its Pyke's revision

Eq. (2.30) to construct the subsequent stress–strain curves under variable amplitude stress:

$$\tau - \tau_a = \frac{G(\gamma - \gamma_a)}{1 + \frac{\gamma - \gamma_a}{n\gamma_f}}, \tag{2.30}$$

where τ_a and γ_a are the shear stress and strain at reversal stress point B, and $n = 1 - \tau_a / (G\gamma_f)$. Additionally, δ_0 and δ in Eq. (2.26) are equal to $|\tau_f - \tau_a|$ and $|\tau_f - \tau|$, respectively (| | is the symbol for the absolute value, and τ_f is a negative number during unloading), as shown in the unloading curve in Fig. 2.8. The tangent modulus of the stress–strain curve can be determined based on Eq. (2.30), which is represented as

$$G_t = \frac{\partial \tau}{\partial \gamma} = \frac{G}{\left(1 + \frac{\tau - \tau_a}{nG\gamma_f - (\tau - \tau_a)}\right)^2} = \frac{G}{\left(1 + \frac{\tau - \tau_a}{\tau_f - \tau}\right)^2} = \frac{G}{\left(1 + \frac{\delta_0 - \delta}{\delta}\right)^2} = \left(\frac{\delta}{\delta_0}\right)^2 G. \tag{2.31}$$

If $\xi = 0$ and $\mu = 2$, Eq. (2.26) is simplified to the following formula:

$$H = \left(\frac{\delta}{\delta_0}\right)^2 H_{\max}. \tag{2.32}$$

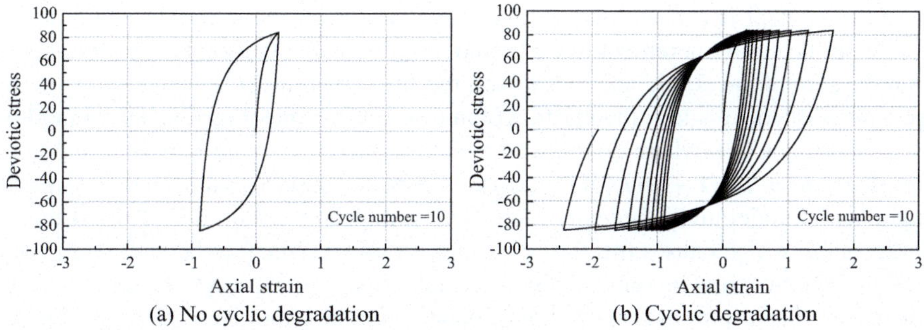

Fig. 2.9 Comparison of stress–strain curves with and without cyclic degradation

It is considered that the expression of the tangent modulus for the initial loading, subsequent unloading and reloading curves in Eqs. (2.29) and (2.31) have the same form as the interpolation function represented as Eq. (2.32). $H_{\max} = G$ for the hyperbolic stress–strain curve of τ versus γ. Therefore, the hyperbolic one-dimensional constitutive model is actually a special simplification of the present three-dimensional single bounding surface model when $\xi = 0$ and $\mu = 2$.

If $\xi = 0$ in Eq. (2.26), the evolution law of elastoplastic shear modulus H is consistent between the unloading and reloading paths for symmetrical two-way cyclic loading, and therefore, the stress–strain hysteresis curve is completely closed without further degradation as the cycle number increases, which is shown in Fig. 2.9a. However, many studies have shown that the soils have significant cyclic degradation characteristics under two-way cyclic loading (Hyodo et al. 1994; Andersen et al. 1980; Vucetic and Dobry 1988). Therefore, $e^{-\xi \int |\bar{\varepsilon}_d^p|}$ is introduced into the interpolation function represented as Eq. (2.26), and its value decreases constantly with the accumulation of equivalent plastic shear strain during cyclic loading, which makes the modulus of reloading smaller than that of unloading, and the modulus of the second unloading smaller than that of reloading. In this way, the cyclic degradation of soils can be described, as shown in Fig. 2.9b.

2.3.4 Incremental Elastoplastic Constitutive Relations

The strain increment includes two parts of the elastic and plastic strain increment. The elastic strain increment is obtained according to the generalized Hooke's law, and the plastic strain increment is obtained according to the hardening rule, the flow rule, and the bounding surface equation. It is assumed that saturated clays have no volume change under undrained conditions because of low permeability when they are subjected to rapid seismic loads. For this case, from the perspective of total stress, the yielding of clays is independent of normal stress and only depends on deviatoric stress (Prevost 1977). Hence, it is assumed that the plastic deviatoric strain increment is caused only by the change in deviatoric stress. In addition, the total volume strain increment of clays under undrained condition is 0, which means that the sum of the elastic and plastic volume strain increment is 0. According to the generalized Hooke's law, the elastic volume strain increment can be determined by $\mathrm{d}\varepsilon_{kk}^{e} = \mathrm{d}\sigma_{kk} / (3K)$, where $\mathrm{d}\sigma_{kk}$ is volume stress increment, and K is Bulk modulus satisfying $K = 2G(1+\nu) / (3(1-2\nu))$. As a special case, Poisson's ratio ν is nearly equal to 0.5 under undrained condition. Hence, the corresponding K is an extremely large value, which causes the elastic volume strain increment to be almost zero at a certain volume stress increment, and then the plastic volume strain increment is also close to 0. Therefore, the expression of plastic volume strain increment is not given in this study. The calculation of deviatoric deformation of undrained clays is mainly introduced in the following.

The associative flow rule is used in the present model. According to the radial mapping rules proposed by Dafalias (1986), the direction of the plastic strain increment at the current stress points (the direction of plastic flow) is assumed to be consistent with the normal direction of the bounding surface at the image stress points, as shown in Fig. 2.7. Hence, the plastic deviatoric strain increment is written as

$$\mathrm{d}e_{ij}^{p} = \langle L \rangle \overline{n}_{ij},\ L = \frac{1}{H'} \overline{n}_{kl} \mathrm{d}s_{kl}, \tag{2.33}$$

where $\mathrm{d}e_{ij}^{p}$ is the plastic deviatoric strain increment; L is the loading index, and the Macauley bracket $\prec \succ$ is defined as follows: $\langle L \rangle = L$ if $L > 0$; otherwise, $\langle L \rangle = 0$. The Macauley bracket ensures that non-positive values of the loading index L lead to $\mathrm{d}e_{ij}^{p} = 0$. It should be noted that if $L < 0$, reverse loading occurs; the mapping center must be updated in this case, and subsequently, $L > 0$ based on the new mapping center after the stress reversal, which means that the reverse plastic flow can be captured. H' is the plastic shear modulus; $\mathrm{d}s_{kl}$ is the deviatoric stress increment at the current stress point; and $\overline{n}_{ij}$ and $\overline{n}_{kl}$ are the exterior normal unit vectors of the bounding surface at the image stress point, which can be obtained as follows:

$$\overline{n}_{ij} = \frac{\partial F}{\partial \overline{s}_{ij}} \Big/ \sqrt{\frac{\partial F}{\partial \overline{s}_{kl}} \frac{\partial F}{\partial \overline{s}_{kl}}} = \sqrt{\frac{3}{2}} \frac{\overline{s}_{ij}}{\alpha \overline{I}_1 + k}. \tag{2.34}$$

After substituting Eq. (2.34) into Eq. (2.33), we obtain Eq. (2.35):

$$\mathrm{d}e_{ij}^{p} = \frac{3}{2H'} \frac{\overline{s}_{ij}}{(\alpha \overline{I}_1 + k)^2} \overline{s}_{kl} \mathrm{d}s_{kl}. \tag{2.35}$$

According to the generalized Hooke's law, the elastic strain increment is obtained:

$$\mathrm{d}e_{ij}^{e} = \frac{\mathrm{d}s_{ij}}{2G}, \tag{2.36}$$

where de^e_{ij} is the elastic deviatoric strain increment, and G is the elastic shear modulus.

The deviatoric strain increment is obtained from Eqs. (2.35) and (2.36):

$$de_{ij} = \frac{1}{2G} ds_{ij} + \frac{3}{2H'} \frac{\bar{s}_{ij}}{(\alpha \bar{I}_1 + k)^2} \bar{s}_{kl} ds_{kl}, \tag{2.37}$$

where de_{ij} is the deviatoric strain increment.

For the undrained simple shear tests without vertical stress (pure shear), the shear strain increment is obtained as follows by simplifying Eq. (2.37):

$$d\varepsilon_{zx} = \left(\frac{1}{2G} + \frac{1}{H'}\right) d\tau_{zx} = \frac{1}{H} d\tau_{zx}, \tag{2.38}$$

where $d\varepsilon_{zx}$ is the shear strain increment, and $d\tau_{zx}$ is the shear stress increment.

For the undrained triaxial tests, the vertical strain increment is obtained using Eq. (2.39) by simplifying Eq. (2.37):

$$d\varepsilon_z = \frac{2}{3}\left(\frac{1}{2G} + \frac{1}{H'}\right) d(\sigma_z - \sigma_x) = \frac{2}{3H} d(\sigma_z - \sigma_x), \tag{2.39}$$

where $d\varepsilon_z$ is the vertical strain increment, σ_z is the vertical stress, and σ_x is the lateral stress.

In Eqs. (2.38) and (2.39), H is the elastoplastic shear modulus, and it obeys the following relation:

$$\frac{1}{H} = \frac{1}{2G} + \frac{1}{H'}. \tag{2.40}$$

At the beginning of loading, $\delta = \delta_0$, and $\int \left|d\bar{\varepsilon}^p_d\right| = 0$ in Eq. (2.36), and then, $H = H_{max} = 2G$. Hence, in this case, the plastic modulus H' is infinite, and only the elastic strain is produced. When the stress point reaches the bounding surface during loading, $\delta = 0$, and then, $H = H' = 0$, which indicates that the plastic strain can be developed infinitely; as a result, failure occurs for the soil element. We can obtain the plastic modulus H' based on Eq. (2.40):

$$H' = \frac{2GH}{2G - H}. \tag{2.41}$$

It is considered that the formula is valid in general three-dimensional stress states. H can be determined by Eq. (2.26), and then, H' can be determined by Eq. (2.41). Therefore, de_{ij} can be determined by Eq. (2.37). Note that the equations are presented in forms that are most suitable for finite element analysis and can easily be implemented in a general finite element program.

2.4 Model Parameters

The proposed model requires five parameters: the internal friction angle φ, cohesion c, elastic shear modulus G, and parameters μ and ξ. The roles of these parameters and the possible methods for determining their values are discussed in the following sections.

The parameters φ and c are the shear strength indices of the soils, which are used to determine α and k in the bounding surface function according to Eq. (2.21). These two parameters can be determined by performing conventional triaxial tests or direct shear tests.

The elastic shear modulus G represents the initial stiffness of the soils. Cyclic triaxial tests (or cyclic simple shear tests) are performed (for a quarter cycle in sinusoidal period loads) to obtain the stress–strain curve of the initial loading. The stress–strain curve of the initial loading can be fitted using the hyperbola curve represented as Eq. (2.42) (or Eq. (2.28) for cyclic simple shear tests).

$$\sigma_d = \frac{E\varepsilon_d}{1 + \frac{\varepsilon_d}{\varepsilon_f}}, \tag{2.42}$$

where E is the elastic compression modulus, σ_d is the axial (vertical) cyclic stress, ε_d is the axial (vertical) cyclic strain, and ε_f is the reference axial strain. Then, the elastic shear modulus G can be determined by $G = 0.5E\big/(1 + \nu)$, where Poisson's ratio ν is 0.5 under an undrained condition, and thus, $G = E\big/3$. G can also be directly determined by Eq. (2.28) for cyclic simple shear tests. To make the constitutive model convenient for engineering applications, G can be determined by the experienced formula $G = a\sigma_c^b$, where G is related closely to the initial consolidation confining pressure σ_c, and a, b are two experienced coefficients that reflect the stiffness of the soils. In addition, G can also be determined by testing the shear wave velocity of the soil layers based on the formula $G = \rho v_s^2$, where ρ is the density of the soils, and v_s is the shear wave velocity of the soils.

Parameter μ captures the shape of the stress–strain curves for the initial loading. It can be determined by the trail-and-error method after obtaining the values of φ, c, and G. We can first assume that μ is equal to 2 (the stress–strain curve is a hyperbola curve in this case), and we compare the test data with the calculation results for the stress–strain curves of the initial loading based on Eq. (2.39). If they are inconsistent, then μ must be adjusted to repeat the above process until the calculation results are relatively consistent with the experimental results. Figure 2.10 shows the comparison between the test results and the calculations for the stress–strain curves of the initial loading with different values for μ. It can be seen from the figure that the decrease in the modulus becomes slower with the increases in μ, and hence, the increase in the strain becomes faster. Here, μ is usually between 0 and 5. It is relatively appropriate to set μ to 1.5 to predict the test data in the figure. It should be noted that the test data are from test results on Newfield clay (Sangrey et al. 1969), and the value of μ will be used to predict the related cyclic test results in the next section.

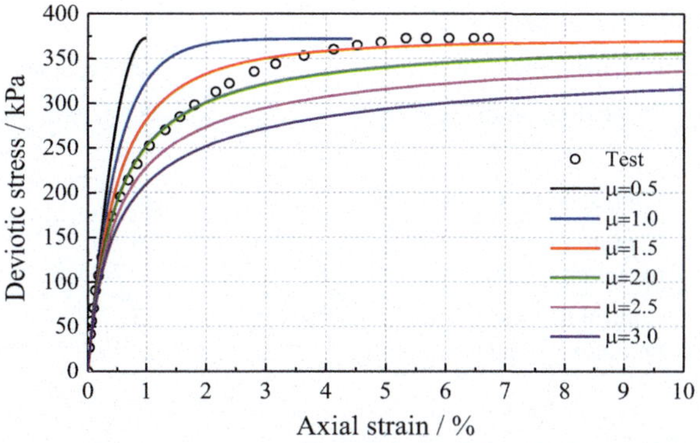

Fig. 2.10 Predicted stress–strain curves for initial loading with different values of μ

The parameter ξ is used to control the cumulative rate of the equivalent plastic strain, which will have a significant effect on the rate of modulus degradation. ξ can be determined by the trial-and-error method after obtaining the values of φ, c, G, and μ. We can first assume that ξ is equal to 1, and we can compare the computation and test results of the stress–strain curve (focusing on the residual plastic strain and maximum strain). If they are inconsistent, then ξ must be adjusted to repeat the above process until the calculation results are relatively consistent with the experimental results. Figure 2.11 shows the cyclic stress–strain curves for one-way and two-way loading under different values of ξ. One can see that the accumulation of strain is becoming faster and the final residual plastic strain is much larger as ξ increases for the one-way cyclic loading, as shown in Fig. 2.11a. The degradation of the modulus is more significant, and the cyclic strain is larger, as ξ increases for the two-way cyclic loading, as shown in Fig. 2.11b. ξ is usually between 0 and 5. It should be noted that the other model parameters are set according to Table 2.2 in the section below, to investigate the effect of ξ on the stress–strain curve.

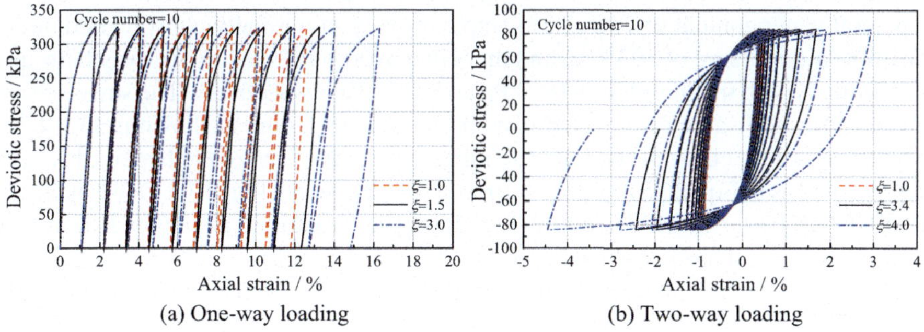

Fig. 2.11 Predicted the stress–strain curves for one-way and two-way loading under different values of ξ

Table 2.2 Values of the model parameters

Model parameter	Parameter function	Newfield clay	Cloverdale clay
φ (°)	Shear strength index of soils	28	30
c (kPa)		13.6	7.7
G(MPa)	Represent the initial stiffness of soils	25	32
μ	Capture the shape of stress–strain curves for initial loading	1.5	2
ξ	Control the rate of strain accumulation and modulus degradation	1.5	3.4

2.5 Verification of the Model Performance

To verify the model capability for simulating the undrained behaviors of clay under cyclic loads, experimental data from Newfield clay (Sangrey et al. 1969) and Cloverdale clay (Zergoun and Vaid 1994) are used to compare with the numerical simulation results obtained from using the model. The predictions of the stress–strain curve for one-way and two-way cyclic triaxial tests are analyzed in detail. The capability of the model for capturing the essential features of the behaviors in saturated clay, including the reverse plastic flow, the evolution of hysteretic loops, the accumulation of plastic deformations, and the soil stiffness degradation, is validated. The related model parameters used for predicting the test results are presented in Table 2.2.

2.5.1 Prediction for One-Way Cyclic Loading Tests

One-way cyclic loading tests were performed by Sangrey et al. (1969). The undisturbed clay used in the tests was obtained from the face of a landslide scarp in Newfield, New York. The clay has a liquid limit of 28%, a plastic limit of 18%, and a natural water content of 24%. The samples were first normally consolidated under an all-round pressure of 393 kPa, and then, they were subjected to one-way half-sine cyclic loads with magnitudes of 180 kPa and 325 kPa, respectively. The stress–strain curve of the soil sample subjected to a single cycle to failure is shown in Fig. 2.12. The stress–strain curve of the initial loading can be fitted using the hyperbola curve represented as in Eq. (2.42), and the fitting curve is shown in Fig. 2.12. One can obtain that $G = E/3 = 25$ MPa according to the fitting equation. The other parameters used for the simulation are listed in Table 2.2.

The predicted and measured stress–strain curve of the Newfield clay at the stress level of 180 kPa and 325 kPa are presented in Fig. 2.13a, b, respectively. The dotted line is the stress–strain curve for monotonic loading to failure. It can be seen from the figures that the nonlinear, hysteretic property and strain accumulation characteristics of soils under cyclic loads can be captured using the constitutive model. The predicted

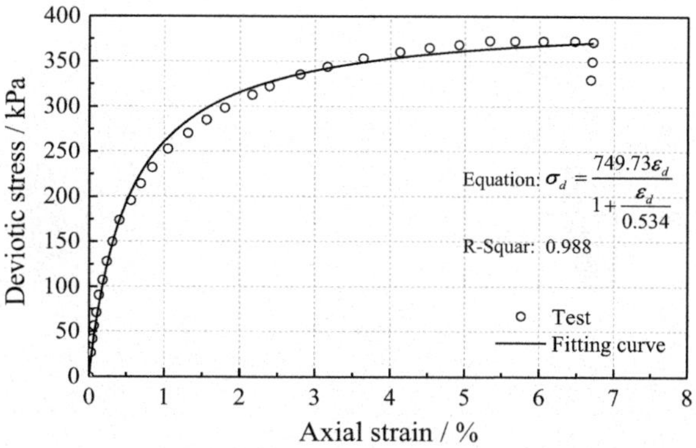

Fig. 2.12 Stress–strain curve of the soil sample subjected to a single cycle to failure

results of the stress–strain curves are consistent with the test results on the overall trend and the final residual plastic deformation, although there is a deviation in the shapes of the individual hysteresis loops. The strain accumulates in the direction of compression for both the tests and predictions, and the residual strain is predominant rather than the cyclic component. At the low stress level, the residual strain is small due to the low rate of strain accumulation. At the high stress level, the failure of the clay occurs quickly after several cycles, due to the high rate of strain accumulation. Unfortunately, the model is unable to capture the cyclic 'shakedown' characteristic of clay at the low stress level due to the gradual degradation of the hardening modulus with the accumulated plastic deviatoric strain. In general, the comparisons indicate that the model can consider reasonably the effect of the stress level on the stress–strain responses of clays.

2.5.2 *Prediction for Two-Way Cyclic Loading Tests*

Two-way cyclic loading tests were performed by Zergoun and Vaid (1994). A soft gray undisturbed marine clay (Cloverdale clay) was used in the tests. The clay has a liquid limit of 50% and a plastic limit of 26% and a natural water content of 50%. The samples were first normally consolidated under a hydrostatic stress of 200 kPa, and then, they were subjected to two-way sine cyclic loads with magnitudes of 78.4 kPa and 84 kPa, respectively. The model parameters used for the simulation are listed in Table 2.2.

The predicted and measured stress–strain curves of the Cloverdale clay at the stress level of 78.4 kPa and 84 kPa are presented in Fig. 2.14a, b, respectively. Generally, the model simulations agree with the measured data. It can be seen that the two-way cyclic behaviors of soil specimens are significantly different from the one-way cyclic behaviors shown in Fig. 2.13. For the two-way cyclic tests without initial static deviatoric stress, the cyclic strain in both the compression and extension is much larger than the average strain, and the extension strain is relatively more significant

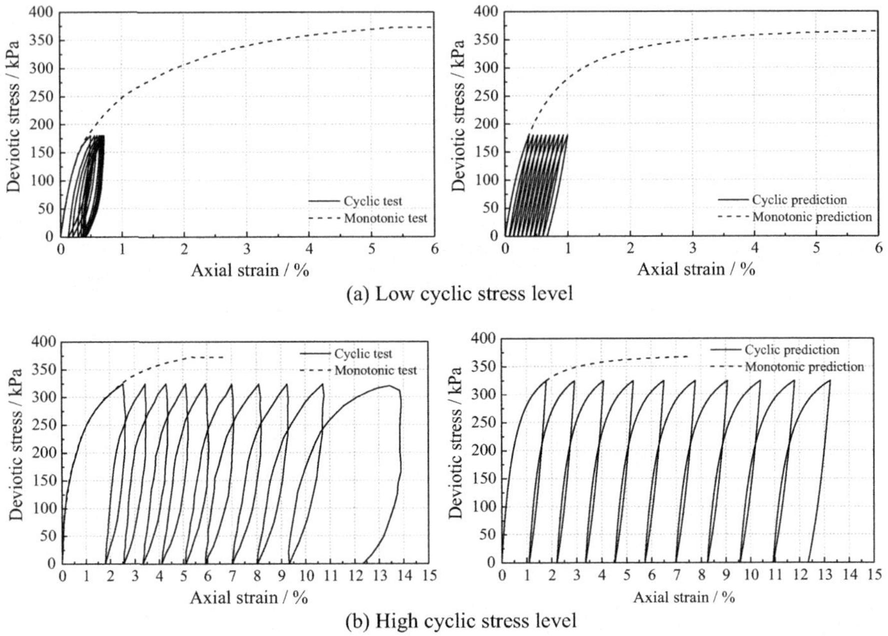

Fig. 2.13 Comparison of the predicted and measured stress–strain curve for Newfield clay at different cyclic stress levels [test data from Sangrey et al. (1969)]

than the compression strain during cyclic loading. At the stress level of 78.4 kPa, the degradation is not significant, and the final cyclic strain is small due to fewer cycles. At the stress level of 84 kPa, the strain amplitude increases significantly as the cycle number increases, and as a result, the failure occurs due to a large cyclic strain in the direction of the extension. The development of the cyclic strain is actually the result of the degradation of soil stiffness, and the degradation becomes more significant with an increase in the cyclic stress level and the cycle number. The comparisons indicate the ability of the model to capture the cyclic degradation characteristics of clays.

2.6 A Simplified Version of the Single Bounding Surface Model

The detailed theoretical framework and performance of the single bounding surface model have been elaborated above (Cheng et al. 2020a). In the model, the undrained clay is regarded as a single-phase medium, and then the deviatoric deformation under cyclic loading can be calculated by using a total stress approach. Compared with the effective stress-based constitutive model (two-phase approach), the total

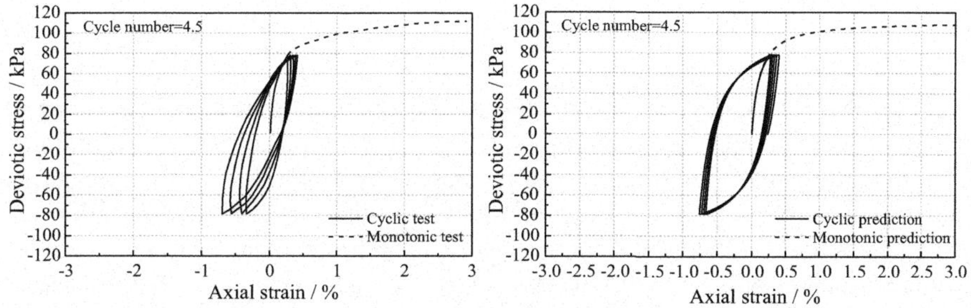

(a) Cyclic stress level of 78.4 kPa

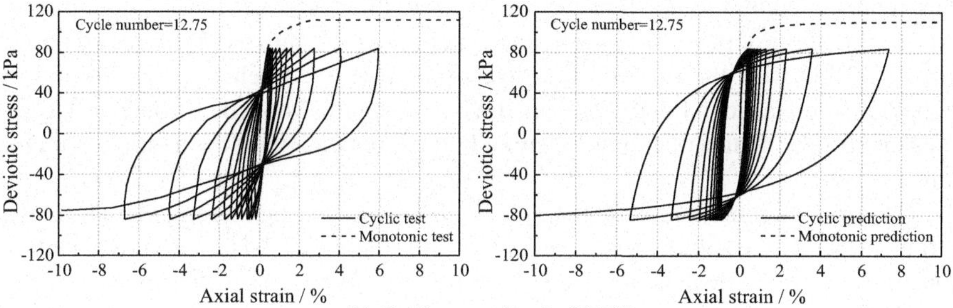

(b) Cyclic stress level of 84 kPa

Fig. 2.14 Comparison of the predicted and measured stress–strain curves for Cloverdale clay at different cyclic stress levels [test data from Zergoun and Vaid (1994)]

stress-based model is simpler with fewer parameters and is easier to implement numerically even though it cannot capture the accumulation and dissipation of pore pressure. The model can capture essential cyclic behavior characteristics of saturated clay, including reverse plastic flow, evolution of hysteretic loops, accumulation of plastic deformations, and degradation of soil stiffness.

A more simplified and practical version of the model will be introduced in this section (Cheng et al. 2021). The shear strength of clay depends on the average principal stress of the soil element; hence, the bounding surface (failure surface) equation is represented by the Drucker-Prager function in Sect. 2.3.1. However, the Von Mises function is usually adopted when the total stress-based model is used to simulate the undrained behaviors of saturated clay (Prevost 1977). It means that, from the perspective of total stress, the yielding of undrained clays is independent of normal stress and depends only on deviatoric stress. Nonetheless, we must keep in mind that the shear strength (failure strength) of soil is closely related to the average principal stress (or normal stress).

In the model, the bounding surface represents the ultimate strength of the soil element, i.e., the failure occurs when the stress point reaches the bounding surface. As a simplified version, the bounding surface equation is described using the Von Mises function, i.e.,

$$F = \frac{3}{2}\bar{s}_{ij}\bar{s}_{ij} - A_0^2 = 0 \tag{2.43}$$

where $\bar{s}_{ij}$ denotes the deviatoric stress tensor of the bounding surface, and A_0 denotes the radius of the bounding surface. It is noted that A_0 represents the soil undrained shear strength, and $A_0 = 2\ S_u$. It must be emphasized that A_0 is not a constant, but increases usually with the soil depth, which is consistent with the variation tend of S_u. A_0 actually reflects the increase in soil undrained shear strength with the average principal stress (or normal stress) as the soil depth increases. In this case, the bounding surface equation represented by the Von Mises function is equivalent to that represented by the Drucker-Prager function in Sect. 2.3.1, that is, A_0 indirectly characterizes the dependency of soil undrained shear strength (soil failure) on normal stress. In this way, the difference between the undrained behavior of soil and the behavior of metals can be distinguished when using the Von Mises function (A_0 is a constant for the metals due to its behavior being independent of the average principal stress).

The strain increment includes two parts: elastic and plastic increments. The elastic strain increment is obtained according to the generalized Hooke's law; and the plastic strain increment is obtained according to the hardening rule, the flow rule, and the bounding surface equation. The model involves only one plastic deviatoric mechanism. The deviatoric strain increment is determined by:

$$\mathrm{d}e_{ij} = \frac{1}{2G}\mathrm{d}s_{ij} + \frac{3}{2H'}\frac{\bar{s}_{ij}}{A_0^2}\bar{s}_{kl}\mathrm{d}s_{kl}, \tag{2.44}$$

where $\mathrm{d}e_{ij}$ is the deviatoric strain increment; G is the elastic shear modulus; $\mathrm{d}s_{ij}$ is the deviatoric stress increment, $\bar{s}_{ij}$ is the deviatoric stress at the image stress point, H' is the plastic shear modulus.

The simplified model includes only four parameters: A_0 and G, μ and ξ, where A_0 and G represent the soil shear strength and initial stiffness, respectively; μ captures the shape of stress–strain curves for initial loading; and ξ controls the rate of strain accumulation and modulus degradation. The values of μ and ξ are usually between 0 and 5. As discussed earlier, the model parameters can be determined by performing dynamic triaxial tests or simple shear tests. A_0 is closely related to the internal friction angle φ and cohesion c. For saturated clays, especially soft clays, the soil strength is usually expressed by the undrained shear strength S_u. In this case, $A_0 = 2S_u$ according to Eq. (2.43) and G can be determined by $G = 0.5E/(1+\nu)$, where E is Young's modulus and ν is Poisson's ratio ($\nu = 0.5$ for undrained clay). E is usually considered to be a multiple of S_u for clay according to engineering experience (e.g., typical range of $E = 300\ S_u$ to $1800\ S_u$) (Robertson and Campanella 1983; Anastasopoulos et al. 2011).

2.7 Numerical Implementation of the Constitutive Model

The user-defined material subroutine UMAT is a secondary development interface provided by ABAQUS for users to define their own material properties. Its main task is to calculate the stress increment that corresponds to the strain increment introduced by the ABAQUS main program according to the user-defined constitutive relationships of the materials, to update the stress state. Thus, the key is how to determine the stress increment that corresponds to the strain increment based on the constitutive relationship, as represented by:

$$\{\Delta\sigma\}_n^i = [D^{\mathrm{ep}}]_n\{\Delta\varepsilon\}_n^i, \tag{2.45}$$

where $\{\Delta\sigma\}_n^i$ and $\{\Delta\varepsilon\}_n^i$ represent the stress increment and strain increment, respectively, that correspond to the i-th iteration for the n-th incremental load step. Here, $[D^{\mathrm{ep}}]_n$ is the elastoplastic matrix at the start of the i-th iteration for the fully explicit integration algorithm or the elastoplastic matrix at the end of the i-th iteration calculation for the implicit integration algorithm. In the study, the modified Euler scheme with substeps of equal size is used to determine the elastoplastic stress state after each iteration for each incremental load step. The modified Euler algorithm is a second-order method; it requires two evaluations of the elastoplastic constitutive matrix $[D^{\mathrm{ep}}]_n$ to update the stresses, and the strain increment $\{\Delta\varepsilon\}_n^i$ is divided into a specific number of subincrements of equal size to improve the calculation's accuracy. The integration algorithm has been introduced in detail by Sloan and Booker (1992). $[D^{\mathrm{ep}}]_n$ corresponds to the constitutive relationship in this study, and it can be determined by derivation of the following formula:

Equation (2.46) can be obtained based on Eq. (2.37):

$$\mathrm{d}s_{ij} = 2G\mathrm{d}e_{ij} - \frac{3}{2}(2G - H)\frac{\overline{s_{ij}}}{(\alpha\overline{I}_1 + k)^2}\overline{s}_{kl}\mathrm{d}e_{kl}. \tag{2.46}$$

Equation (2.47) can be obtained based on the generalized Hooke's law:

$$\mathrm{d}\sigma_{kk} = 3K\mathrm{d}\varepsilon_{kk}, \tag{2.47}$$

where $\mathrm{d}\sigma_{kk}$ is the volume stress increment. Additionally, $\mathrm{d}\varepsilon_{kk}$ is the volume strain increment. K is the bulk modulus, and $K = 2G(1+\nu)\big/(3(1-2\nu))$. As a special case, Poisson's ratio ν is set to 0.49 under the undrained condition during the finite element calculation.

In addition, $\mathrm{d}\varepsilon_{ij} = \mathrm{d}e_{ij} + \frac{1}{3}\delta_{ij}\mathrm{d}\varepsilon_{kk}$, and we can obtain the following incremental stress–strain relationship by combining Eq. (2.46) with Eq. (2.47):

$$\mathrm{d}\sigma_{ij} = 2G\mathrm{d}\varepsilon_{ij} + \frac{2G\nu}{1-2\nu}\delta_{ij}\mathrm{d}\varepsilon_{kk} + \frac{(H-2G)}{2(\alpha\overline{I}_1 + k)^2}\overline{s}_{ij}\overline{s}_{kl}\mathrm{d}\varepsilon_{kl}, \tag{2.48}$$

where $d\sigma_{ij}$ is the stress increment, $d\varepsilon_{ij}$ is the strain increment, and δ_{ij} is the Kronecker delta.

It is assumed that $F = \frac{2G\nu}{1-2\nu}$, and $M = \frac{(H-2G)}{2(\alpha\bar{I}_1+k)^2}$, and then, $[D^{ep}]_n$ can be represented as the following matrix based on Eq. (2.48):

$$[D^{ep}]_n = \begin{bmatrix} 2G+F+M\bar{s}_{11}\bar{s}_{11} & F+M\bar{s}_{11}\bar{s}_{22} & F+M\bar{s}_{11}\bar{s}_{33} & M\bar{s}_{11}\bar{s}_{12} & M\bar{s}_{11}\bar{s}_{23} & M\bar{s}_{11}\bar{s}_{31} \\ F+M\bar{s}_{22}\bar{s}_{11} & 2G+F+M\bar{s}_{22}\bar{s}_{22} & F+M\bar{s}_{22}\bar{s}_{33} & M\bar{s}_{22}\bar{s}_{12} & M\bar{s}_{22}\bar{s}_{23} & M\bar{s}_{22}\bar{s}_{31} \\ F+M\bar{s}_{33}\bar{s}_{11} & F+M\bar{s}_{33}\bar{s}_{22} & 2G+F+M\bar{s}_{33}\bar{s}_{33} & M\bar{s}_{33}\bar{s}_{12} & M\bar{s}_{33}\bar{s}_{23} & M\bar{s}_{33}\bar{s}_{31} \\ M\bar{s}_{12}\bar{s}_{11} & M\bar{s}_{12}\bar{s}_{22} & M\bar{s}_{12}\bar{s}_{33} & G+M\bar{s}_{12}\bar{s}_{12} & M\bar{s}_{12}\bar{s}_{23} & M\bar{s}_{12}\bar{s}_{31} \\ M\bar{s}_{23}\bar{s}_{11} & M\bar{s}_{23}\bar{s}_{22} & M\bar{s}_{23}\bar{s}_{33} & M\bar{s}_{23}\bar{s}_{12} & G+M\bar{s}_{23}\bar{s}_{23} & M\bar{s}_{23}\bar{s}_{31} \\ M\bar{s}_{31}\bar{s}_{11} & M\bar{s}_{31}\bar{s}_{22} & M\bar{s}_{31}\bar{s}_{33} & M\bar{s}_{31}\bar{s}_{12} & M\bar{s}_{31}\bar{s}_{23} & G+M\bar{s}_{31}\bar{s}_{31} \end{bmatrix}. \tag{2.49}$$

The model has no elastic domains, and it includes only an outer bounding surface. The bounding surface describes the failure of the soils, and the stress states are not located beyond it; therefore, the calculated stress states that fall outside of the bounding surface must be brought back to the bounding surface. The radial return method (Cheng et al. 2016; Wu 1990) is used to modify those stress states beyond the bounding surface. In addition, the mobile mapping center is used to describe the evolution of the hardening modulus, which means that the mapping center must be updated to determine the stress states after the stress reversal. The stress reversal is judged according to the angle between the deviatoric stress increment vector and the exterior normal vector at the image stress point on the bounding surface, as shown in Fig. 2.7. The key of the UMAT program is to determine the stress increment and to update the stress states based on Eq. (2.48). Some state variables are set to store and update the mapping center and the equivalent plastic strain in the UMAT program. The main steps of updating the stress states based on the proposed constitutive model are given next, and the corresponding flow chart of the stress updating is shown in Fig. 2.15.

(1) The current tress point and the mapping center can be obtained from the last incremental step.
(2) Next, we judge whether the current tress point is the mapping center. If that is real, then the hardening modulus H is the maximum elastoplastic shear modulus $H_{\text{max},}$ and the stress increment $d\sigma_{ij}$ can be determined based on the elastic stress–strain relationship. If not, then it is necessary to determine the stress increment according to the elastoplastic stress–strain relationship.
(3) The image stress point that corresponds to the current stress point can be determined based on the radial mapping rule after obtaining the position of the mapping center and the current stress point.
(4) The distance between the current stress point, the mapping center and the image stress point δ, δ_0 can be determined. The elastoplastic shear modulus H of the current stress point is determined according to Eq. (2.26).
(5) The stress increment $d\sigma_{ij}$ can be determined based on Eq. (2.48), and then, the stress states at the end of the increment step can be determined.

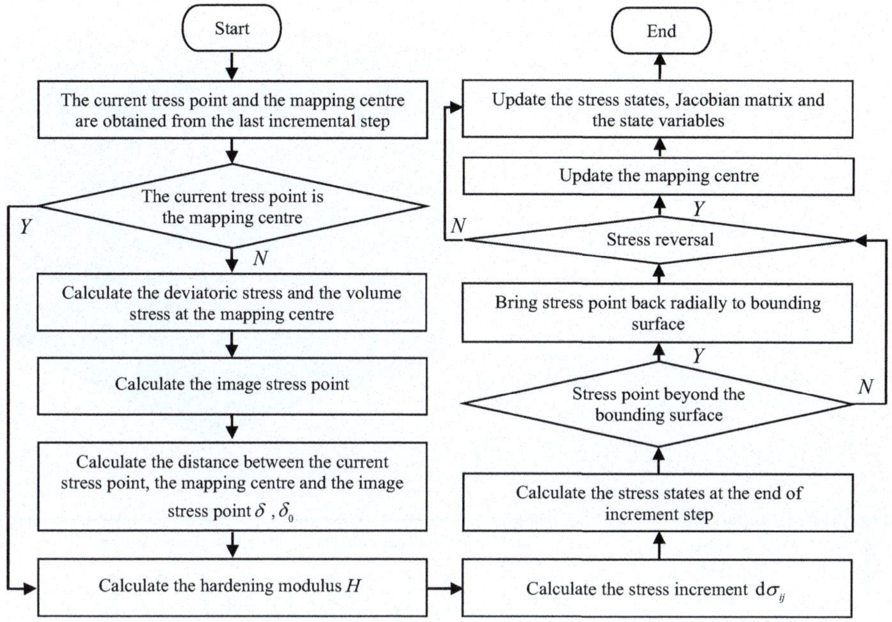

Fig. 2.15 Flowchart of stress updating based on the proposed constitutive model

(6) Next, we judge whether the stress state at the end of the increment step is beyond the bounding surface. If it goes beyond the bounding surface, then it is necessary to bring the stress point back radially to the bounding surface using the radial return method.
(7) Next, we judge whether a stress reversal occurs. In the case of a stress reversal, the mapping center must be updated.
(8) We next update the stress states, Jacobian matrix, and state variables. Then, we go to the next incremental step.

A finite element model with only one element was established to simulate the soil sample to verify the subroutine UMAT. The vertical displacement of the bottom boundary is constrained, and the other boundaries are free in the model, as shown in Fig. 2.16. The half-sine cyclic axial deviator stress with the peak value of 325 kPa is applied to the top of the sample to simulate the stress state of the Newfield clay, and the related model parameters can be seen in Table 2.2. The comparison between the constitutive model prediction and the finite element results for the stress–strain curve is shown in Fig. 2.17. The finite element result of the stress–strain curve is in agreement with the result predicted by the constitutive model, which indicates that the integration algorithm used in this study has an acceptable accuracy and the subroutine UMAT has been correctly implemented in ABAQUS.

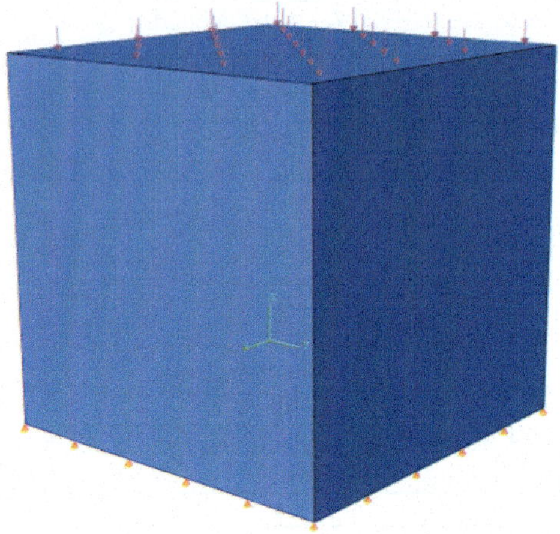

Fig. 2.16 Finite element model of the soil sample

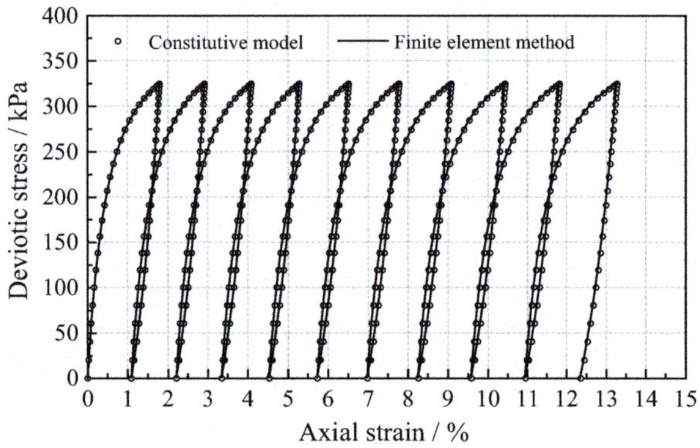

Fig. 2.17 Comparison of the stress–strain curves calculated by the finite element method and the constitutive model

2.8 Seismic Response Analysis of the Homogeneous Horizontal Site

The seismic response of a homogeneous clay layer on bedrock is analyzed. The thickness of the clay layer is 40 m. The internal friction angle of the clay is $\varphi = 24°$ and the cohesion $c = 10$ kPa. The soil unit weight $\gamma = 19$ kN/m^3. The shear wave velocity $v = 160$ m/s. The elastic shear modulus $G = 48.6$ MPa, and the model parameter μ and ξ are set to 1.5 and 2, respectively. A three-dimensional finite element model is established based on ABAQUS software. The model is discretized by 40 8-node linear brick elements with the thickness of 1 m. The seismic action is applied in only one direction, and the movement of the soil layer in the other two directions is restricted to simulate a one-dimensional seismic response. Implicit dynamic analysis of the seismic responses is performed by using the proposed nonlinear constitutive model. In addition, a computer program for Equivalent-linear Earthquake site Response

Analysis (EERA) (Bardet et al. 2000) is used for comparative analysis. To ensure the comparability of the results of the two analysis methods, the shear modulus and the damping ratio for the different levels of shear strain in EERA were determined based on the stress–strain hysteretic curves of the proposed nonlinear constitutive model. In other words, the stress–strain hysteretic curves of the clays for different levels of shear strain are determined by the proposed nonlinear constitutive model based on the above model parameters. Then, the shear modulus and damping ratio can be inversely calculated according to the stress–strain hysteretic curves, as shown in Fig. 2.18. The clay site is subjected to a seismic accelerogram from the Loma Prieta earthquake record in the USA in 1989, which had a scaled peak acceleration (PGA) of 0.1 g, as shown in Fig. 2.19.

The acceleration time histories and the acceleration response spectra at the ground surface computed by the two analysis methods are shown in Figs. 2.20 and 2.21, respectively. It can be seen that the EERA predicts PGA = 0.177 g, the nonlinear elastoplastic analysis predicts 0.174 g, and the corresponding time of the PGA for the two are both approximately 11.2 s. The acceleration time histories computed by the two analysis methods agree with each other in the overall trend; although the two methods are essentially different, one is nonlinear analysis and the other

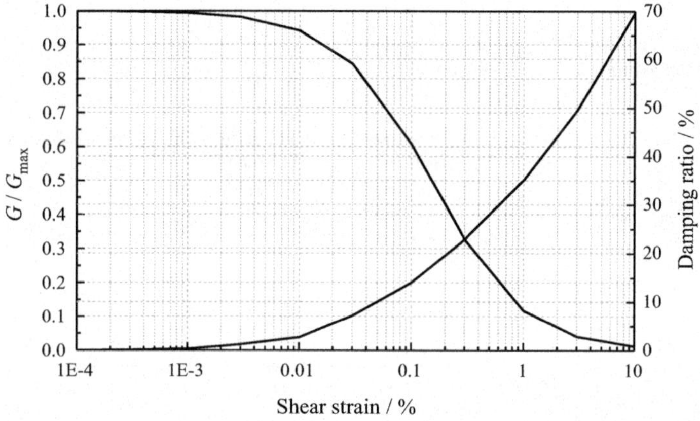

Fig. 2.18 Shear modulus and damping ratio versus the shear strain

Fig. 2.19 Acceleration recorded in the Loma Prieta earthquake

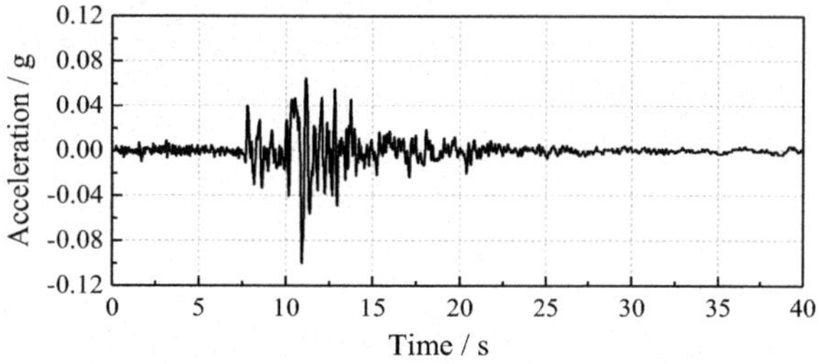

is equivalent linear analysis. In a strict sense, the acceleration value computed by ABAQUS is larger than that computed by EERA over the whole time history. A similar relation occurs in the comparison of the acceleration response spectrum of the ground surface in Fig. 2.21. The main reason is that the damping for ABAQUS is smaller than that for EERA during the computation. Stress–strain loops computed by the two methods at depths of 10 m and 30 m, respectively, are also compared in Fig. 2.22. It can be seen that the level of stress and strain calculated by ABAQUS agrees with that calculated by EERA, overall. However, the hysteresis loop for the former is more 'thin' than that for the latter; in other words, the damping of the latter is greater. This finding arises mainly as a result of the inherent defects of the equivalent linear analysis because of ignoring the true nature of the soil's nonlinearity and the change in the soil stiffness during the earthquake action. In addition, the level of stress and strain of the soil layer increases with the increase in the depth, while the nonlinear hysteretic characteristics and the loss of energy increase. It can be seen from the stress–strain curve of the soil at a depth of 30 m that the nonlinear constitutive model proposed in this study can capture the strain accumulation characteristics of the soil under seismic loads, but the equivalent linear model has no such ability. The above comparative analysis verifies the ability of the constitutive model to simulate the seismic response of the soil layer.

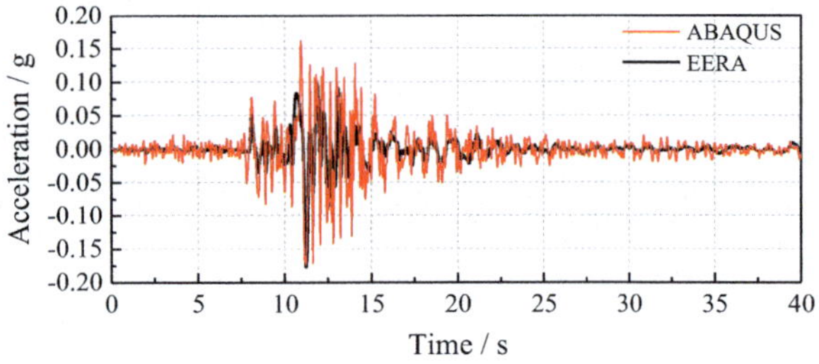

Fig. 2.20 Acceleration time histories of the ground surface

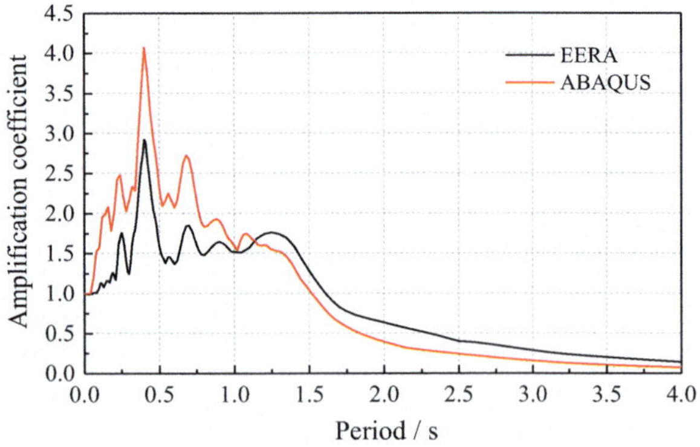

Fig. 2.21 Amplification coefficient of the acceleration response spectrum of the ground surface

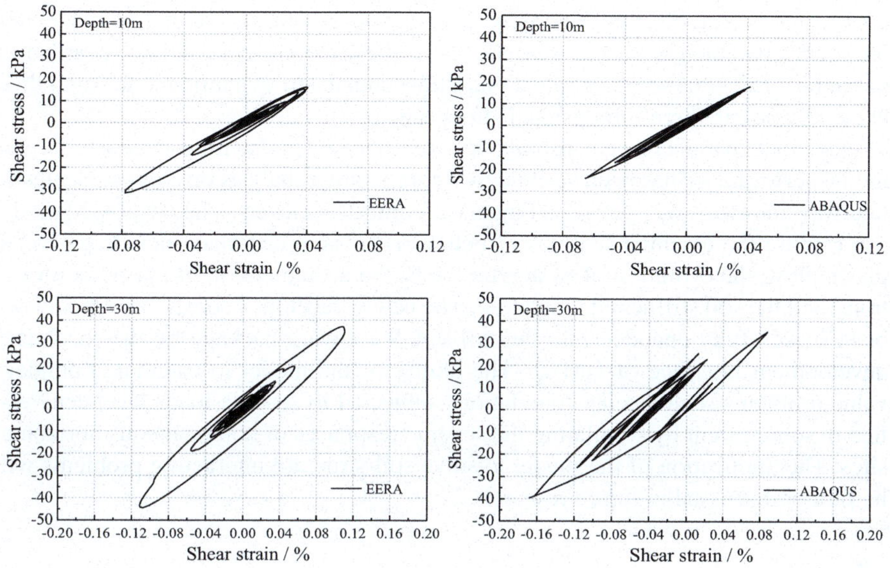

Fig. 2.22 Stress–strain curves at different depths of the soil layer

2.9 Summary

A simple single bounding surface constitutive model has been developed to predict the undrained cyclic behaviors of saturated clays. The model includes only one outer bounding surface without any inner loading surface or yield surface, which is based on the concept of a 'vanishing elastic domain'. The model does not involve changes in the multiple yield surfaces in terms of the size and position in stress space. Therefore, it does not involve complex kinematic hardening rules; it is only required to memorize important stress reverse events, which makes the model easy to implement numerically, and have a high computational efficiency in solving boundary value problems. In other words, the simplicity should be the largest advantage of the model. However, the model is based on the total stresses and is thus unable to predict pore-pressure buildup due to changes in the mean effective stress. Problems that require prediction of the pore pressure buildup during cyclic loading are best formulated in terms of the effective stresses. Furthermore, without additional enhancements, the model is unable to replicate the strain softening behavior.

This research proposed a new interpolation function of the elastoplastic shear modulus based on the bounding surface theories. The constitutive model can capture the cyclic degradation characteristics of clays by introducing the cumulative equivalent plastic shear strain into the interpolation function. The evolution of the hardening modulus is described in the deviatoric stress space by the movement and updating of the mapping center based on the new interpolation function, which enables the model to describe the stress–strain hysteretic responses of clays under cyclic loading. The present model can be simplified into some classical one-dimensional models when the model parameters take certain special values; therefore, it can be regarded as the improvement of some classical one-dimensional soil dynamic models and their generalization in three-dimensional stress space.

The model parameters can usually be determined by performing triaxial tests. The model performance has been verified by a comparative analysis on clays subjected to one-way and two-way cyclic loading at different stress levels. The developed model can capture the essential features of behaviors in saturated clay, including reverse plastic flow, the evolution of hysteretic loops, the accumulation of plastic deformations, and the soil stiffness degradation. The newly developed constitutive model has been successfully encoded into the ABAQUS software package by the secondary development interface of UMAT. The ability of the model to calculate boundary value problems such as clay foundations subjected to seismic loads has been verified to some extent by simulating the seismic responses of homogeneous horizontal sites. The application of the model to solve soil-structure interacting problems will be further discussed in future research.

References

Anastasopoulos, I., F. Gelagoti, R. Kourkoulis, et al. 2011. Simplified constitutive model for simulation of cyclic response of shallow foundations: Validation against laboratory tests. *Journal of Geotechnical and Geoenvironmental Engineering* 137 (12): 1154–1168.

Andersen, K.H., W.F. Rosenbrand, S.F. Brown, et al. 1980. Cyclic and static laboratory tests on Drammen clay. *Journal of Geotechnical and Geoenvironmental Engineering* 106 (5): 499–529.

Bardet, J.P. 1986. Bounding surface plasticity model for sands. *Journal of Engineering Mechanics* 112 (11): 1198–1217.

Bardet, J.P., K. Ichii, C.H. Lin. 2000. *EERA: A Computer Program for Equivalent Linear Earthquake Site Response Analysis of Layered Soil Deposits*. Report of University of Southern California, Department of Civil Engineering.

Borja, R.I., and A.P. Amies. 1994. Multiaxial cyclic plasticity model for clays. *International Journal Geotechnical Engineering* 120 (6): 1051–1070.

Cheng, X.L., and J.H. Wang. 2016. An elastoplastic bounding surface model for the cyclic undrained behaviour of saturated soft clays. *Geomechanics and Engineering* 11 (3): 325–343.

Cheng, X.L., J.H. Wang, and Z.X. Wang. 2016. Incremental elastoplastic FEM for simulating the deformation process of suction caissons subjected to cyclic loads in soft clays. *Applied Ocean Research* 59: 274–285.

Cheng, X.L., X.L. Du, D.C. Lu, et al. 2020a. A simple single bounding surface model for undrained cyclic behaviours of saturated clays and its numerical implementation. *Soil Dynamics and Earthquake Engineering* 139: 106389.

Cheng, X.L., N. Li, and Z.H. Yang. 2020b. A simple anisotropic bounding surface model for saturated clay considering the cyclic degradation. *European Journal Environmental Civil Engineering* 24 (12): 2094–2115.

Cheng, X.L., T.J. Wang, J.X. Zhang, et al. 2021. Finite element analysis of cyclic lateral responses for large diameter monopiles in clays under different loading patterns. *Computers and Geotechnics* 134: 104104.

Crouch, R.S., and J.P. Wolf. 1994. Unified 3D critical state bounding-surface plasticity model for soils incorporating continuous plastic loading under cyclic paths. *International Journal for Numerical and Analytical Methods in Geomechanics* 18 (11): 735–784.

Dafalias, Y.F. 1986. Bounding surface plasticity Part I: Mathematical foundation and hypoplasticity. *Journal of Engineering Mechanics* 112 (9): 966–987.

Dafalias, Y.F., and E.P. Popov. 1975. A model of nonlinearly hardening materials for complex loading. *Acta Mechanica* 21 (3): 173–192.

Dafalias, Y.F., and E.P. Popov. 1977. Cyclic loading for materials with a vanishing elastic domain. *Nuclear Engineering and Design* 41 (2): 293–302.

Dafalias, Y.F., and L.R. Herrmann. 1986. Bounding surface plasticity. Part II: Application to isotropic cohesive soils. *Journal of Engineering Mechanics* 112(12): 1263–1291.

Fike, J.A. 2013. *Multi-Objective Optimization Using Hyper-Dual Numbers*, Ph.D. Thesis. Stanford University, Palo Alto.

Gerolymos, N., and G. Gazetas. 2005. Constitutive model for 1-D cyclic soil behavior applied to seismic analysis of layered deposits. *Soils and Foundations* 45 (3): 147–159.

Hardin, B.O., and V.P. Drnevich. 1972. Shear modulus and damping in soils: Design equations and curves. *Geotechnical Special Publication* 98 (7): 667–692.

Hong, P.Y., J.M. Pereira, Y.J. Cui, et al. 2014. An elastoplastic model with combined isotropic-kinematic hardening to predict the cyclic behavior of stiff clays. *Computers and Geotechnics* 62: 193–202.

Huang, M.S., and Y. Liu. 2014. Numerical analysis of axial cyclic degradation of a single pile in saturated soft soil based on nonlinear kinematic hardening constitutive model. *Chinese Journal of Geotechnical Engineering* 36 (12): 2170–2178.

Huang, M.S., Y.H. Liu, and D.C. Sheng. 2011. Simulation of yielding and stress-strain behavior of shanghai soft clay. *Computers and Geotechnics* 38: 341–353.

Hyodo, M., Y. Yamamoto, and M. Sugiyama. 1994. Undrained cyclic shear behaviour of normally consolidated clay subjected to initial static shear stress. *Soils and Foundations* 34 (4): 1–11.

Kimoto, S., B. Shahbodagh Khan, M. Mirjalili, et al. 2015. Cyclic elastoviscoplastic constitutive model for clay considering nonlinear kinematic hardening rules and structural degradation. *International Journal of Geomechanics* 15 (5): A4014005.

Krieg, R.D. 1975. A practical two surface plasticity theory. *Journal of Applied Mechanics* 42 (3): 641–646.

Li, T., and H. Meissner. 2002. Two-surface plasticity model for cyclic undrained behavior of clays. *Journal of Geotechnical and Geoenvironmental Engineering* 128 (7): 613–626.

Li, X.Q., D.C. Lu, Q.T. Lin, X.Q. Dong, and X.L. Du. 2023. A hydro-mechanical constitutive model for unsaturated soils over a wide saturation range. *Computers and Geotechnics* 159: 105475.

Liang, R.Y., and F.G. Ma. 1992. Anisotropic plasticity model for undrained cyclic behavior of clays. *Journal of Geotechnical Engineering* 118 (2): 229–265.

Lu, D.C., J.Y. Liang, X.L. Du, C. Ma, and Z.W. Gao. 2019a. Fractional elastoplastic constitutive model for soils based on a novel 3d fractional plastic flow rule. *Computers and Geotechnics.* 105: 277–290.

Lu, D.C., X. Zhou, X.L. Du, and G.S. Wang. 2019b. A 3d fractional elastoplastic constitutive model for concrete material. *International Journal of Solids and Structures.* 165: 160–175.

Lu, D.C., X. Zhou, X.L. Du, and G.S. Wang. 2020. 3D dynamic elastoplastic constitutive model of concrete within the framework of rate-dependent consistency condition. *Journal of Engineering Mechanics.* 146 (11): 4020124.

Lu, D.C., F.P. Meng, X. Zhou, Y.H. Zhuo, Z.W. Gao, and X.L. Du. 2023a. A dynamic elastoplastic model of concrete based on a modeling method with environmental factors as constitutive variables. *Journal of Engineering Mechanics.* 149 (12): 4023102.

Lu, D.C., Y.N. Zhang, X. Zhou, C.C. Su, Z.W. Gao, and X.L. Du. 2023b. A robust stress update algorithm for elastoplastic models without analytical derivation of the consistent tangent operator and loading/unloading estimation. *International Journal for Numerical and Analytical Methods in Geomechanics.* 47 (6): 1022–1050.

Mroz, Z. 1967. On the description of anisotropic work hardening. *Journal of the Mechanics and Physics of Solids* 15 (3): 163–175.
Mroz, Z., V.A. Norris, and O.C. Zienkiewicz. 1978. An anisotropic hardening model for soils and its application to cyclic loading. *International Journal for Numerical and Analytical Methods in Geomechanics* 2 (3): 203–221.
Mroz, Z., V.A. Norris, and O.C. Zienkiewicz. 1981. An anisotropic critical state model for soils subject to cyclic loading. *Géotechnique* 31 (4): 451–469.
Ni, J., B. Indraratna, X.Y. Geng, et al. 2015. Model of soft soils under cyclic loading. *International Journal of Geomechanics* 15 (4): 04014067.
Prevost, J.H. 1977. Mathematical modelling of monotonic and cyclic undrained clay behaviour. *International Journal for Numerical and Analytical Methods in Geomechanics* 1 (2): 195–216.
Prevost, J.H. 1978. Anisotropic undrained stress-strain behavior of clays. *Journal of the Geotechnical Engineering Division, ASCE* 104 (GT8): 1075–1090.
Pyke, R.M. 1979. Nonlinear soil models for irregular cyclic loading. *Journal of Geotechnical Engineering* 105 (6): 715–726.
Robertson, P.K., and R.G. Campanella. 1983. Interpretation of cone penetration tests. Part II: Clay. *Canadian Geotechnical Journal* 20 (4): 734–745.
Sangrey, D.A., D.J. Henkel, and M.I. Esrig. 1969. The effective stress response of a saturated clay soil to repeated loading. *Canadian Geotechnical Journal* 6: 241–252.
Seidalinov, G., and M. Taiebat. 2014. Bounding surface SANICLAY plasticity model for cyclic clay behaviour. *International Journal for Numerical and Analytical Methods in Geomechanics* 38 (7): 702–724.
Simo, J.C., and T.J. Hughes. 2006. *Computational Inelasticity*. New York: Springer Science & Business Media.
Sloan, S.W., and J.R. Booker. 1992. Integration of Tresca and Mohr-Coulomb constitutive relations in plane strain elastoplasticity. *International Journal for Numerical Methods in Engineering* 33 (1): 163–196.
Su, C.C., D.C. Lu, X. Zhou, G.S. Wang, X.Y. Zhuang, and X.L. Du. 2023. An implicit stress update algorithm for the plastic nonlocal damage model of concrete. *Computer Methods in Applied Mechanics and Engineering*. 414: 116189.
Tabbaa, A.A., and D.M. Wood. 1989. An experimentally based "bubble" model for clay. In *Numerical Models in Geomechanics*, 91–99, ed. by S Pietruszczak, and G.N. Pande. Elsevier Science Publishers Ltd.
Vucetic, M., and R. Dobry. 1988. Degradation of marine clays under cyclic loading. *Journal of Geotechnical Engineering* 114 (2): 133–149.
Wang, J.H., and M.I. Yao. 1996. Elastoplastic simulation of the cyclic undrained behaviour of soft clays. *Chinese Journal of Geotechnical Engineering* 18 (3): 11–18.
Wu, W. 1990. A unified numerical integration formula for the perfectly plastic von Mises model. *International Journal for Numerical Methods in Engineering* 30 (3): 491–504.
Yang, M., G. Seidalinov, and M. Taiebat. 2019. Multidirectional cyclic shearing of clays and sands: Evaluation of two bounding surface plasticity models. *Soil Dynamics and Earthquake Engineering* 124: 230–258.
Yao, H., and J. Wang. 2018. Elastoplastic simulation of stress-strain response considering the cyclic degradation of saturated clay. *Journal of Offshore Mechanics and Arctic Engineering* 140 (4).
Yin, Z.Y., Q. Xu, and P.Y. Hicher. 2013. A simple critical-state-based double-yield-surface model for clay behavior under complex loading. *Acta Geotechnica* 8 (5): 509–523.
Yu, H.S., C. Khong, and J. Wang. 2007. A unified plasticity model for cyclic behaviour of clay and sand. *Mechanics Research Communications* 34 (2): 97–114.
Zergoun, M., and Y.P. Vaid. 1994. Effective stress response of clay to undrained cyclic loading. *Canadian Geotechnical Journal* 31: 714–727.
Zhou, X., D.C. Lu, X.L. Du, G.S. Wang, and F.P. Meng. 2020. A 3d non-orthogonal plastic damage model for concrete. *Computer Methods in Applied Mechanics and Engineering* 360: 112716.

Zhou, X., D.C. Lu, C.C. Su, Z.W. Gao, and X.L. Du. 2022a. An unconstrained stress updating algorithm with the line search method for elastoplastic soil models. *Computers and Geotechnics* 143: 104592.

Zhou, X., D.C. Lu, Y.N. Zhang, X.L. Du, and T. Rabczuk. 2022b. An open-source unconstrained stress updating algorithm for the modified cam-clay model. *Computer Methods in Applied Mechanics and Engineering* 390: 114356.

Zhou, X., A. Shi, D. Lu, Y. Chen, X. Zhuang, X. Lu, and X. Du. 2023. A return mapping algorithm based on the hyper dual step derivative approximation for elastoplastic models. *Computer Methods in Applied Mechanics and Engineering.* 417: 116418.

Zhuang, H.Y., G.X. Chen, and D.H. Zhu. 2006. Dynamic visco-plastic memorial nested yield surface model of soil and its verification. *Chinese Journal of Geotechnical Engineering* 28 (10): 1267–1272.

Chapter 3
A Cyclic p-y Elastoplastic Model Applied to Lateral Loaded Pile in Soft Clays

The p-y method as a simplified analysis tool has been widely used to analyze the behavior of laterally loaded piles. This chapter develops a novel cyclic p-y elastoplastic model within the framework of the single-surface bounding surface theory. The model can capture the soil stiffness degradation during cyclic loading by incorporating the cumulative plastic displacement to an interpolation function of the elastoplastic resistance coefficient. The model is relatively simple with only four parameters that can be determined from standard soil properties and stress–strain responses measured in cyclic direct simple shear (DSS) tests. The performance of developed model is validated by predicting the cyclic lateral response of piles installed in soft clay during field and centrifuge tests published in the literature. The model can reliably simulate monotonic and cyclic responses of piles under different lateral loading patterns, and capture main characteristics of the pile head load–displacement curve, such as nonlinearity, hysteresis, displacement accumulation, and stiffness degradation. It can also predict the evolution of the lateral deflection and sectional bending moment along the pile during cyclic loading.

3.1 Introduction

In harsh offshore environmental conditions, offshore platforms and wind turbines (OWT) supported by pile foundations are subjected to cyclic lateral loads due to wind, sea waves, and currents. Consequently, the piles may experience large sectional bending moments and significant cumulative displacement due to the soil stiffness degradation caused by long-term lateral cyclic loading, especially for piles installed in soft clay. For wind turbine foundations, serviceability limit state (SLS) requirements typically dominate the whole structure design (Det Norske Veritas (DNV) 2013), and foundation response to environmental loads pose a big challenge in their design. This

X. Cheng et al., *Dynamic Analysis of Offshore Wind Turbine Foundations in Soft Clays*,
https://doi.org/10.1007/978-981-97-9454-6_3

motivated extensive research to develop reliable methods to analyze cyclic lateral response of piles.

The load-transfer approach, often referred to as the p-y method, is widely used to analyze the behavior of laterally loaded pile. In this approach, the pile is simulated as an elastic beam and the soil is idealized as a series of discrete nonlinear springs along the pile shaft. The p-y springs describe the relationship between the local lateral soil resistance p and pile lateral relative displacement y. Although discrete p-y springs do not rigorously capture the soil continuum behavior, they are still widely utilized because of their simplicity and computational efficiency compared to continuum modeling. The properties of p-y springs are usually characterized by a specific p-y model, which depends on soil strength, stiffness, and pile properties. Selection of an appropriate p-y model is therefore very important for accurate simulation of the pile lateral behavior.

Matlock (1970) proposed a p-y model for piles installed in soft clay, which was adopted by the API RP 2GEO (American Petroleum Institute (API) 2014) and is widely used in industry. The model was developed based on limited pile load tests at the Sabine River site in the 1950s, and its limitations have been widely discussed. The model is too conservative and underestimates the initial stiffness and ultimate soil reaction (Jeanjean 2009; Zhu et al. 2017). In addition, for cyclic loading cases, an empirical reduction factor is usually introduced to modify the ultimate lateral soil resistance to account for cyclic degradation effect. This reduction factor is too crude and cannot properly account for the cyclic load level or pattern or loading history. To account for number of load cycles and their amplitude, a cyclic degradation factor was proposed to improve the monotonous p-y models (Fan and Long 2005; Dewaikar and Patil 2006; Zhu et al. 2017). However, these cyclic degradation models are empirical, derived based on a few cyclic load tests, and cannot simulate lateral cyclic hysteretic responses of piles.

The pile cyclic lateral response was also analyzed by developing cyclic p-y hysteresis curves based on the beam on nonlinear Winkler foundation (BNWF) method. In these models, some pile-soil interaction characteristics are simulated using specific components such as spring, dashpot, drag cell, and gap cell to capture the dynamic hysteretic behavior of the soil-pile interface (El Naggar and Novak 1996; Gerolymos and Gazetas 2005a; Allotey and El Naggar 2008; Heidari et al. 2014a, b; Liang et al. 2018). However, the introduction of too many components complicates the implementation of the model. In addition, the finite element method (Basack 2010a, 2015; Rathod et al. 2016; Hong et al. 2017; Zhang et al. 2022; Nimbalkar and Basack 2022), the boundary element method (Basack 2010b; Basack et al. 2022), the strain wedge method (Heidari and El Naggar 2018; Yang et al. 2021) and the energy-based method (Arvan and Arockiasamy 2022) were also employed to analyze the cyclic behavior of laterally loaded piles.

The bounding surface theory developed by Dafalias (1986) has been widely used to describe the cyclic stress–strain response of soil. In recent years, this theory has also been innovatively applied to describe cyclic p-y response. Several p-y models were developed within the framework of the bounding surface elastoplastic theory to simulate the lateral sand-pile interaction (Su and Yan 2013; Choi 2015) and clay-pile interaction (McCarron 2015, 2021; Yu et al. 2020). The theoretically robust p-y models that involve only spring elements based on bounding surface theory offer the promise of good numerical implementation and applicability.

Only a few studies investigated the cyclic p-y elastoplastic models for clays based on bounding surface theory. Single-surface bounding surface model (Yu et al. 2020) degenerates the elastic domain to a loading stress point, and hence, unlike the two-surface bounding surface model (McCarron 2015) does not involve the evolution of the size and position of the kinematic hardening yield surface. Therefore, it is relatively simple and easy to implement numerically. Yu et al. (2020) employed the softening model of soil resistance to simulate the cyclic degradation of undrained clay, which requires in-situ T-bar testing to determine model parameters. The model applicability was validated only under two-way symmetric cyclic load. However, piles in clay may exhibit vastly different responses under different lateral cyclic loading patterns (i.e., one-way loading, two-way symmetric, and asymmetric loading, Cheng et al. 2021). In addition, the pile may exhibit cyclic stability or instability responses under different lateral load levels.

To address these shortcomings, this chapter proposes a new method to describe the cyclic degradation of soft clay stiffness and strength utilizing an elastoplastic resistance coefficient (Cheng et al. 2023). This method gives rise to a relatively simple incremental cyclic p-y elastoplastic model developed by constructing a novel interpolation function of the elastoplastic resistance coefficient within the framework of the single-surface bounding surface theory. The model performance in simulating both monotonic and cyclic responses of piles under different cyclic loading patterns is validated by simulating several existing field and centrifugal model tests.

3.2 Cyclic p-y Model Based on Single Bounding Surface Theory

3.2.1 Model Framework

Cyclic p-y curves often exhibit significant nonlinear hysteresis characteristics as shown in Fig. 3.1, which are closely related to the soil stress–strain response. Monotonic p-y curve may be constructed by scaling the soil stress–strain curve (Zhang and Andersen 2017; Jeanjean et al. 2017). Cyclic stress–strain response can usually be described by an incremental stress–strain relationship. Similarly, cyclic p-y response can be described by an incremental p-y relationship. Based on the incremental elastic–plastic constitutive theory, the incremental displacement dy induced by the incremental resistance dp can be decomposed into two parts, i.e.,

$$\mathrm{d}y = \mathrm{d}y^e + \mathrm{d}y^p, \tag{3.1}$$

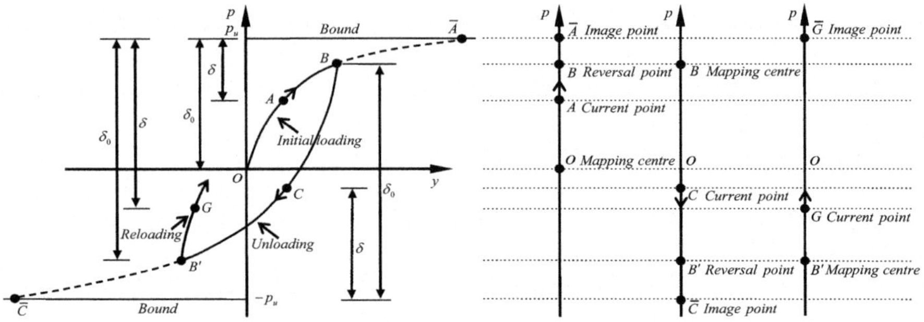

(a) A typical cyclic p-y curve (b) Initial loading (c) Unloading (d) Reloading

Fig. 3.1 Cyclic p-y curve and radial mapping rule for the initial loading–unloading-reloading path

where $\mathrm{d}y^e$ is the incremental elastic displacement and $\mathrm{d}y^p$ is the incremental plastic displacement. The incremental resistance $\mathrm{d}p$ can be obtained from incremental elastic displacement, incremental plastic displacement, and incremental total displacement, based on the one-dimensional expression:

$$\mathrm{d}p = k^e \mathrm{d}y^e, \tag{3.2a}$$

$$\mathrm{d}p = k^p \mathrm{d}y^p, \tag{3.2b}$$

$$\mathrm{d}p = k^{ep} \mathrm{d}y, \tag{3.2c}$$

where k^e is the elastic resistance coefficient; k^p is the plastic resistance coefficient; k^{ep} is the elastoplastic resistance coefficient (tangent modulus of p-y curve). Based on Eqs. (3.1) and (3.2), one can obtain:

$$k^{ep} = \frac{k^e k^p}{k^e + k^p}, \tag{3.3}$$

$$k^p = \frac{k^e k^{ep}}{k^e - k^{ep}}. \tag{3.4}$$

Based on Eq. (3.2c), the key to establish the relationship between $\mathrm{d}p$ and $\mathrm{d}y$ is to determine the elastoplastic resistance coefficient k^{ep} on the cyclic loading path. In the present model, the bounding surface constitutive theory (Dafalias 1986) can be used to establish one-dimensional incremental p-y relationship. It is assumed that the elastic domain enclosed by the bounding surface is reduced to a point and that incremental plastic displacement $\mathrm{d}y^p$ occurs immediately for any incremental resistance $\mathrm{d}p$ within the bounding surface, which is inspired by the concept of 'vanishing elastic domain' proposed by Dafalias and Popov (1977). The size of bounding surface represents the ultimate soil resistance p_u, which means that the failure occurs when the soil resistance p reaches the bounding surface.

A typical cyclic p-y curve for the initial loading–unloading-reloading path is shown in Fig. 3.1a. The radial mapping rule proposed by Dafalias (1986) is adopted herein, as shown schematically in Fig. 3.1b–d. In the initial loading path (Fig. 3.1b), O, A, and B denote the initial resistance, current resistance, and resistance reversal points, respectively. According to the radial mapping rule, O is the mapping center, and $\overline{A}$ denotes the image point, which is intersection of the bounding surface and a line that passes through the mapping center and the current resistance point. In one-dimensional case, the image point becomes the ultimate soil resistance point in the same direction as the current soil resistance. In the unloading path after resistance reversal (Fig. 3.1c), the resistance reversal point B becomes the new mapping center, C denotes the current resistance point, and $\overline{C}$ denotes the image point for unloading path. In the reloading path (Fig. 3.1d), upon the resistance reversal again, the reversal point B' is updated as the mapping center, G denotes the current resistance point, and $\overline{G}$ denotes the image point for the reloading path. During cyclic loading, the soil resistance reversal point where the incremental resistance dp abruptly changes direction is updated as the new mapping center upon resistance reversal.

Figure 3.1a shows that the elastoplastic resistance coefficient k^{ep} has a maximum value at the beginning of loading (the mapping center), and $k^{\mathrm{ep}} = k^e$ ($k^p = +\infty$) according to Eqs. (3.3) and (3.4). When the soil resistance point reaches the bounding surface (the image point), k^{ep} has its minimum value, and $k^{\mathrm{ep}} = 0$ ($k^p = 0$). When the current resistance point moves from the mapping center to the resistance reversal point, k^{ep} gradually decreases to a certain value from the maximum value. It is assumed that k^{ep} reaches the maximum value again upon soil resistance reversal and gradually decreases again to a certain value at the next resistance reversal point. After reaching the maximum and minimum values of k^{ep}, the method of constructing modulus interpolation function in bounding surface theory can be used to determine k^{ep} for any current resistance point in the cyclic loading path.

To determine the elastoplastic resistance coefficient k^{ep} for the current soil resistance point in the cyclic loading path, with reference to the expression of interpolation function of elastoplastic modulus in the bounding surface model (Cheng et al. 2020), a new interpolation function of k^{ep} is constructed based on the radial mapping rule (Dafalias 1986), i.e.,

$$k^{\mathrm{ep}} = \left(\frac{\delta}{\delta_0}\right)^{\mu} k^e \exp\left(-\xi \frac{\int |\mathrm{d}y^p|}{D}\right), \tag{3.5}$$

where δ denotes the distance between the current soil resistance point and image point, and δ_0 denotes the distance between the mapping center and the image point as shown in Fig. 3.1; μ is a parameter that reflects the shape of p-y curves for the initial loading; dy^p is the incremental plastic displacement determined by combining Eqs. (3.2.b) and (3.4); $\int |\mathrm{d}y^p|$ is the cumulative plastic displacement during cyclic loading, which increases monotonically with time and reflects the degradation of soil stiffness as its plastic displacement accumulates; and ξ is a coefficient that controls the rate of degradation of stiffness. Obviously, k^{ep} is maximum when $\delta = \delta_0$ (the mapping center) and is minimum when $\delta = 0$ (the image point).

3.2.2 *Model Parameters*

The cyclic p-y model requires four parameters: ultimate soil resistance p_{u}, elastic resistance coefficient k^e, and parameters μ and ξ. The roles of these parameters and how to determine their values are discussed in the following sections.

3.2.2.1 Ultimate Soil Resistance p_{u}

The ultimate lateral soil resistance p_{u} is calculated assuming shear strength profiles approximately linearly increasing with depth (Murff and Hamilton 1993), i.e.,

$$
\begin{aligned}
p_{\mathrm{u}} &= N_p \cdot s_{\mathrm{u}} + \gamma' z \\
N_{\mathrm{p}} &= N_1 - N_2 \cdot \exp\left(\frac{-\zeta \cdot z}{D}\right) \\
\zeta &= 0.25 + 0.05 \cdot \lambda \quad \text{for } \lambda < 6 \\
\zeta &= 0.55 \quad \text{for } \lambda \geq 6 \\
\lambda &= s_{\mathrm{u0}} / (k \cdot D)
\end{aligned}
\tag{3.6}
$$

where s_{u} is soil undrained shear strength; N_p is non-dimensional lateral bearing factor; γ' is soil submerged unit weight; z is soil depth below seafloor; D is pile diameter; s_{u0} is shear strength intercept at seafloor; k is rate of increase in shear strength with depth; N_1 is limiting value of bearing factor; $N_1 - N_2$ is intercept of bearing factor at soil surface. Factors N_1 and N_2 are closely related to the characteristics of pile-soil interface and soil properties. Murff and Hamilton (1993) suggested that $N_1 = 9$ and $N_2 = 7$ for smooth pile; same values were used in determining p_{u} for the field tests of offshore driven pile in soft clay carried out by Zhu et al. (2017). However, Jeanjean (2009) suggested that $N_1 = 12$ and $N_2 = 4$ based on a series of centrifuge testing and finite element modeling results for piles in soft clays. For these different values, the non-dimensional lateral bearing factor N_p with the normalized soil depth z/D is plotted for upper limit case ($\lambda \geq 6$) and lower limit case ($\lambda = 0$) as shown in Fig. 3.2. For comparison, the widely used American Petroleum Institute (API) (2014) expression of N_p is also represented as

$$
\begin{aligned}
N_p &= 3 + J \cdot \frac{z}{D} \quad \text{for } \frac{z}{D} < \frac{6}{J}, \\
N_p &= 9 \quad \text{for } \frac{z}{D} \geq \frac{6}{J},
\end{aligned}
\tag{3.7}
$$

where J is a dimensionless empirical constant with values ranging from 0.25 to 0.5 determined from field testing. The profile of N_p for upper limit case ($J = 0.5$) and lower limit case ($J = 0.25$) is also plotted in Fig. 3.2.

Figure 3.2 demonstrates that the maximum of N_p proposed by API is equal to the value proposed by Murff and Hamilton for smooth piles, both ≤ 9. However, before reaching the maximum value, value of N_p proposed by API is significantly smaller than the latter except for shallow depth near the seafloor. Test results from Jeanjean (2009) and Zhu et al. (2017) demonstrated that API p-y curve significantly underestimated the value of ultimate soil resistance. In addition, Jeanjean (2009) found that the maximum value of N_p exceeded 9 and even reached 12.7 for numerical simulation results and 13.4 for centrifugal test results. Hence, Jeanjean (2009) limits the maximum value of N_p to 12 to be consistent with plasticity solutions proposed by Randolph and Houlsby (1984) for a rough pile (see Fig. 3.2). Thus, maximum N_p has lower and upper limits of 9 and 12 for smooth and rough pile-soil interface, respectively. Correspondingly, the present study suggests $N_1 = 12$ and $N_2 = 7$, which renders N_p within the values recommended by Murff and Hamilton (1993) and Jeanjean (2009) as demonstrated in Fig. 3.2. In practice, values of N_1 and N_2 in Eq. (3.6) should be determined from actual pile-soil interface and soil properties by performing pile load test. In the absence of test data, one can refer to the values proposed in this study.

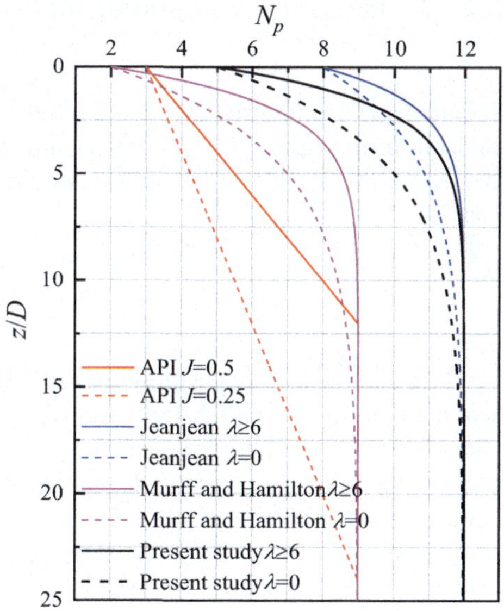

Fig. 3.2 Comparison of non-dimensional lateral bearing factor N_p

3.2.2.2 Elastic Resistance Coefficient k^e

The coefficient k^e is the initial tangent modulus (initial stiffness) of the p-y curve and can expressed as (Vesic 1961):

$$k^e = 0.65 \sqrt[12]{\frac{E_s D^4}{E_p I_p}} \frac{E_s}{1 - \nu^2}, \tag{3.8}$$

where E_s and ν are the soil Young's modulus and Poisson's ratio; $E_p I_p$ is the flexural rigidity of pile section. For clay soils, E_s is usually considered to be a multiple of undrained shear strength s_u (i.e., $E_s = 300\, s_u$ to $1800\, s_u$) (Anastasopoulos et al. 2011), and s_u is generally considered to increase linearly with depth z. In ensuing analyses, k^e is determined employing Eq. (3.8). Alternatively, k^e can be approximately assumed to increase with depth (Poulos and Davis 1980), i.e.,

$$k^e = \eta_h z, \tag{3.9}$$

where η_h represents the variation of soil modulus with depth, taken as 160–3450 kN/m^3 for soft normally consolidated clays (Reese and Matlock 1956).

3.2.2.3 Parameter μ Capturing the Shape of Monotonic p-y Backbone Curve

Parameter μ captures the shape of monotonic p-y backbone curve (the initial loading) and the term $\exp\left(-\xi \int |\mathrm{d}y^p| / D\right)$ is introduced in Eq. (3.5) to capture the cyclic degradation of soil. For monotonic loading, $\xi = 0$ and Eq. (3.5) can be simplified to:

$$k^{ep} = \left(\frac{\delta}{\delta_0}\right)^{\mu} k^e. \tag{3.10}$$

Based on Eq. (3.10), the monotonic p-y backbone curve can be obtained by numerical integration of Eq. (3.2c). Figure 3.3 shows the normalized monotonic p-y curves for different μ (p is normalized by p_u, and y is normalized by D). The decrease in tangent modulus slows as the displacement increases, and the variation of p-y from linear to nonlinear becomes more gradual and starts earlier as μ increases. It is found that μ ranges from 0 to 5 to fit the actual p-y curves. It is worth noting that the current model can be simplified into hyperbolic p-y model (Kondner 1963) when $\mu = 2$, i.e.,

$$p = \frac{y}{\frac{1}{k^e} + \frac{y}{p_u}}. \tag{3.11}$$

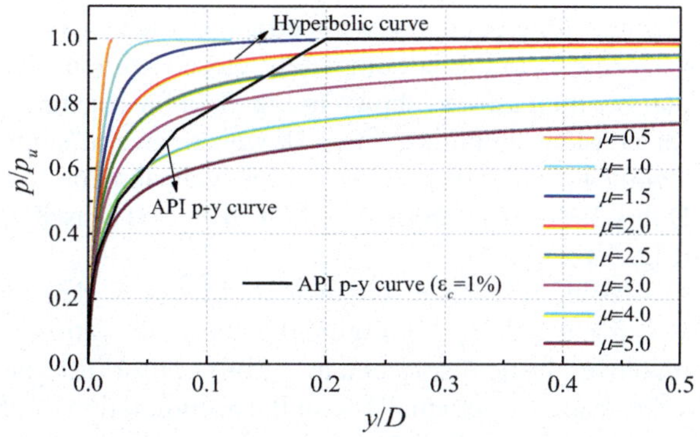

Fig. 3.3 Predicted monotonic p-y curves with different μ ($p_u = 100$ kPa, $k^e = 10$ MPa, $D = 1$ m)

The proposed incremental p-y model with variable μ is more versatile and offers wider applicability to represent different soft clays and pile conditions. Also, the shape of p-y curves bears strong similarities with the soil stress–strain curves (Bransby 1999). The stiffer the soil stress–strain response, the stiffer is the p-y response, and vice versa. Zhang and Andersen (2017) constructed p-y curves by simply scaling the soil stress–strain response measured in direct simple shear (DSS) tests. In this study, parameter μ can be determined by curve fitting the shear stress–strain curve (τ / s_u versus γ) measured in DSS test, based on the incremental stress–strain relationship corresponding to Eq. (3.10) (Cheng et al. 2020).

The normalized API p-y curve with ε_c of 1% is also plotted in Fig. 3.3 (ε_c is strain at one-half the maximum stress on laboratory undrained compression tests of undisturbed soil samples). The shape of API p-y curve is completely different from that for various μ values. For the API p-y curve, the soil resistance *p* reaches the limit value when the lateral displacement *y* reaches 0.2 times the pile diameter *D*, i.e., *y*/*D* = 0.2. If *y*/*D* = 0.2 is taken as the displacement failure criterion, the p-y curve for $\mu = 1.5$ meets this criterion. However, the stiffness (tangent slope) of p-y curve in the initial loading for $\mu = 1.5$ is significantly larger than API p-y curve, which illustrates the rationality of the current p-y curve. Both model tests and finite element analysis indicated that the API p-y curve is too conservative as it underestimates soil stiffness (Jeanjean 2009; Zhu et al. 2017).

3.2.2.4 Parameter ξ Controlling the Rate of Cyclic Degradation of Stiffness

The term $\exp\left(-\xi \int |\mathrm{d}y^p| / D\right)$ is introduced into the interpolation function (Eq. 3.5) to capture the degradation of soil stiffness with the accumulation of plastic displacement during cyclic loading. The parameter ξ controls the rate of soil stiffness degradation by adjusting the rate of plastic displacement accumulation. If $\xi = 0$ in Eq. (3.5), the evolution of elastoplastic resistance coefficient k^{ep} is consistent between unloading and reloading paths for symmetrical two-way cyclic loading, and therefore, the p-y hysteresis curve is completely closed without further degradation as the cycle number increases, as shown in Fig. 3.4a for constant amplitude of p and in Fig. 3.4d for constant amplitude of y. However, several studies revealed that clays exhibit significant cyclic degradation under cyclic loading (Andersen et al. 1980; Vucetic and Dobry 1988). Therefore, $exp\left(-\xi \int |\mathrm{d}y^p| / D\right)$ is used to capture the cyclic degradation (Fig. 3.4b, c, e, f).

Elastic resistance coefficient k_d^e after degradation is given by: $k_d^e = k^e \exp\left(-\xi \int |\mathrm{d}y^p| / D\right)$. Figure 3.5 shows the ratio k_d^e / k^e with the normalized accumulative plastic displacement $\int |\mathrm{d}y^p| / D$ for different ξ. Figure 3.5 reveals that its value decreases gradually with the accumulation of plastic displacement, especially as ξ increases; for $\xi = 5$, the value approaches 0 when accumulative plastic displacement equals the pile diameter ($\int |\mathrm{d}y^p| / D = 1$), which means that the soil stiffness approaches 0. Empirically, ξ ranges from 0 to 5 to fit observed cyclic p-y curves. Cyclic p-y curves with different ξ are presented in Fig. 3.4, where Fig. 3.4a–c shows that the cyclic degradation of soil stiffness is more significant as ξ increases for constant amplitude p; and Fig. 3.4d–f shows the same trend for constant amplitude y. Hence, parameter ξ controls the rate of cyclic degradation of soil stiffness. The parameter ξ can be determined by curve fitting the cyclic shear stress–strain curve measured in DSS test based on the incremental stress–strain relationship (Cheng et al. 2020a).

Under one-way cyclic loading (no resistance reversal), the soil exhibits more significant displacement accumulation rather than stiffness degradation for each hysteretic cycle, as shown in Fig. 3.4g–i; as ξ increases the displacement accumulation increases. However, the rate of increase is insignificant, compared with the increase in stiffness degradation under two-way cyclic loading. Based on numerical simulations, ξ can be taken as 0 for one-way cyclic loading.

The developed model is especially relevant to response of offshore pile foundation subjected to cyclic environmental loads such as winds and waves. The frequency range of wind and wave loads is usually 0.05–0.3 Hz (Arany et al. 2016), i.e., low-frequency (long period) loads. When studying the long-term lateral cyclic behavior of offshore piles, these loads are usually regarded as reciprocating quasi-static loads, i.e., the associated inertia and damping forces can be ignored. Therefore, the p-y model proposed in this chapter is mainly applicable to low-frequency loads such as winds and waves, where the influence of load frequency on model parameters can be ignored.

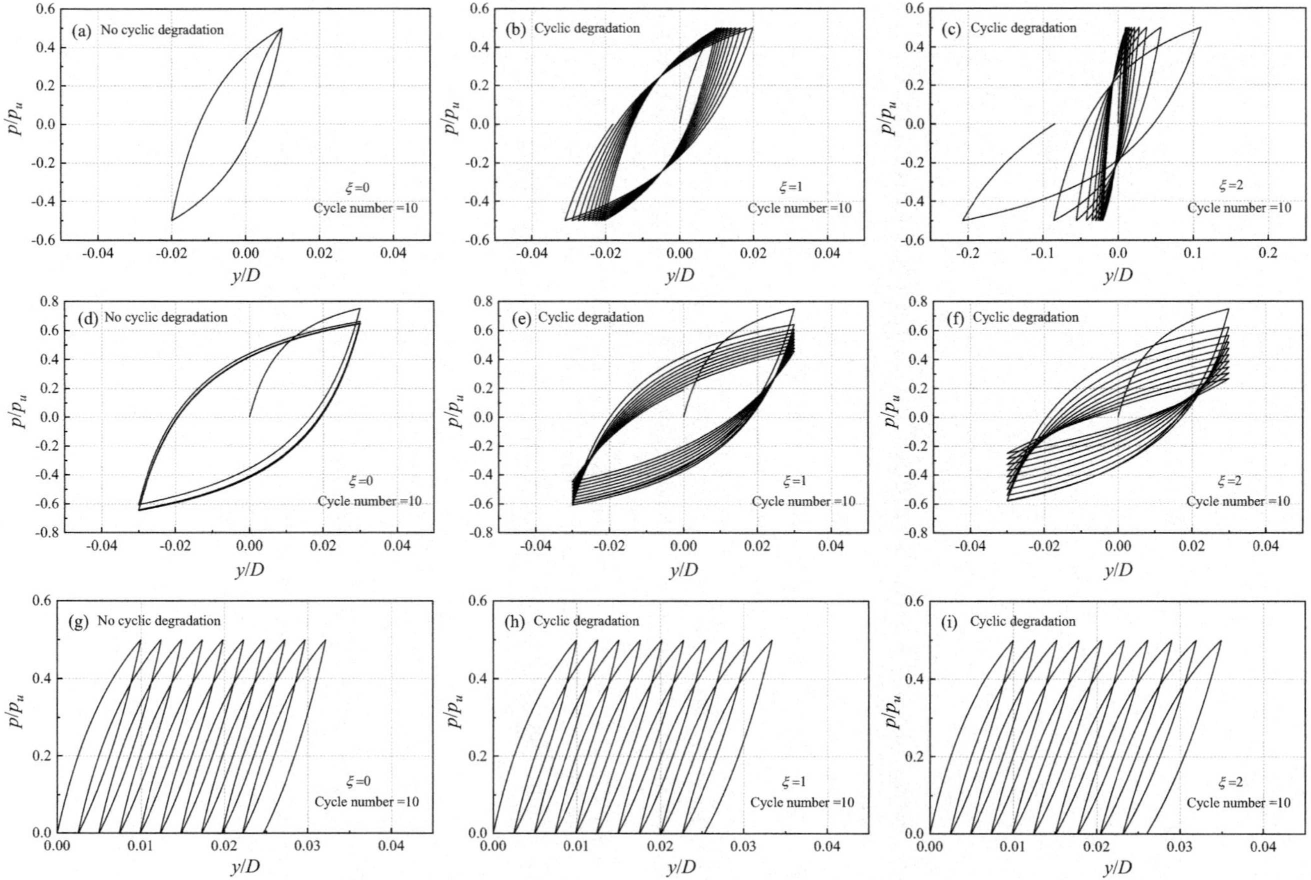

Fig. 3.4 Normalized cyclic p-y curves for different ξ: **a–c** for constant amplitude of p (two-way loading); **d–f** for constant amplitude of y (two-way loading); **g–i** for constant amplitude of p (one-way loading); ($p_u = 100$ kPa, $k^e = 10$ MPa, $\mu = 2$, $D = 1$ m)

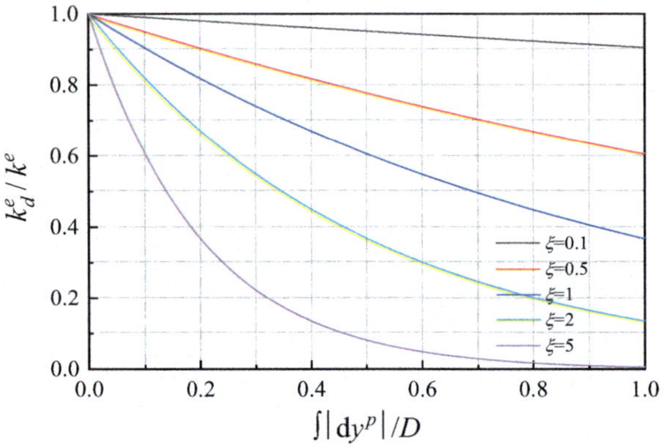

Fig. 3.5 Degradation of elastic resistance coefficient with accumulative plastic displacement

3.3 Incremental Finite Element Calculation for Laterally Loaded Piles

The cyclic p-y elastoplastic model is numerically implemented in a beam on nonlinear Winkler foundation (BNWF) method. The pile and adjoining soil are discretized into elements, with each soil node connected to one spring as shown in Fig. 3.6. The pile is simulated by elastic beam elements, and the pile-soil interaction is simulated by nonlinear spring. The incremental load vector is expressed as ΔF; the incremental soil resistance vector is expressed as ΔP; the incremental displacement vector is expressed as ΔU with two components: horizontal displacement Δv and angular displacement $\Delta\theta$; the global beam stiffness matrix is expressed as K_p. The beam equilibrium equation is given by:

$$K_\text{p} \cdot \Delta U = \Delta F - \Delta P. \tag{3.12}$$

In Eq. (3.12), the incremental soil resistance vector ΔP can be determined by

$$\Delta P = K_\text{s} \cdot \Delta U, \tag{3.13}$$

where K_s represents the global stiffness matrix of the nonlinear spring. Combining Eq. (3.12) with Eq. (3.13), the equilibrium equation is rewritten as:

$$\left(K_\text{p} + K_\text{s}\right) \cdot \Delta U = \Delta F. \tag{3.14}$$

In Eq. (3.14), K_p is obtained by integrating the beam element stiffness matrix K_p^e, i.e.,

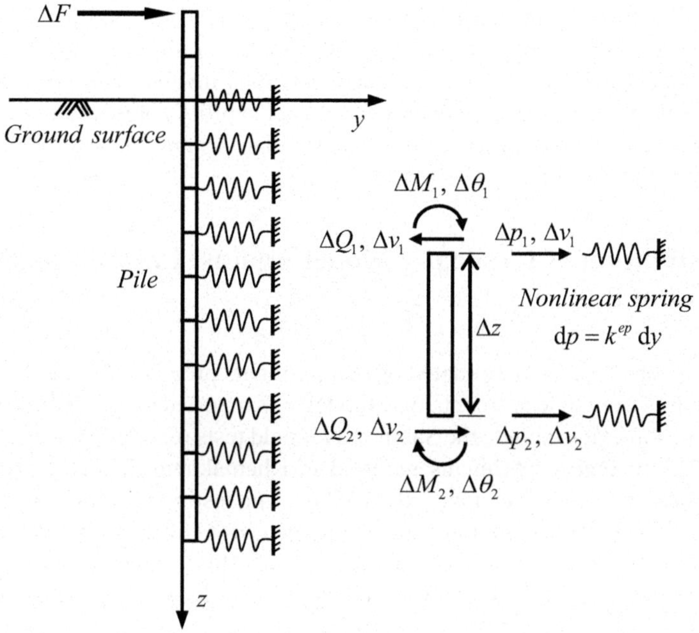

Fig. 3.6 Incremental finite element analysis for laterally loaded piles by the beam on nonlinear Winkler foundation (BNWF) method

$$K_{\mathrm{p}}^{e} = \frac{E_{\mathrm{p}} I_{\mathrm{p}}}{\Delta z^{3}} \begin{bmatrix} 12 & 6\Delta z & -12 & 6\Delta z \\ 6\Delta z & 4\Delta z^{2} & -6\Delta z & 2\Delta z^{2} \\ -12 & -6\Delta z & 12 & -6\Delta z \\ 6\Delta z & 2\Delta z^{2} & -6\Delta z & 4\Delta z^{2} \end{bmatrix}, \tag{3.15}$$

where E_{p} is pile Young's modulus; I_{p} is second moment of area of pile cross-section; Δz is length of beam element. Obviously, K_{p}^{e} remains unchanged during the calculation.

In Eq. (3.14), K_{s} can be obtained by integrating the stiffness matrix of each discrete spring K_{s}^{e} represented as

$$K_{\mathrm{s}}^{e} = k^{\mathrm{ep}} \cdot D \cdot \Delta z. \tag{3.16}$$

It should be noted that K_{s}^{e} changes with the change of the elastoplastic resistance coefficient k^{ep} due to the nonlinearity of the p-y curve, which can be determined using Eq. (3.5).

The incremental displacement vector ΔU can be obtained from Eq. (3.14). Then, the incremental internal force at each element node, including incremental bending moment ΔM and shear force ΔQ, can be determined by $K_{\mathrm{p}} \cdot \Delta U$. Finally, the total displacement, bending moment, and shear force can be obtained by numerical integration. The whole calculation process is realized by compiling a Matlab program. The p-y model based on bounding surface theory only involves the spring cell, which simplifies its numerical implementation and applicability compared to models with the gap and dashpot cells; however, it cannot simulate separation at pile-soil interface.

3.4 Validation of Cyclic p-y Model Against Lateral Loading Test of Pile

The model performance is validated by comparing its predictions with the measured pile responses from field and centrifuge model tests reported in the literature. Selected pile tests in soft clays include the Sabine River field tests reported in Matlock (1970), SOLCYP centrifuge experiments reported in Khemakhem (2012), and centrifuge experiments reported in Yu et al. (2018, 2020). The testing programs are first introduced briefly, and then monotonic and cyclic behaviors of laterally loaded piles are predicted by the developed incremental p-y model. Pile geometries, soil properties, and model parameters for the analyses of test piles are summarized in Tables 3.1 and 3.2.

3.4.1 Matlock's Field Pile Tests

Matlock (1970) conducted lateral monotonic and cyclic load tests on steel pipe piles installed in soft clay soil at a site near the Sabine River. Based on the test results, he proposed the p-y curves for soft clays that were adopted by API RP 2GEO (2014). The test pile had an outer diameter of 0.324 m with a wall thickness of 12.7 mm. The pile total length was 13.1 m with 12.8 m embedded depth. The lateral load was applied at 0.3 m above the ground surface. The test site consisted of marine-deposited, high plasticity clay, interbedded with thin silty sand and sand layers. Zhang et al. (2020) analyzed the results of site investigation and proposed a representative

Table 3.1 Geometry of test piles used for model validation

Tests	Pile head condition	Pile diameter D (m)	Embedded depth L (m)	Loading height H (m)	Wall thickness T (mm)	Elastic modulus E_p (GPa)	Flexural rigidity $E_p I_p$ (MNm^2)
Matlock's field test	Free	0.324	12.8	0.3	12.7	210	31.65
Khemakhem's centrifuge test	Free	0.954	16	2	40	74	889
Yu et al.'s centrifuge test	Free	2	12	2	110	103	30,000

Table 3.2 Soil properties and model parameters

Tests	Loading type	Parameters related to ultimate soil resistance p_u and elastic resistance coefficient k^e			Parameter capturing the shape of p-y backbone curve	Parameter controlling the rate of cyclic degradation
		Undrained shear strength s_u (kPa)	Elastic modulus E_s (kPa)	Effective unit weight γ' (kN/m^3)	μ	ξ
Matlock's field test	Monotonic	$1.54\,z + 9.92$	$600\,s_u$	7	3	0
	Cyclic (two-way)					0.03
Khemakhem's centrifuge test	Monotonic	$1.17\,z + 1.27$	$1000\,s_u$	7	1.4	0
	Cyclic					0 (one-way) 0.25 (two-way)
Yu et al.'s centrifuge test	Monotonic	$1.39\,z$	$1400\,s_u$	6.4	2	0
	Cyclic (two-way)					2

undrained shear strength profile. Correspondingly, a simplified linear shear strength profile described by $s_u = 1.54z+9.92$ (z is soil depth) is used herein. One monotonic loading test and one cyclic loading test (both free-head pile) are analyzed employing the developed model. The relevant parameters of the current analysis are given in Tables 3.1 and 3.2.

3.4.1.1 Monotonic Loading Tests

Figure 3.7a compares the calculated pile head load–displacement curve with the test results (lateral displacement is normalized by pile diameter D). Figure 3.7a shows a good match between the calculated and measured responses for the monotonic load test. The parameter $\xi = 0$ is used (i.e., no cyclic degradation) for the monotonic loading. Figure 3.7b, c compares the calculated lateral deflection and bending moment profiles along the pile and those evaluated from the field measurements for five different pile head load levels. The calculated responses agree well with the test results for all load levels, which demonstrates that the developed p-y model can well predict the increase of lateral deflection and bending moment profiles with increasing the load level, and the location of the maximum bending moment for different load levels. It also demonstrates the ability of the developed model to predict lateral monotonic behavior of the pile in soft clays.

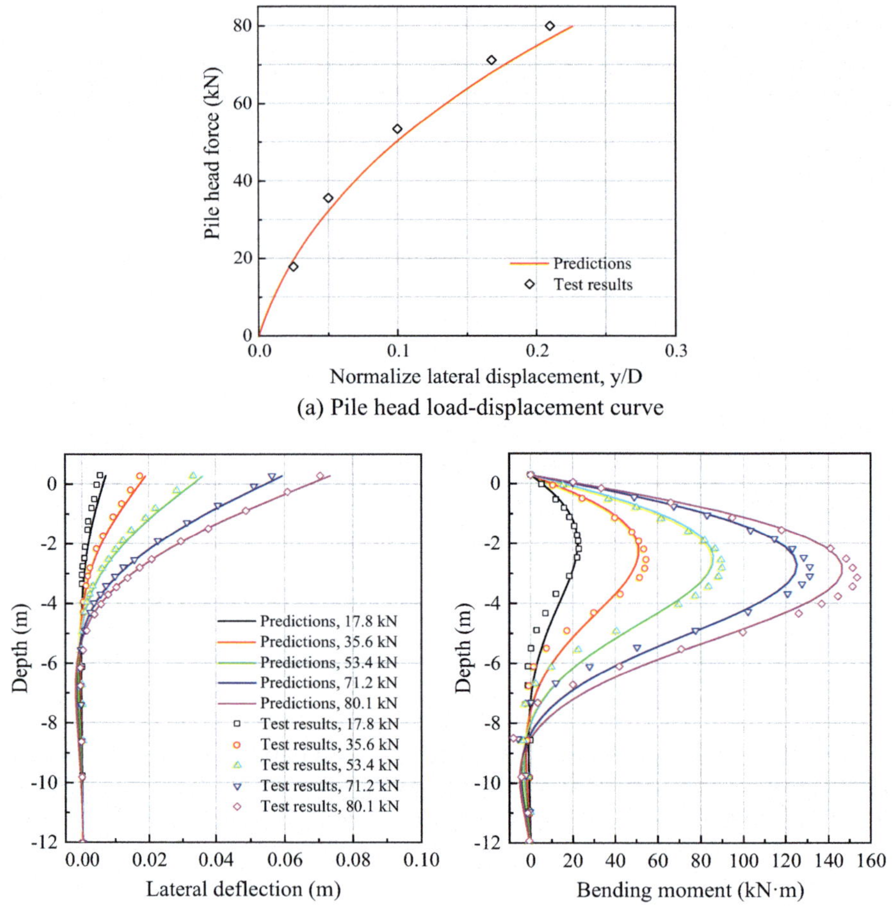

(a) Pile head load-displacement curve

(b) Lateral deflection profile

(c) Bending moment profile

Fig. 3.7 Calculated and measured test results of pile head load–displacement curve, lateral deflection, and bending moment profiles along the pile

3.4.1.2 Cyclic Loading Tests

The load history of cyclic lateral load test was summarized by Zhang et al. (2020) as shown in Table 3.3. The load history consisted of four sinusoidal load epochs, and the cyclic loading level remained constant within each epoch. The cyclic tests were performed in load-controlled mode. The pile was subjected to nonzero average load and cyclic load, and the minimum load was the same for all four load epochs. The minimum load with a negative value means the load reversal direction during cycling. The load level increased with the epoch number. Except epoch 1 with 400 load cycles, the other three epochs comprised 200 load cycles. The period for each load cycle was 20 s. Figure 3.8a, b compares the predictions with test results including

Table 3.3 Cyclic load history of the free-head cyclic test (Zhang et al. 2020)

Epoch	No. of cycles	Min. load (kN)	Max. load (kN)	Ave. load (kN)	Cyc. load (kN)
1	400	– 8.9	17.8	4.45	13.35
2	200	– 8.9	35.6	13.35	22.25
3	200	– 8.9	53.4	22.25	31.15
4	200	– 8.9	60	25.55	34.45

the evolution of the lateral deflection and sectional bending moment along the pile with cyclic loading. Zhang et al. (2020) emphasized that the test results were based on readings of peak strain gauge values during continuous cyclic loading, and hence the results along the depth do not correspond to a given time but for a given period, as indicated by the legend. The first and last load cycles predictions are shown in Fig. 3.8a, b.

The comparison reveals that the calculated lateral deflection and sectional bending moment profiles are in general good agreement with the measured data. Maximum bending moments and their locations agree with the experimental data. This agreement reflects the ability of the developed p-y model to simulate the gradual increase in lateral deflection and bending moment along the pile as cyclic loading progressed due to soil stiffness degradation. It also enabled accurate prediction of the position of the maximum moment, which shifted downward gradually as the cycle number increased. Figure 3.8c shows the predicted pile head load–displacement curve for Epoch 3. Unfortunately, the corresponding experimental data was not available for comparison. The predicted curve exhibits the characteristics of nonlinearity, hysteresis, displacement accumulation, and stiffness degradation. The displacement accumulation stabilizes gradually with cyclic loading, illustrating the model ability to capture lateral cyclic behaviors of the pile in soft clays.

3.4.2 *Khemakhem's Centrifuge Tests*

Khemakhem (2012) reported a set of centrifuge model tests of free-head piles in Kaolin clay under monotonic and cyclic lateral loading. The tests formed part of the SOLCYP project (Puech and Garnier 2017). The model tests were carried out at 50 g using the IFSTTAR centrifuge facility. In prototype scale, the model pile had an outer diameter of 0.954 m with a wall thickness of 40 mm (accounting for increase in pile diameter caused by the coating on strain gauges). The pile total length was 18 m with an embedded depth of 16 m. The lateral load was applied 2 m above the ground surface. The monotonic and cyclic tests were carried out in slightly overconsolidated Kaolin clay. The selected tests were described by Zhang et al. (2020) to validate their models. In present research, the representative soil profile suggested by Khemakhem (2012) is used, i.e., $s_u = 1.17z + 1.27$. The specific parameters of

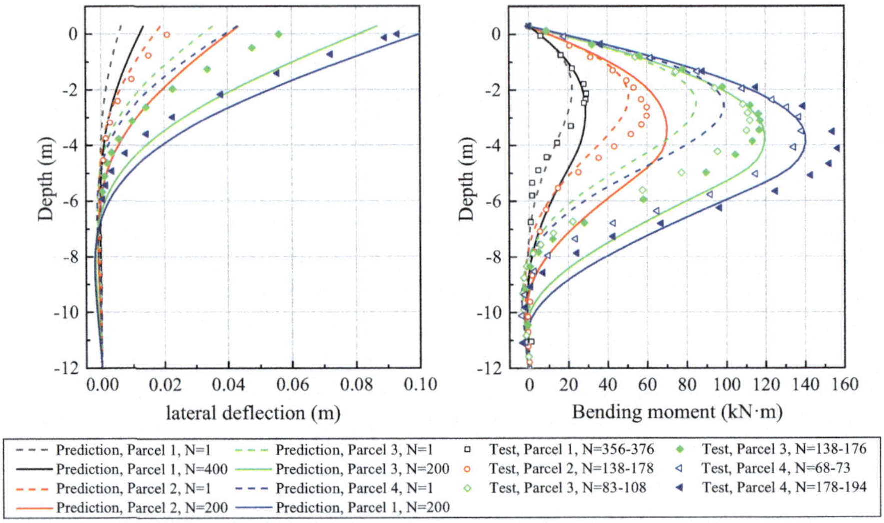

(a) Evolution of lateral deflection (b) Evolution of bending moment

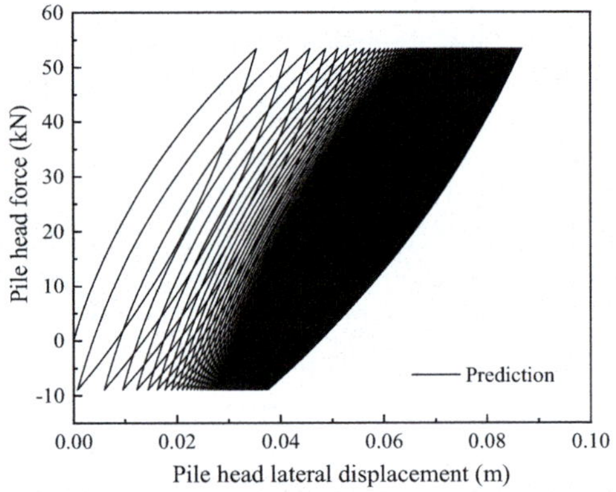

(c) Predicted pile head load-displacement curve for Parcel 3

Fig. 3.8 Calculated and measured test results for the evolution of lateral deflection and sectional bending moment during cyclic loading, and predicted pile head load–displacement curve

pile geometry and soil properties used in the analyses are presented in Tables 3.1 and 3.2.

3.4.2.1 Monotonic Loading Tests

The monotonic tests were carried out in displacement-controlled mode at a constant velocity of 20 mm/s at prototype scale. Figure 3.9a compares the predicted pile head

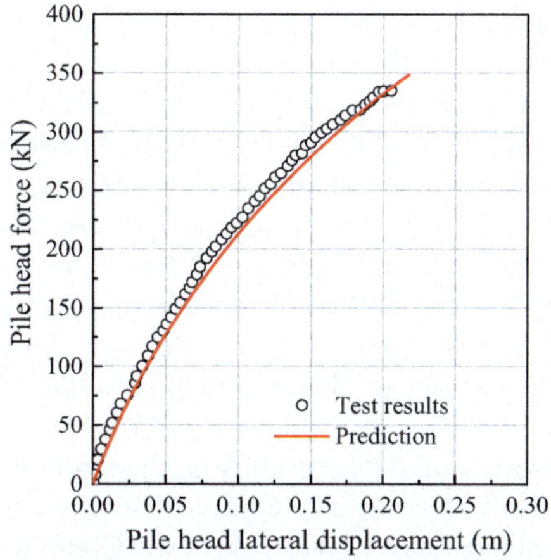

(a) Pile head load-displacement curve

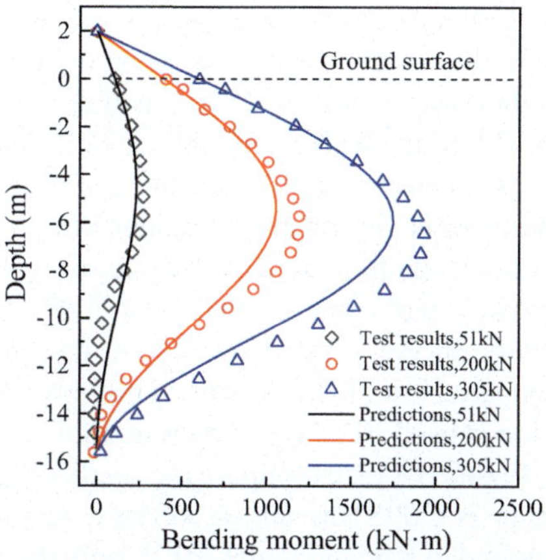

(b) Bending moment profile along the pile

Fig. 3.9 Calculated and measured test results of pile head load–displacement curve and bending moment profile

load–displacement curve against the measured data, which demonstrates good agreement between the two sets. Figure 3.9b compares the calculated bending moment profile along the pile with that established from the strain gauges measurements during the tests at different load levels. The predictions agree with the physical measurements in the general trend despite some deviation.

3.4.2.2 Cyclic Loading Tests

The piles were subjected to cyclic load-controlled tests, in which sinusoidal load with a frequency of 0.25 Hz was applied to the pile head. Two cyclic loading tests are selected to validate the proposed p-y model. Both tests had the same average lateral load of 150 kN, but different cyclic amplitudes: test C09S1ins had a cyclic amplitude of 200 kN, and the load direction was reversed during each cycle (cycle number = 40); and test C10S1ins had a cyclic amplitude of 50 kN, and the lateral load did not reverse direction during cycling (cycle number = 1000).

Figure 3.10a, b compares predictions and test results of pile head load–displacement curves for test C09S1ins. The measured load–displacement curve exhibited a distinctive response pattern in different load cycles; the hysteretic loop for first 20 cycles ($N \leq 20$) was 'thin' and the following cycles were 'fat'. This trend can be seen more clearly from the evolution of pile head displacement with the number of load cycles, where the increase of displacement was first slow and then accelerated as the cyclic loading continued as shown in Fig. 3.11e. It is interesting to note that the calculated response could capture such complex trend. Figure 3.10c shows the calculated pile head load–displacement curve for test C10S1ins. Regrettably, the corresponding experimental results were not available for comparison. The predictions for test C10S1ins without loading reversal exhibit different trends from test C09S1ins with loading reversal. For test C10S1ins, the increase of displacement slows down gradually with the cycle number, and the cumulative displacement of pile head is significantly smaller due to the relatively small cyclic amplitude. The parameter ξ that controls the rate of cyclic degradation of stiffness is taken as 0 for one-way cyclic loading (i.e., test C10S1ins). For test C09S1ins, the loading reversal occurs during asymmetric cyclic loading, however, the reversal is not significant compared with the two-way symmetrical cyclic loading. Hence, a relatively small ξ value is used (0.25) as shown in Table 3.2.

Figure 3.11a, b compare the predicted and measured pile bending moment profiles for tests C09S1ins and C10S1ins, respectively. Due to availability of test results reported in Khemakhem (2012), only the comparisons for 10th cycle of test C09S1ins and 100th cycle of test C10S1ins are presented. The predictions agree with the

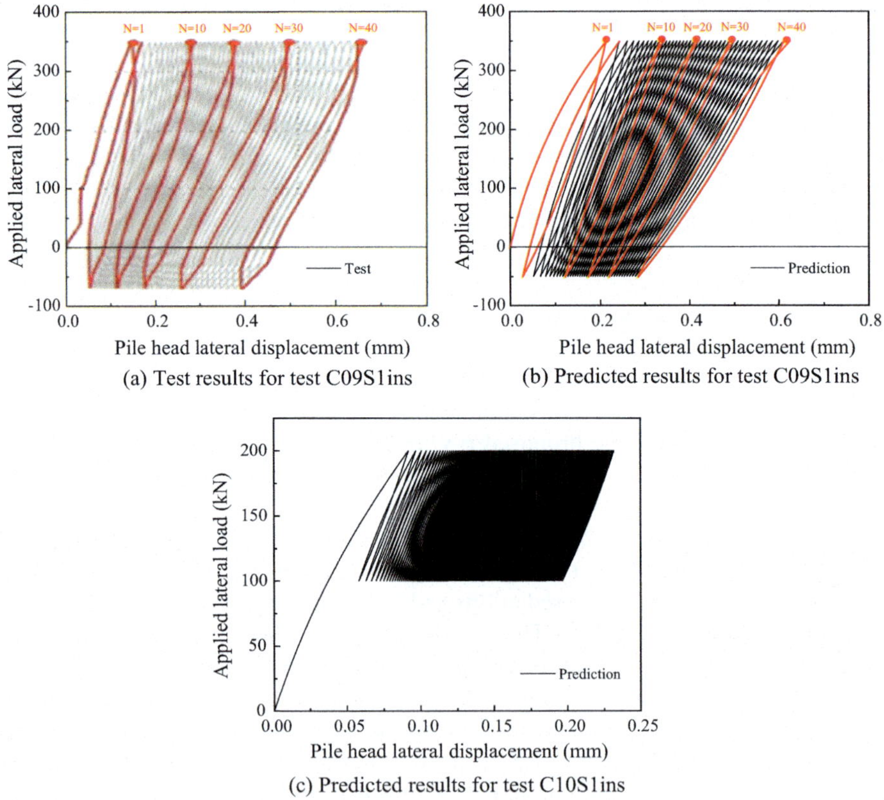

Fig. 3.10 Comparison between predicted and measured pile head load–displacement curves for Khemakhem's centrifuge tests

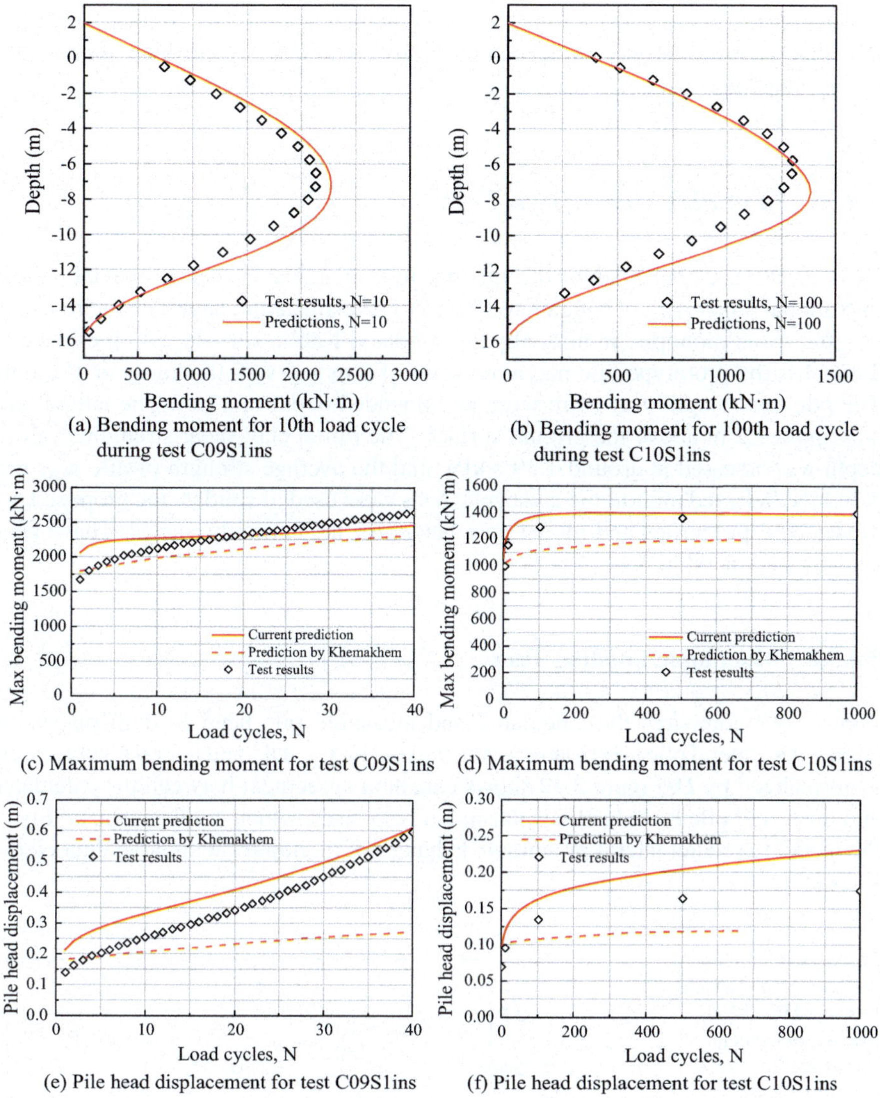

(a) Bending moment for 10th load cycle during test C09S1ins
(b) Bending moment for 100th load cycle during test C10S1ins
(c) Maximum bending moment for test C09S1ins
(d) Maximum bending moment for test C10S1ins
(e) Pile head displacement for test C09S1ins
(f) Pile head displacement for test C10S1ins

Fig. 3.11 Comparison between predicted and measured bending moment profiles, and evolution of maximum bending moment and pile head displacement with the load cycle during tests

measured results in both cases, although the maximum bending moment is slightly overestimated. Figure 3.11c–f compares the evolution of pile maximum bending moment and the pile head deflection with the number of load cycles for tests C09S1ins and C10S1ins, respectively, along with the predictions by Khemakhem (2012). The comparisons demonstrate that the calculated responses agree well with the test results in the general trend despite some deviations. The comparisons illustrate that the proposed p-y model can generally capture the cyclic behavior revealed by the centrifuge tests.

3.4.3 Yu et al.'s Centrifuge Tests

Yu et al. (2018, 2020) conducted centrifuge tests at 50 g at the centrifugal laboratory of National University of Singapore (NUS) to investigate the lateral cyclic response of a free-head monopile in normally consolidated Kaolin soft clay. As presented in Table 3.1, the prototype pile had a diameter of 2 m and wall thickness of 110 mm. The pile total length was 14 m with an embedded depth of 12 m. The lateral load was applied 2 m above the ground surface. The initial clay shear strength s_u along depth was assessed at around 1.39 z kPa, and the average strength profile $s_{u,av}$ was 8.34 kPa. Selected monotonic and cyclic tests were used to validate the proposed p-y model. Soil properties and model parameters for the currently analyses have been summarized in Table 3.2.

3.4.3.1 Monotonic Loading Tests

Figure 3.12 compares the calculated and measured pile head load–displacement curves. The lateral pile head load P is normalized by $s_{u,av}$ DL, and lateral displacement is normalized by D. Figure 3.12 shows excellent agreement between the calculated and measured pile head load–displacement responses, which confirms the ability of the model to predict lateral monotonic behavior of monopiles with large diameter in soft clays.

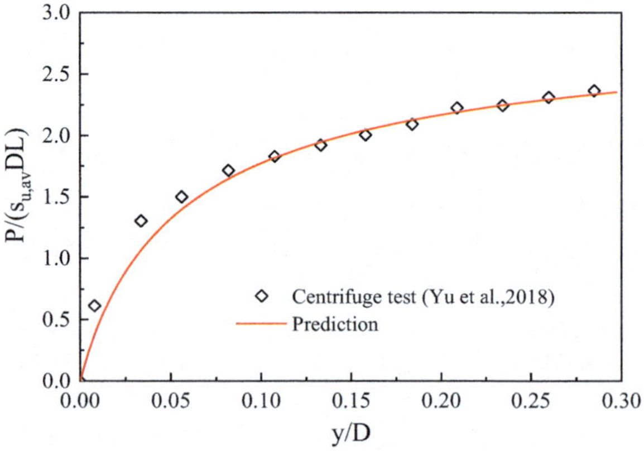

Fig. 3.12 Comparison between predicted and measured pile head load–displacement curves for Yu et al.'s centrifuge tests

3.4.3.2 Cyclic Loading Tests

The selected two-way cyclic tests were performed in displacement-controlled mode, which involved applying 20 load cycles at three different sinusoidal displacement amplitudes of 0.05D, 0.1D, and 0.2D. Figure 3.13 compares the calculated and measured load–displacement curves at the pile head. In general, the calculated and measured responses are in good agreement. Both the calculated and measured curves exhibit significant stiffness degradation during cyclic loading, and the nonlinear hysteretic characteristics became more evident as the displacement amplitude increased. It should be noted that the cyclic degradation stabilized after a certain number of cycles. Einav and Randolph (2005) and Yu et al. (2020) employed the fully remolded ratio between fully remolded and intact state soil resistance to characterize the cyclic degradation phenomenon of the undrained clay. The cyclic degradation stabilizes when the soil resistance is degraded to the fully remolded resistance, i.e., fully remolded soil resistance represents the lower limit of soil resistance after experiencing degradation. Similarly, for the cyclic tests, soil stiffness has a lower bound on degraded value. In present research, the ratio between the elastic resistance coefficient k_{d}^{e} after degradation and initial value k^{e} as shown in Fig. 3.5 is limited to 0.2 to avoid infinite degradation. Figure 3.14 compares the calculated and measured degradation of the pile head secant stiffness with the number of cycles for the three tests with different displacement amplitudes. The secant stiffness is determined by $P/(s_{\mathrm{u,av}}Dy_{\mathrm{max}})$, where y_{max} is the cyclic displacement amplitude. The calculated responses agree with the test results, which demonstrate the rationality of the developed model in simulating the stiffness degradation of the pile in soft clays.

3.5 Summary

Summary: A simplified cyclic p-y elastoplastic model is developed within the framework of the single-surface bounding surface theory. A novel interpolation function is proposed to determine the elastoplastic resistance coefficient for any current resistance point in the cyclic loading path, and the cumulative plastic displacement is introduced to the interpolation function to capture the degradation of the soil stiffness during cyclic loading. The p-y hysteretic curves can be constructed based on the radial mapping rule and movable mapping centers. The cyclic p-y model requires only four parameters with clear physical significance. The model is numerically implemented into the BNWF method, and its performance is validated by calculating the responses of lateral monotonic and cyclic load tests of full scale and model piles installed in soft clays. The good match between predictions and test results has demonstrated the model ability to predict the essential cyclic behavior of laterally loaded piles in soft clays.

Model novelty: The proposed cyclic p-y elastoplastic model is relatively easy to implement numerically because it does not involve the evolution of the size and position of the kinematic hardening yield surface. The cyclic degradation of soft clay

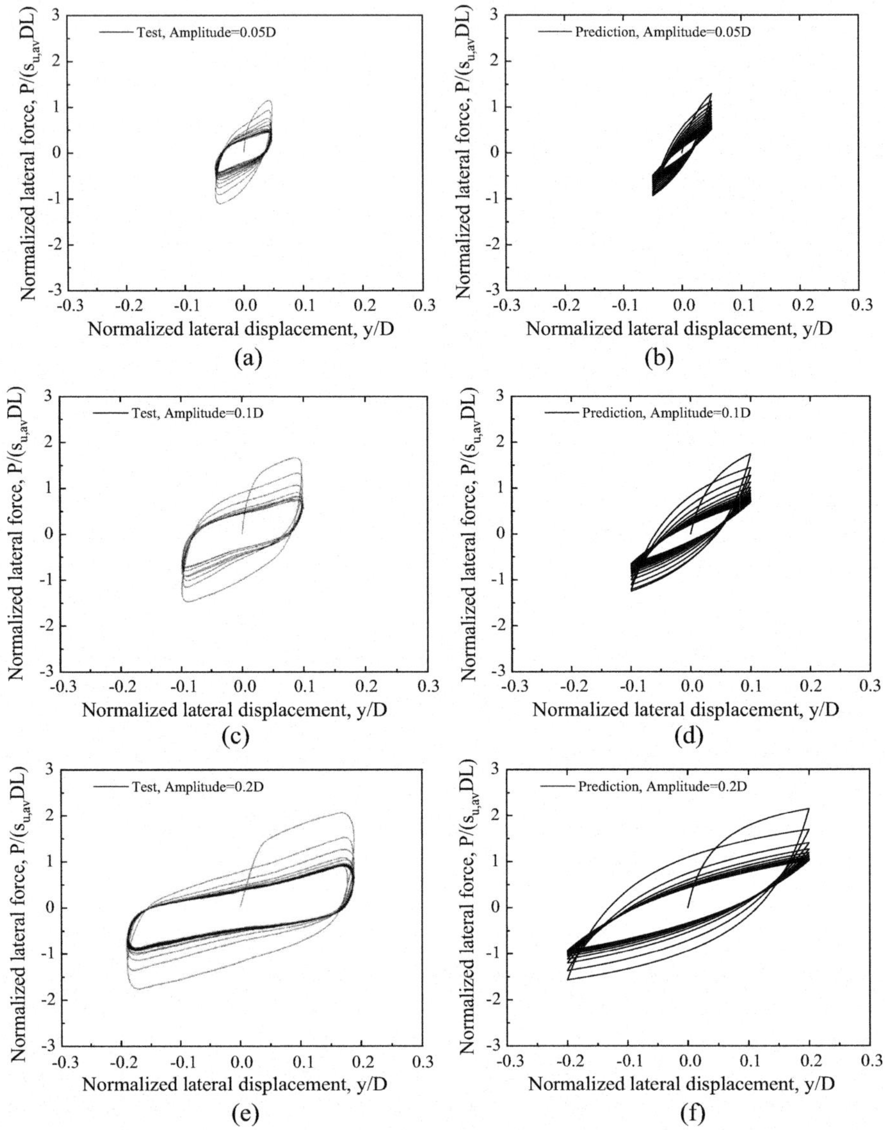

Fig. 3.13 Load–displacement hysteresis curves at pile head with different displacement amplitudes: **a**, **c**, **e** centrifuge test results (Yu et al. 2018); **b**, **d**, **f** predictions

is described using a coefficient of degradation of elastoplastic resistance, which is used to reflect the degradation of soil stiffness and resistance. The developed model can analyze the response of piles in soft clay under different lateral loading patterns including one-way loading, two-way symmetric, and asymmetric loading. The model can capture the main characteristics of pile head load–displacement curve, such as

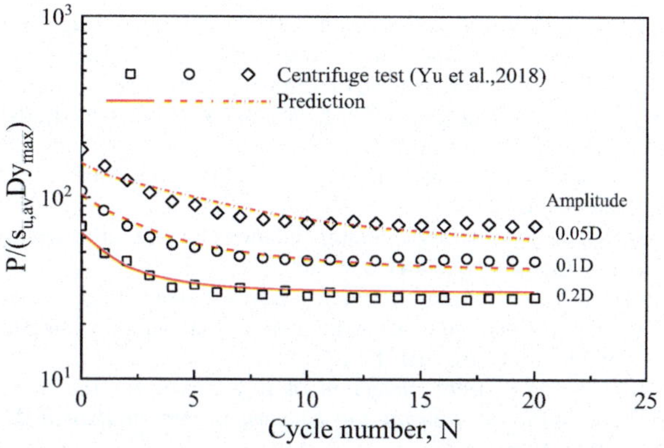

Fig. 3.14 Comparison between the calculated and measured degradations of the pile head secant stiffness with the number of cycles

nonlinearity, hysteresis, displacement accumulation, and stiffness degradation, and predict the evolution of the lateral deflection and section bending moment along the pile with cyclic loading.

Significance and Application: The simulation of proper cyclic p-y hysteretic response is the most challenging aspect in the analysis of piles lateral cyclic responses employing the BNWF method. This chapter proposes a relatively simple and practical cyclic p-y model within the framework of the single-surface bounding surface theory. It offers computational efficiency compared with the 3D finite element method based on advanced dynamic constitutive model. The model parameters can be easily determined from standard soil properties and stress–strain responses measured in direct simple shear (DSS) tests. When applied to engineering practice, the model can improve the design of horizontally loaded piles. However, the proposed p-y model has its limitations. The current version is applicable to soft clay only as it does not capture the strain softening response of stiff clay and can be used for low-frequency loading (e.g., wind and waves) as it ignores inertia and damping forces.

References

Allotey, N., and M.H. El Naggar. 2008. Generalized dynamic Winkler model for nonlinear soil-structure interaction analysis. *Canadian Geotechnical Journal* 45 (4): 560–573.

Anastasopoulos, I., F. Gelagoti, R. Kourkoulis, et al. 2011. Simplified constitutive model for simulation of cyclic response of shallow foundations: Validation against laboratory tests. *Journal of Geotechnical and Geoenvironmental Engineering* 137 (12): 1154–1168.

Andersen, K.H., W.F. Rosenbrand, S.F. Brown, et al. 1980. Cyclic and static laboratory tests on Drammen clay. *Journal of Geotechnical and Geoenvironmental Engineering* 106 (5): 499–529.

API Recommended Practice 2GEO. 2014. *Geotechnical and Foundation Design Considerations*, 1st Edition Addendum 1. American Petroleum Institute.

Arany, L., S. Bhattacharya, J.H. Macdonald, et al. 2016. Closed form solution of Eigen frequency of monopile supported offshore wind turbines in deeper waters incorporating stiffness of substructure and SSI. *Soil Dynamics and Earthquake Engineering* 83: 18–32.

Arvan, P.A., and M. Arockiasamy. 2022. Energy-based approach: Analysis of a laterally loaded pile in multi-layered non-linear elastic soil strata. *Geotechnics* 2 (3): 570–598.
Basack, S. 2010a. Response of vertical pile group subjected to horizontal cyclic load in soft clay. *Latin American Journal of Solids and Structures* 7: 91–103.
Basack, S. 2010b. A boundary element analysis on the influence of Krc and e/d on the performance of cyclically loaded single pile in clay. *Latin American Journal of Solids and Structures* 7: 265–284.
Basack, S. 2015. Design recommendations for pile subjected to cyclic load. *Marine Georesources & Geotechnology* 33 (4): 356–360.
Basack, S., M. Karami, and M. Karakouzian. 2022. Pile-soil interaction under cyclic lateral load in loose sand: Experimental and numerical evaluations. *Soil Dynamics and Earthquake Engineering* 162: 107439.
Bransby, M.F. 1999. Selection of p–y curves for the design of single laterally loaded piles. *International Journal for Numerical and Analytical Methods in Geomechanics* 23 (15): 1909–1926.
Cheng, X.L., X.L. Du, D.C. Lu, et al. 2020. A simple single bounding surface model for undrained cyclic behaviours of saturated clays and its numerical implementation. *Soil Dynamics and Earthquake Engineering* 139: 106389.
Cheng, X.L., T.J. Wang, J.X. Zhang, et al. 2021. Finite element analysis of cyclic lateral responses for large diameter monopiles in clays under different loading patterns. *Computers and Geotechnics* 134: 104104.
Cheng, X.L., M.H. El Naggar, D.C. Lu, et al. 2023. A cyclic p-y elastoplastic model applied to lateral loaded pile in soft clays. *Canadian Geotechnical Journal* 60 (6): 885–901.
Dafalias, Y.F. 1986. Bounding surface plasticity Part I: Mathematical foundation and hypoplasticity. *Journal of Engineering Mechanics* 112 (9): 966–987.
Dafalias, Y.F., and E.P. Popov. 1977. Cyclic loading for materials with a vanishing elastic domain. *Nuclear Engineering and Design* 41 (2): 293–302.
Det Norske Veritas (DNV). 2013. *DNV-OS-J101 Design of Offshore Wind Turbine Structures*. Det Norske, Veritas AS.
Dewaikar, D.M., and P.A. Patil. 2006. Analysis of a laterally loaded pile in cohesion-less soil under static and cyclic loading. *Indian Geotechnical Journal* 36: 181.
Einav, I., and M.F. Randolph. 2005. Combining upper bound and strain path methods for evaluating penetration resistance. *International Journal for Numerical Methods in Engineering* 63 (14): 1991–2016.
El Naggar, M.H., and M. Novak. 1996. Nonlinear analysis for dynamic lateral pile response. *Soil Dynamics and Earthquake Engineering* 15 (4): 233–244.
Fan, C.C., and J.H. Long. 2005. Assessment of existing method for predicting soil response of laterally loaded piles in sand. *Computers and Geotechnics* 32 (4): 274–289.
Gerolymos, N., and G. Gazetas. 2005. Constitutive model for 1-D cyclic soil behavior applied to seismic analysis of layered deposits. *Soils and Foundations* 45 (3): 147–159.
Heidari, M., and M.H. El Naggar. 2018. Analytical approach for seismic performance of extended pile-shafts. *Journal of Bridge Engineering* 23 (10): 04018069.
Heidari, M., M.H. El Naggar, M. Jahanandish, et al. 2014a. Generalized cyclic p–y curve modeling for analysis of laterally loaded piles. *Soil Dynamics and Earthquake Engineering* 63: 138–149.
Heidari, M., M. Jahanandish, M.H. El Naggar, et al. 2014b. Nonlinear cyclic behavior of laterally loaded pile in cohesive soil. *Canadian Geotechnical Journal* 51 (2): 129–143.
Hong, Y., B. He, L.Z. Wang, et al. 2017. Cyclic lateral response and failure mechanisms of semi-rigid pile in soft clay: Centrifuge tests and numerical modelling. *Canadian Geotechnical Journal* 54 (6): 806–824.
Jeanjean, P., Y. Zhang, A. Zakeri, et al. 2017. A framework for monotonic p-y curves in clays. In *Keynote Lecture, OSIG SUT Conference*. Society for Underwater Technology, London.

Jeanjean, P. 2009. Re-assessment of P-Y curves for soft clays from centrifuge testing and finite element modelling. In *Proceedings of the Offshore Technology Conference*, Paper number: OTC-20158-MS.

Khemakhem, M. 2012. *Etude expérimentale de la réponse aux charges latérales monotones et cycliques d'un pieu foré dans l'argile*. Doctoral dissertation. Ecole centrale de Nantes.

Kondner, R.L. 1963. Hyperbolic stress-strain response: Cohesive soils. *Journal of the Soil Mechanics and Foundation Engineering* 89 (1): 115–144.

Liang, F., H. Chen, and Y. Jia. 2018. Quasi-static p-y hysteresis loop for cyclic lateral response of pile foundations in offshore platforms. *Ocean Engineering* 148: 62–74.

Matlock, H. 1970. Correlations for design of laterally loaded piles in soft clay. *Proceedings of the Offshore Technology Conference, Houston, Paper OTC* 1204: 577–594.

McCarron, W.O. 2015. Bounding surface model for soil resistance to planar cyclic lateral pile displacements. *Computers and Geotechnics* 65: 285–290.

McCarron, W.O. 2021. Efficient analysis of cyclic laterally loaded piles. *Results in Engineering* 10: 100213.

Murff, J.D., and J.M. Hamilton. 1993. P-ultimate for undrained analysis of laterally loaded piles. *International Journal of Geotechnical Engineering* 119 (1): 91–107.

Nimbalkar, S., and S. Basack. 2022. Pile group in clay under cyclic lateral loading with emphasis on bending moment: Numerical modelling. *Marine Georesources & Geotechnology* 41 (3): 269–284.

Poulos, H.G., and E.H. Davis. 1980. *Pile Foundation Analysis and Design*, vol. 397. New York: Wiley.

Puech, A., and J. Garnier. 2017. *Design of Piles Under Cyclic Loading: Solcyp Recommendations*. John Wiley & Sons.

Randolph, M.F., and G.T. Houlsby. 1984. The limiting pressure on a circular pile loaded laterally in cohesive soil. *Geotechnique* 34 (4): 613–623.

Rathod, D., K. Muthukkumaran, and T.G. Sitharam. 2016. Response of laterally loaded pile in soft clay on sloping ground. *International Journal of Geotechnical Engineering* 10 (1): 10–22.

Reese, L.C., and H. Matlock. 1956. Non-dimensional solutions for laterally loaded piles with soil modulus assumed proportional to depth. In *Proceedings of the 8th Texas Conference on Soil Mech. and Foundation Engg.*, Special Publication No. 29, Bureau of Engg. Research, University of Texas, Austin.

Su, D., and W.M. Yan. 2013. A multidirectional p–y model for lateral sand–pile interactions. *Soils and Foundations* 53 (2): 199–214.

Vesic, A.S. 1961. Bending of beams resting on isotropic elastic solid. *Journal of Engineering Mechanics* 87 (2): 35–53.

Vucetic, M., and R. Dobry. 1988. Degradation of marine clays under cyclic loading. *Journal of Geotechnical Engineering* 114 (2): 133–149.

Yang, Y., X. Gao, W. Wu, et al. 2021. A simplified method for analysis of laterally loaded piles considering cyclic soil degradation. *Advances in Civil Engineering* 2021 (1): 9096540.

Yu, J., C.F. Leung, and M. Huang. 2018. Response of monopile foundations under cyclic lateral loading in normally consolidated clay. *International Journal of Offshore and Polar Engineering* 28 (04): 411–418.

Yu, J., J. Zhu, K. Shen, et al. 2020. Bounding-surface-based p-y model for laterally loaded piles in undrained clay. *Ocean Engineering* 216 (2): 107997.

Zhang, J., and H. Wang. 2022. Development of offshore wind power and foundation technology for offshore wind turbines in China. *Ocean Engineering* 266: 113256.

Zhang, Y., and K.H. Andersen. 2017. Scaling of lateral pile p-y response in clay from laboratory stress-strain curves. *Marine Structures* 53: 124–135.

Zhang, Y., K.H. Andersen, and P. Jeanjean. 2020. Verification of a framework for cyclic p-y curves in clay by Hindcast of Sabine River, SOLCYP and centrifuge laterally loaded pile tests. *Applied Ocean Research* 97 (102085).

Zhu, B., Z. Zhu, T. Li, et al. 2017. Field tests of offshore driven piles subjected to lateral monotonic and cyclic loads in soft clay. *Journal of Waterway, Port, Coastal, and Ocean Engineering* 143 (5): 05017003.

Chapter 4
Lateral Cyclic Responses of OWT Monopile in Soft Clays

It is very important to analyze the lateral cyclic response of monopiles for the design of offshore wind turbines (OWTs). A three-dimensional finite element method for lateral cyclic responses of large-diameter monopiles is developed based on a simplified constitutive model of clays that has been successfully encoded into the ABAQUS software package (as detailed in Chap. 2). The applicability of this method is validated by simulating the existing centrifugal model test. The method can predict the nonlinear hysteresis responses of monopiles in clays during cyclic loading, and the evolution of bending moment and lateral deflection profile of monopile with loading cycles. Various lateral cyclic loads with different loading patterns including one-way and two-way loading, symmetric and asymmetric loading, variable-amplitude and constant-amplitude loading are applied to monopiles. The impact of different cyclic loading patterns on lateral responses of large-diameter monopiles is systematically investigated. The research results can provide some reference for the engineering design of large-diameter monopiles under lateral cyclic loads.

4.1 Introduction

Offshore wind power has become a promising solution for global energy development. Monopiles are widely used for supporting offshore wind turbines (OWTs), which account for more than 75% of currently installed OWTs (Hamilton et al. 2013; Esteban et al. 2015). Differing from the conventional piles, the piles supporting OWTs usually have the diameters ranging between 3 and 8 m, which can be regarded as large-diameter monopiles. In marine environment, these monopiles mainly are subjected to lateral cyclic loads due to winds, waves, and tides. In current guidelines for offshore wind turbines, requirements of serviceability limit state (SLS) typically dominate the whole design (Arany et al. 2015; Liao et al. 2018). In terms of SLS, monopiles are often designed to not exceed 0.25° of tilt or rotation at the mudline (or other similar

X. Cheng et al., *Dynamic Analysis of Offshore Wind Turbine Foundations in Soft Clays*,
https://doi.org/10.1007/978-981-97-9454-6_4

value as dictated by the turbine manufacturer) which is typically induced by lateral cyclic loads during the design life of the OWTs (Det Norske Veritas (DNV) 2013). Therefore, it is necessary to propose a method to analyze lateral cyclic responses of monopiles for the design of OWTs foundation.

At present, the simplified *p–y* curve methods proposed by available design standards are generally used to calculate lateral responses of piles subjected to cyclic loads (Det Norske Veritas (DNV) 2013; American Petroleum Institute (API) 2007). However, the *p–y* curves in design standards are usually obtained by model tests of small diameter piles (usually less than 2 m), hence the size effect of large-diameter monopiles is not considered, which may lead to inaccurate analysis results. In addition, an empirical reduction factor is introduced usually to modify the ultimate lateral soil resistance to account for cyclic effect, which is too crude to consider the cyclic load level, cyclic load pattern, cyclic loading history, etc. As a result, conventional *p–y* methods from codes have been questioned, and new computational methods have been developed by many researchers.

The cyclic degradation factor related to the number of cycles and the level of cyclic loads was proposed to improve the conventional *p–y* methods (Rajashree and Sundaravadivelu 1996; Fan and Long 2005; Dewaikar and Patil 2006; Zhu et al. 2017). The cyclic lateral responses were also analyzed by developing new cyclic *p–y* hysteresis curves based on nonlinear Winkler model. In this way, many factors including soil nonlinearity, pile-soil contact nonlinearity, radiation damping, and soil stiffness degradation can be considered (Gerolymos et al. 2009; Memarpour et al. 2010; Heidari et al. 2014a, b; Choi et al. 2015; Liang et al. 2018). However, the p-y curve method does not consider the continuity of soil, and it is difficult to apply to special boundary conditions.

The degradation stiffness model was proposed to assess the long-term performance of OWT monopiles under cyclic lateral loading, the impacts of a series of factors such as pile geometry, loading conditions, embedment length, and the loading cycles number on the pile cyclic performance were investigated (Achmus et al. 2009; Kuo et al. 2011; Depina et al. 2015; Ma et al. 2017; Yang et al. 2018; Chen et al. 2020a, b). This method is actually using the degradation of soil modulus to reflect the effect of cyclic loading on strength and stiffness of soils. It cannot track the time history of cyclic deformation for monopiles, so it can't reveal the mechanism of pile-soil interaction during cyclic loading.

Lateral cyclic responses of monopiles have also been analyzed by using some advanced soil constitutive models and establishing finite element model of interaction between piles and soils. Bourgeois et al. (2010) performed three-dimensional finite element computations for lateral cyclic responses of piles using a constitutive model that combines a linear isotropic elastic law with only one single deviatoric plastic mechanism. Lateral cyclic responses of a vertical pile embedded in a dry dense sand were studied by Giannakos et al. (2012) by performing three-dimensional finite element analyses with a nonlinear kinematic hardening soil model. Hong et al. (2017) investigated the cyclic lateral response and failure mechanisms of semi-rigid pile in soft clay using a hypoplastic clay model considering the cyclic degradation effects. Shao et al. (2019) studied the degradation of the lateral bearing capacity of piles in soft clay subjected to cyclic lateral loading by using a modified kinematic hardening constitutive model capturing the cyclic degradation of soft clay.

Generally speaking, there are still many obstacles in using advanced constitutive model to analyze cyclic lateral response of piles although it has many advantages. For example, the dynamic constitutive model is usually complicated because of various hardening rules involved, which leads to the difficulty of numerical implementation and the relative low calculation efficiency. Therefore, it is necessary to use simple and applicable constitutive model to describe the cyclic behavior of soil. In addition, the lateral load applied to OWT monopiles is usually not a regular one-way or two-way cyclic load, but a variable-amplitude irregular load. How the difference of load action mode will affect the lateral responses of piles is also lack of systematic research. As a result, main objectives of this chapter focus on: (a) develop and validate a three-dimensional finite element method for lateral cyclic responses of large-diameter monopiles based on a simplified constitutive model of clay; (b) investigate the impact of different load action modes on lateral responses of large-diameter monopiles.

4.2 Numerical Simulation Method and Verification

4.2.1 Brief Introduction of Centrifuge Tests

Centrifuge model tests of monopiles in Malaysia kaolin clay were reported by Yang et al. (2019). The open-ended model pile was made of 6061 aluminum alloy pipe with an elastic modulus of 68.9 GPa. The outer diameter and the total length of the model pile were 0.059 m, 0.83 m, respectively. The centrifuge acceleration of 100*g* is adopted, so it represents a prototype pile of 5.9 m diameter and 83 m length, as shown in Fig. 4.1. The pile was deemed semi-rigid with a relative pile-soil stiffness of 0.0043 based on the definition proposed by Poulos and Hull (1989). The Malaysia kaolin clay has a liquid limit of 80% and plastic limit of 35%. The unit weight is 15.5–16.4 kN/m^3, and internal friction angle is 23°. The undrained shear strength S_u varies almost linearly with depth of soil layer, which is represented by $S_u = 1.26\,Z$ (Z denotes the depth of soil layer).

The lateral static monotonic loading test was firstly performed for the model pile with a loading velocity of 0.003 m/s. Then three lateral cyclic loading episodes numbered C1, C2, and C3 were applied in turn to the pile, and the loading position is 0.23 m (prototype size) away from the mud surface. Test C1, C2, C3 had a loading amplitude from 0.25MN to 1MN, from 0.45 MN to 1.75 MN, from 0.45 MN to 3.7 MN, respectively. The cycle number of loading for Tests C1, C2, and C3 was 100, 100, and 180, respectively. The loading frequency was 0.2 Hz for all tests. The reconsolidation was allowed during the interval between different cyclic episodes.

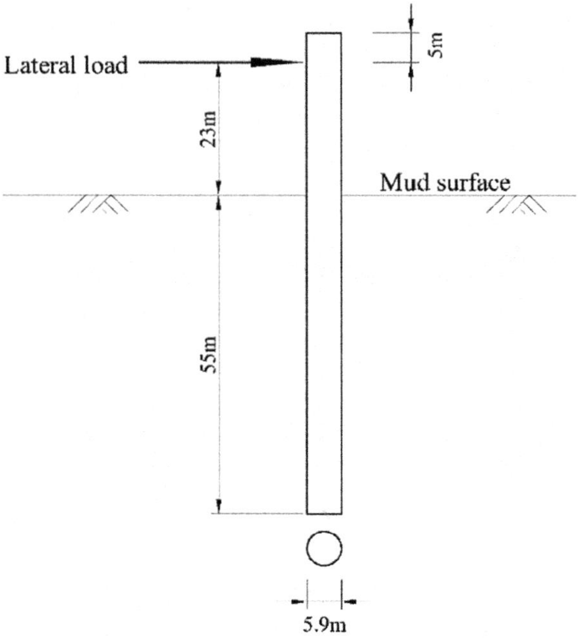

Fig. 4.1 Diagrammatic sketch of the model pile

4.2.2 Numerical Simulation of Centrifuge Tests

The 3D finite element model was established using the ABAQUS software package to simulate the above centrifuge model tests. Considering the symmetries of the geometries and loading conditions during the tests, only a half portion of the pile and soil domain was meshed to improve the calculation efficiency, as shown in Fig. 4.2. Trial computation results indicate that a model of size $20D \times 14D$ (the cross section is a semi-circle, and D is the pile diameter) is sufficient to avoid boundary effects on simulation results. The pile and the soils were all simulated using 14,512 8-node linear brick elements (C3D8). Normal horizontal constraints were applied to the vertical boundaries, and fixed constraints were applied to the bottom boundary. The top boundary was fully free. The pipe section of the monopile was replaced by a solid section pile with equivalent bending stiffness (Achmus et al. 2009). Lateral loads are applied to the pile according to the loading condition in the above centrifugal tests.

The interaction between the pile and the surrounding soil was modeled by the contact-pairs in ABAQUS. In this approach, the master surface is defined as a surface that belongs to the material that either is relatively stiff or has a finer mesh geometry, and the slave surface corresponds to the softer material or material with a coarser mesh. In this study, the pile surface was defined as the master surface, and the soil surface in contact with the pile was defined as the slave surface. The separation between the interface elements was allowed when they were subjected to tension.

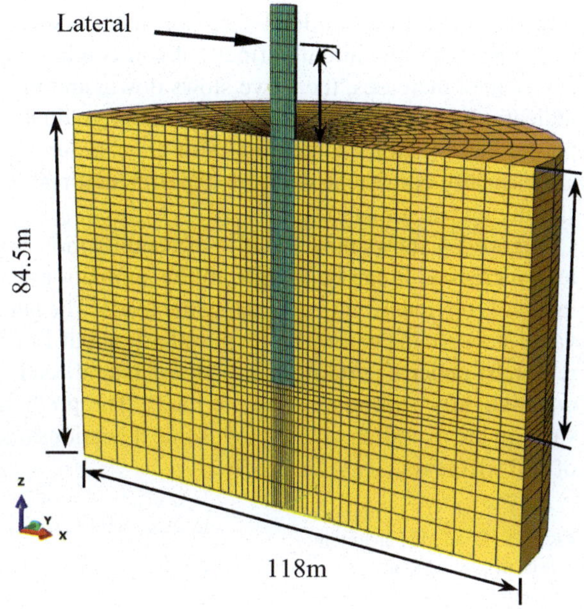

Fig. 4.2 Three-dimensional finite element model of large-diameter monopile

When the two surfaces were in contact, the interface behavior was governed by Coulomb's friction theory, and the friction coefficient was assumed to be 0.3.

The material behavior of the monopile with an equivalent solid section is simulated using linear elastic constitutive model. The elastic compression modulus E is 18.44 GPa, the material density is 2700 kg/m^3, and Poisson's ratio is 0.3. The cyclic behavior of soil was simulated by the simplified version of single bounding surface model proposed by authors mentioned in Chap. 2. Parameter $A_0 = 2\ S_u$, and $S_u = 1.26\ Z$, hence A_0 increases linearly along the depth Z of soil layer. According to the soil properties in the above centrifugal tests, the elastic compression modulus $E = 400\ S_u$, and elastic shear modulus $G = 0.5E \big/ (1 + \nu)$, where Poisson's ratio ν is 0.5 under an undrained condition, and at this case $G = E \big/ 3$. It means that G increases also linearly along the depth Z of soil layer. Empirically, parameters μ and ξ are usually between 0 and 5. μ and ξ are set as 2 and 1 respectively in the simulation.

4.2.3 Comparisons of Predictions and Centrifuge Model Test Results

4.2.3.1 Monotonic Loading Tests

Figure 4.3 shows the comparison between the current simulation and centrifugal tests 'results for load–displacement curves of pile head under lateral monotonic loading. The displacement has been normalized by the pile diameter. It can be seen from the figure that the load–displacement curve is more steep at initial loading, which indicates the initial stiffness of soil is relatively large during loading. As the lateral load increases, the curve slows down, and the displacement increases faster, indicating the degradation of soil stiffness gradually. It reflects the nonlinear and plastic deformation characteristics of clays around the pile. The simulation results are basically consistent with the experimental results although the soil stiffness is slightly underestimated.

Figure 4.4 shows the evolution of bending moment along pile shaft as the lateral load increases. It can be seen from the simulation results that the bending moment increases obviously with increasing the loads. The bending moment reaches the peak value at a certain position along pile shaft (about 17 m below the mud surface in this research) under a certain magnitude of lateral load, and the position point moves down slightly with increasing the loads. The predictions agree with the measured value in the overall trend although some deviations due to the slight underestimation of soil stiffness.

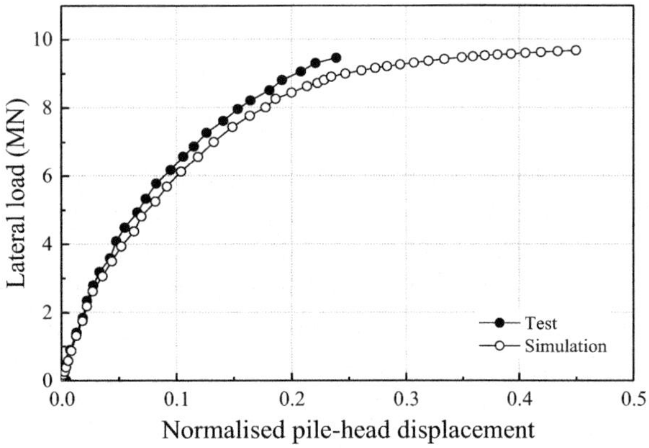

Fig. 4.3 Load–displacement curves of pile head for monotonic loading

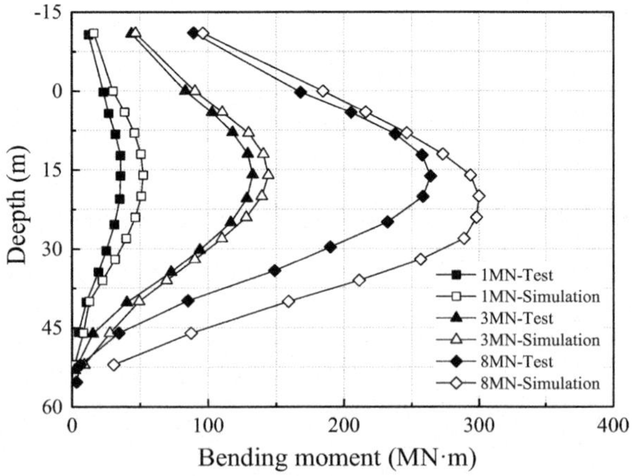

Fig. 4.4 Distribution of bending moment along pile shaft with increasing loads

4.2.3.2 Cyclic Loading Tests

The comparison between the current simulation and centrifugal tests results for load–displacement curves of pile head under lateral one-way cyclic loading is shown in Fig. 4.5. It can be seen from the figure that significant cumulative displacement of pile head occurs under cyclic loading, and it increases gradually as the cyclic load level increases from Test C1 to C3. The predicted hysteresis loops reveal the nonlinear responses of the monopile foundation in clays, and it becomes more significant as the cyclic load level increases. The predicted results are generally in agreement with the test results in every episode although some deviations in Fig. 4.5b ($\xi = 1$). The predicted cumulative displacement during Test C2 and C3 is relatively larger than measured value. The reason may be that the reconsolidation is allowed during the interval between three different cyclic episodes during tests, which enhances the strength and stiffness of soils around pile. However, the effect of soil reconsolidation is not considered in the numerical simulation; hence, the strength and stiffness of soil are underestimated to a certain extent. In addition, in order to investigate the influence of parameter ξ on the numerical simulation, the predicted load–displacement curves with different ξ are also shown in Fig. 4.5c and d ($\xi = 2$ and 3). It can be seen that the predicted displacement of pile head becomes larger as ξ increases, which is mainly because the increase of ξ accelerates the soil stiffness degradation and strain accumulation.

Figure 4.6 shows the comparison between measured and computed bending moment profiles at the 1st and 90th cycles of loading. In general, the predictions are in agreement with the measured results. The numerical model predicts well the shape of the moment distribution and the increase of the bending moments with increasing number of cycles. The numerical model is also capable of predicting the depth of the maximum bending moment.

Lateral cyclic responses of large-diameter monopiles under two-way cyclic loading are also simulated to validate the three-dimensional finite element method furtherly. Three lateral two-way cyclic loading episodes are applied in turn to the pile, and the peak load for every episode is the same as the previous one-way loading centrifuge model tests reported by Yang et al. (2019). The parameter values of the soil constitutive model are completely consistent with the above numerical simulation for one-way cyclic loading. However, due to the lack of centrifugal test results under lateral two-way cyclic loading, only the predicted load–displacement curves of pile head are shown in Fig. 4.7. Despite the lack of quantitative comparison, it can be seen that the numerical method can qualitatively predict the basic behaviors of monopiles under lateral two-way cyclic loading, such as the hysteretic behavior and stiffness degradation.

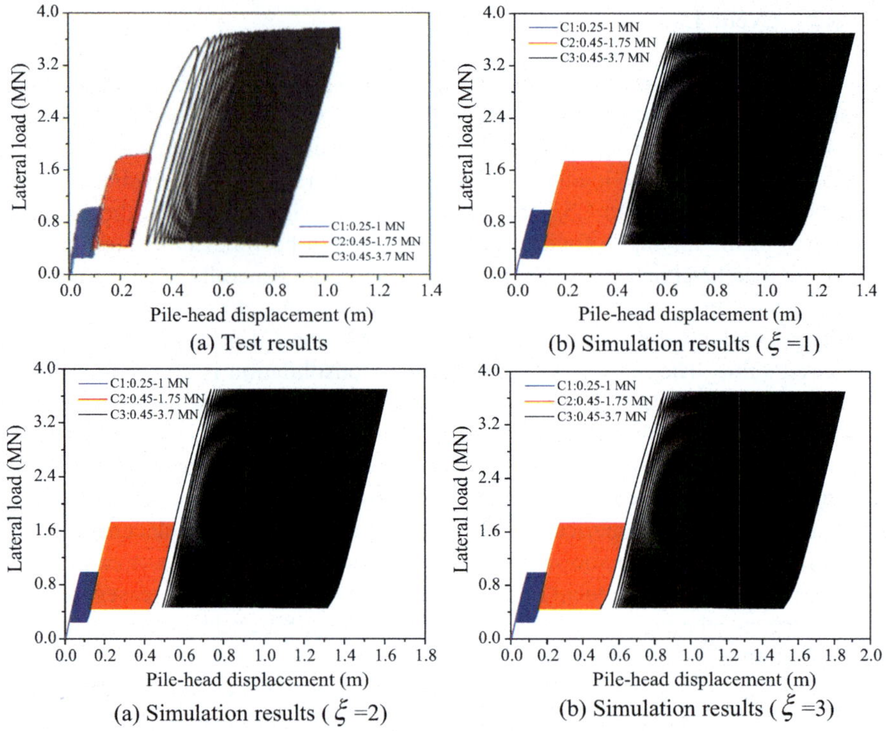

Fig. 4.5 Load–displacement curves of pile head for lateral cyclic loading

Fig. 4.6 Evolution of bending moment profile with increasing the number of loading cycles *N*

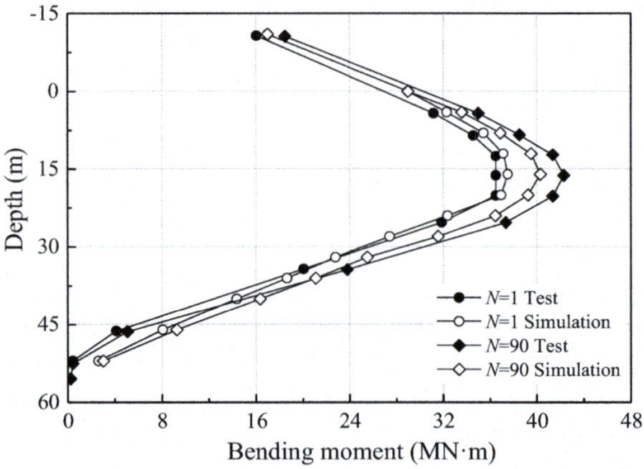

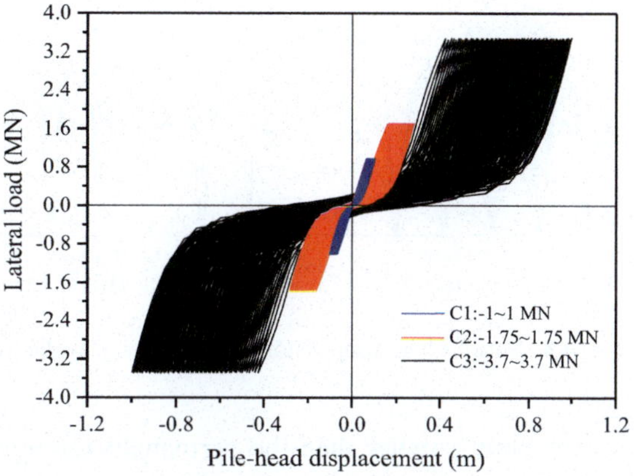

Fig. 4.7 Predicted load–displacement curves of pile head for lateral two-way cyclic loading

4.3 Predicting Lateral Cyclic Responses of Monopiles Under Different Loading Patterns

F_{SLS} is defined as the magnitude of lateral load at the serviceability limit state (SLS) of OWTs when the rotation of monopiles at the mudline reaches the 0.25 degree. Here, $F_{SLS}/F_{ult} = 0.32$ (F_{SLS} is 3.06 MN and F_{ult} is 9.64 MN). In the following section, different patterns of lateral cyclic loads will be applied to the monopiles, and the load magnitude will be set with reference to F_{SLS}. The three-dimension finite element model in Sect. 4.2 above is still used to investigate the impact of different loading patterns on lateral responses of large-diameter monopiles.

4.3.1 One-Way Cyclic Loading

Different patterns of load-time histories for one-way cyclic loading are shown in Fig. 4.8. Five semi-sinusoidal cyclic loads with different amplitudes of 0.1, 0.3, 0.5, 0.7, and 0.9 times F_{SLS} is set, as shown in Fig. 4.8a. In Fig. 4.8b, the loads' magnitude after unloading is 0, 0.1, and 0.2 times F_{SLS}, respectively. In Fig. 4.8c, the variable-amplitude cyclic load is set with the amplitude ranging from 0.1 to 0.9 times F_{SLS}.

Figure 4.9 shows load-rotation and rotation-time curves of monopiles under various levels of one-way loading shown in Fig. 4.8a. The load-rotation curves show significant nonlinear hysteretic characteristics during cyclic loading. For any given loading level, the rotation angle increases with the number of loading cycles. We define the rotation that can be recovered after unloading as cyclic rotation, the unrecoverable rotation as permanent rotation, and the maximum value of rotation as the peak rotation, as shown in Fig. 4.10. Obviously, the peak rotation equals to the sum of cyclic rotation and permanent rotation. It can be seen that the cyclic rotation is

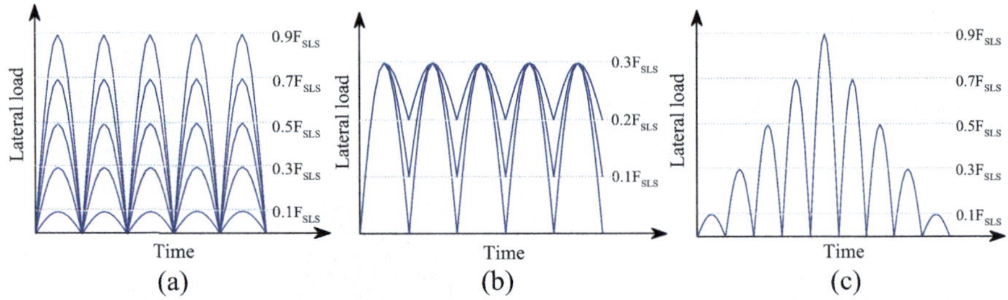

Fig. 4.8 Various load-time histories for one-way cyclic loading

significantly smaller than the permanent rotation in Fig. 4.9b; hence, the gradual increasing permanent rotation is the main reason for reaching the allowable value of serviceability limit state during the design life of the OWT. In addition, both cyclic rotation and permanent rotation increases with the level of loading, and the permanent rotation increases more significantly than the cyclic rotation. Obviously, the higher the cyclic load level, the less the number of loading cycles for reaching the serviceability limit state. For the case of soil and pile in this study, when the load level exceeds 0.3 F_{SLS}, the rotation will reach the serviceability limit state within 100 loading cycles. The accumulation of rotation is very slow within the load level of 0.1 F_{SLS}. The number of loading cycles due to winds and waves could exceed 10^8 over the lifetime of the OWT (Achmus et al. 2009). However, it is almost unrealistic to complete the calculation of 10^8 cycles using the current three-dimensional finite element method because of the huge calculation cost. Despite such limitations, the method based on advanced soil model proposed in this chapter can help people better understand the lateral cyclic behaviors of large-diameter monopiles.

Figure 4.11 shows the evolution of the bending moment and lateral deflection profiles of monopile with increasing the number of loading cycles for the load level of 0.3 F_{SLS}. It can be seen from Fig. 4.11a that the bending moment increases

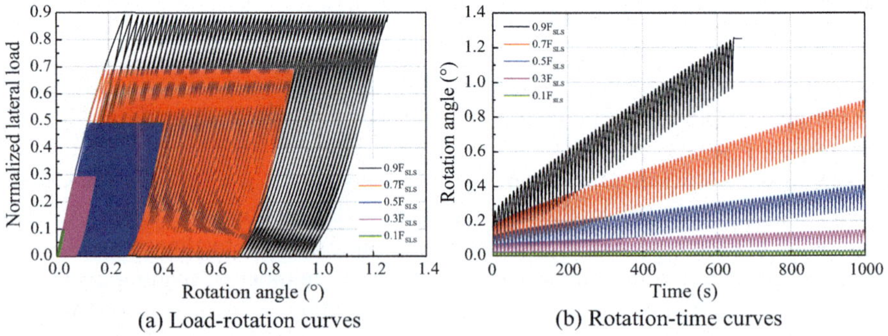

Fig. 4.9 Load-rotation and rotation-time curves at the mud surface under load-time histories shown in Fig. 4.8a

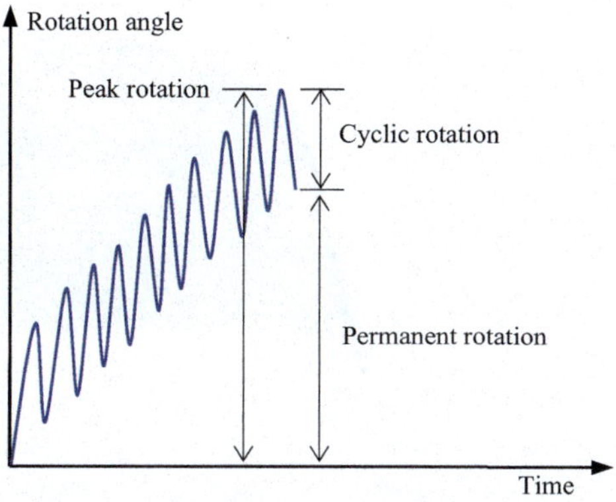

Fig. 4.10 Illustration of rotation angle of monopile at the mud surface

significantly when the loading cycles is less than 25, however, it tends to be stable when the loading cycles are more than 25. The main reason is that lateral peak soil resistance tends to be stable after a certain number loading cycles although the deflection is still increasing, as shown in the cyclic *p*–*y* curve in Fig. 4.12. The lateral deflection increases continuously with the loading cycles in Fig. 4.11b. There is a zero deflection point and it gradually moves down with cyclic loading. The lateral deflection increases inversely below this point. Similar results can be found at other load levels. The predictions agree well with the findings of Achmus et al. (2009) and Hong et al. (2017) for the semi-rigid pile exhibiting an intermediate behavior between the flexible pile and the rigid pile.

Figure 4.13 shows load-rotation and rotation-time curves of monopiles under the load-time history shown in Fig. 4.8b. It can be seen that the peak rotation is different although the same peak load. In the first unloading, the greater the load reduction, the smaller the permanent rotation. However, as the number of cycles increases,

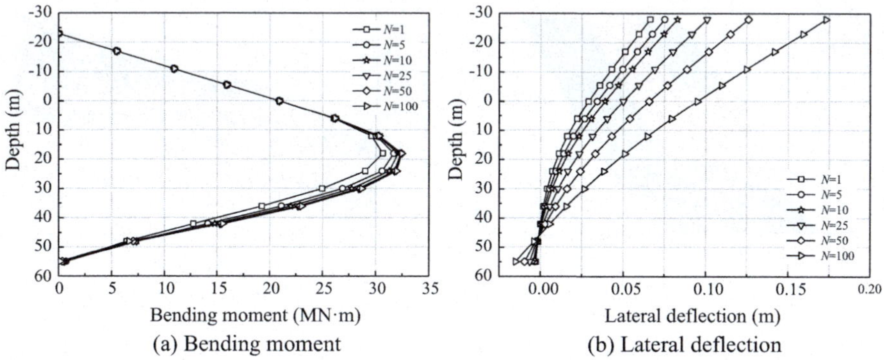

Fig. 4.11 Evolution of bending moment and lateral deflection with loading cycles (load level of 0.3 F_{SLS}) for one-way cyclic loading

Fig. 4.12 Typical cyclic *p–y* curves

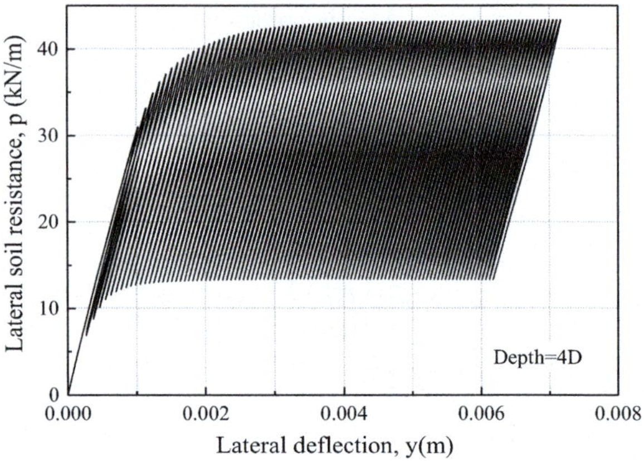

the rotation accumulation is faster for the greater load reduction. When the number of cycles reaches 100 (Time = 1000 s), the permanent rotation is almost the same for these three cases, but the peak rotation is larger for greater the load reduction. The main reason is that the reloading after each unloading will cause larger soil cumulative plastic deformation and then lead to the greater cyclic rotation for the greater load reduction.

Figure 4.14 shows load-rotation and rotation-time curves of monopiles under the variable-amplitude one-way loading shown in Fig. 4.8c. It can be seen that nonlinear hysteretic characteristics become more significant with the amplitude increasing from 0.1 to 0.9 times F_{SLS}. The peak rotation increases more significant rather than permanent rotation. The gradual increasing peak rotation will cause the monopile to reach the serviceability limit state during the design life of the OWT. It should be noted that previous loading history has a certain influence on the rotation caused by subsequent loading. The same load amplitude will cause greater peak rotation after

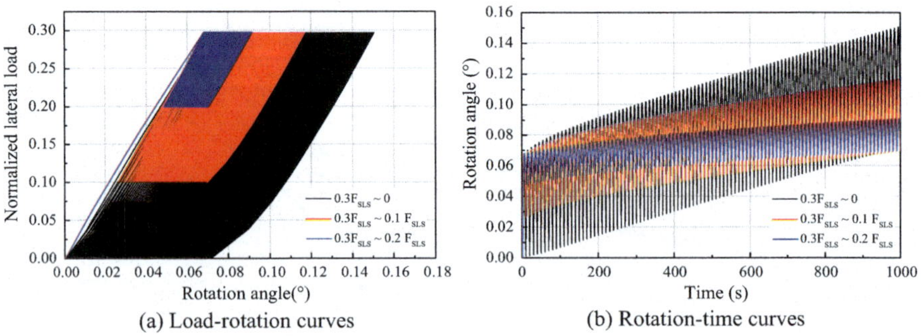

Fig. 4.13 Load-rotation and rotation-time curves at the mud surface under load-time histories shown in Fig. 4.8b

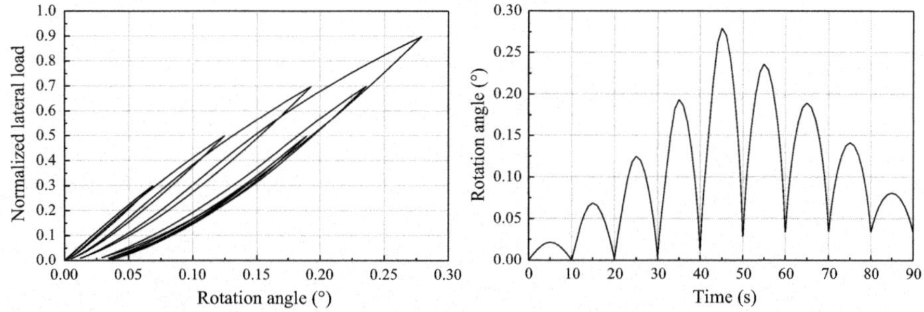

(a) Load-rotation curves (b) Rotation-time curves

Fig. 4.14 Load-rotation and rotation-time curves at the mud surface under load-time histories shown in Fig. 4.8c

experiencing the maximum load amplitude of 0.9 F_{SLS}. However, the permanent rotation will hardly increase as the load amplitude decreases gradually.

4.3.2 Two-Way Cyclic Loading

Different patterns of load-time histories for two-way cyclic loading are shown in Fig. 4.15. Symmetric cyclic loads are set with five different amplitudes of 0.1, 0.3, 0.5, 0.7, and 0.9 times F_{SLS}, as shown in Fig. 4.15a. Five loading cases are set for asymmetric cyclic loading, the positive load amplitudes are 0.3 F_{SLS} for all cases, and negative load amplitudes are set as 0, 0.1, 0.2, 0.3, and 0.4 times F_{SLS}, respectively, as shown in Fig. 4.15b. The variable-amplitude cyclic loading is set with the amplitude ranging from 0.1 to 0.9 times F_{SLS} in Fig. 4.15c.

Figure 4.16 shows load-rotation and rotation-time curves of monopiles under various levels of symmetric two-way loading shown in Fig. 4.15a. Being different from the one-way cyclic loading, the rotation accumulates almost symmetric at both the front and rear faces of the pile shaft during cyclic loading. When the load is reduced to 0 during unloading, the cumulative rotation returns to 0. Hence, the peak rotation almost equals to the cyclic rotation. The gradual increasing cyclic rotation is the main reason for reaching the allowable value of serviceability limit state during the design life of the OWT. The increasing cyclic rotation attributes to the degradation of soil strength and stiffness with increasing the cycle number. Meanwhile, it also indicates the ability of the soil constitutive model to simulate the cyclic degradation properties. As the number of loading cycles increases, the nonlinear response of monopiles becomes more and more significant. The straight line-shape of the cyclic loops gradually changes to the S-shape, which is consistent with the computed results of Heidari et al. (2014a, b). The main reason is that progressive gap forms at the front and rear faces of the pile shaft during cyclic loading, which may lead to lateral load deformation loops that are 'pinched' with relatively little hysteretic energy

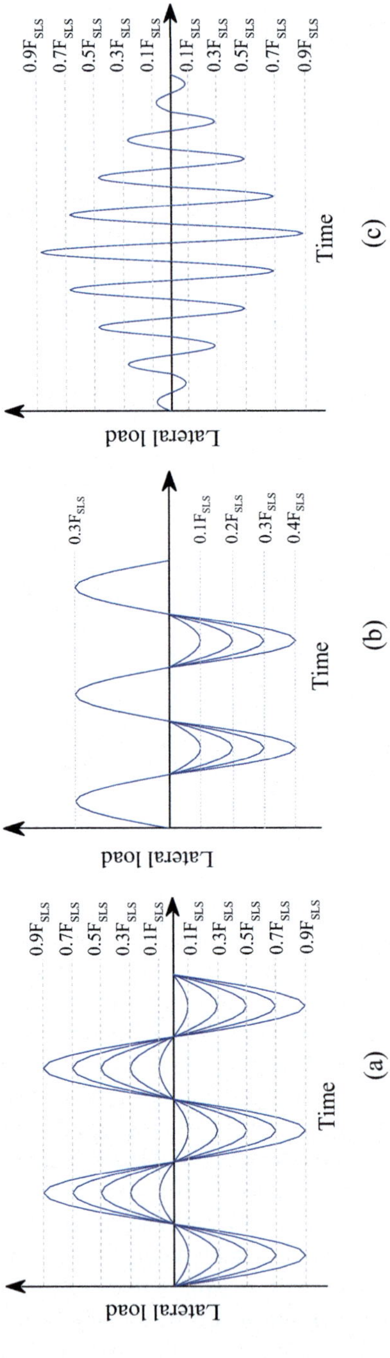

Fig. 4.15 Different patterns of load-time histories for two-way cyclic loading

dissipation. Figure 4.17 shows that the bending moment and lateral deflection at both the front and rear faces of the pile increase almost symmetrically with the number of loading cycles (load amplitude of 0.3 F_{SLS}). The magnitude of bending moment and lateral deflection will increase significantly as the cyclic load level increases, but the evolution law of them with the number of cycles has no significant changes.

Figure 4.18 shows load-rotation and rotation-time curves of monopiles under various levels of two-way asymmetric loading shown in Fig. 4.15b (only the last 5 cycles for the rotation-time curves are plotted for clarity). It can be seen that the accumulation of rotation is also asymmetric at the front and rear faces of the pile shaft. The rotation accumulation is more significant at the side of larger load. The load-rotation response on the negative side is almost linear when the negative load amplitude equals to 0.1 F_{SLS}. The nonlinear hysteretic characteristics become more significant gradually with increasing the load level on the negative side. When the negative load amplitude equals to 0.4 F_{SLS}, the rotation accumulation on the negative side is even more significant than that on the positive side. It is worth noting that the negative loading seems to have a certain effect on the positive load-rotation responses. The larger the negative load amplitude is, the smaller the positive peak rotation is. In addition, the permanent rotation corresponding to lateral load of 0 gradually develops to the negative side from the positive side as the negative load amplitude increases.

Figure 4.19 shows load-rotation and rotation-time curves of monopiles under variable-amplitude two-way symmetric loading shown in Fig. 4.15c. The load-rotation and rotation-time curves are basically symmetrical in the figures. Compared with variable-amplitude one-way loading case shown in Fig. 4.14, the maximum peak rotation is basically the same as the former, but the permanent rotation is less than the former. Due to the influence of previous loading history, the same load amplitude before and after the maximum load amplitude of 0.9 FSLS causes different peak rotations, which is similar to the one-way loading case. The reason may be that previous loading causes the degradation of soil stiffness and strength, which will cause greater peak rotation for the same load amplitude.

4.4 Summary

Cyclic lateral responses for large-diameter monopiles in clays under different loading patterns are analyzed by three-dimensional finite element method. The key conclusions can be summarized as follows:

(1) A three-dimensional finite element method for lateral cyclic responses of large-diameter monopiles is proposed based on a simplified bounding surface constitutive model of clay that has been embedded in the ABAQUS software package. The applicability of this method is validated by simulating the existing centrifugal model test. The method can predict the nonlinear hysteresis responses of the monopile in clays during cyclic loading, and the evolution of bending moment and lateral deflection profile of monopile with loading cycles.

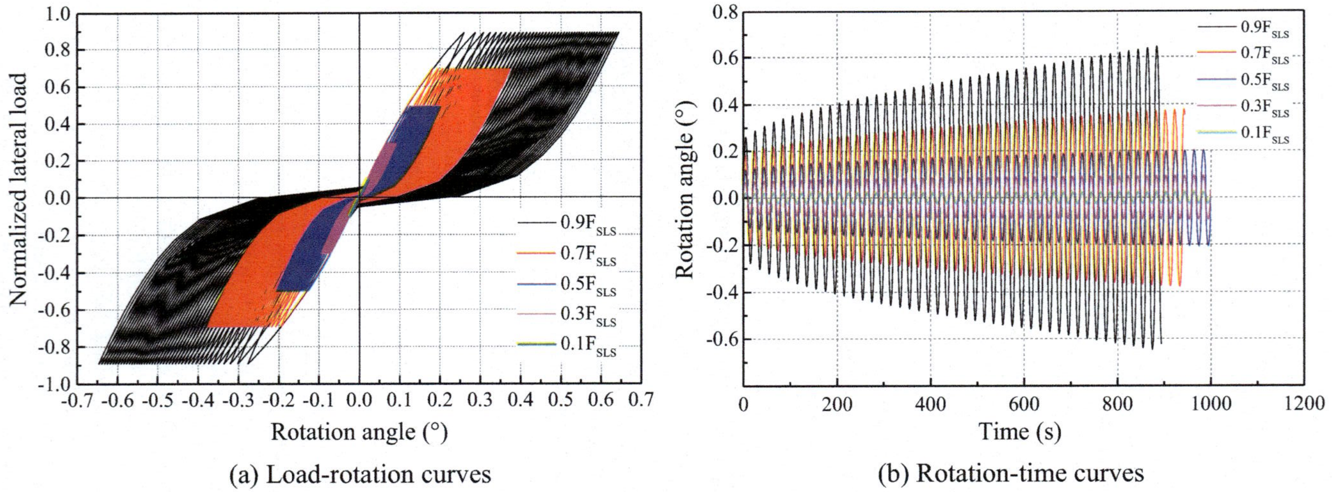

Fig. 4.16 Load-rotation and rotation-time curves at the mud surface under load-time histories shown in Fig. 4.15a

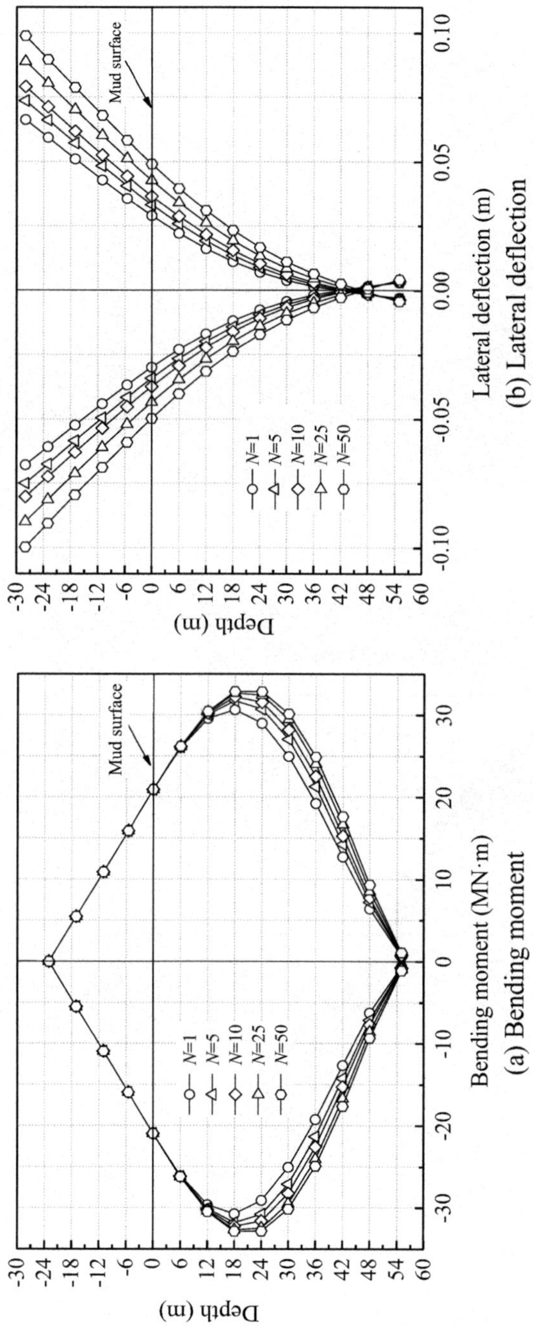

Fig. 4.17 Evolution of bending moment and lateral deflection with loading cycles (load level of 0.3 F_{SLS}) for two-way cyclic loading

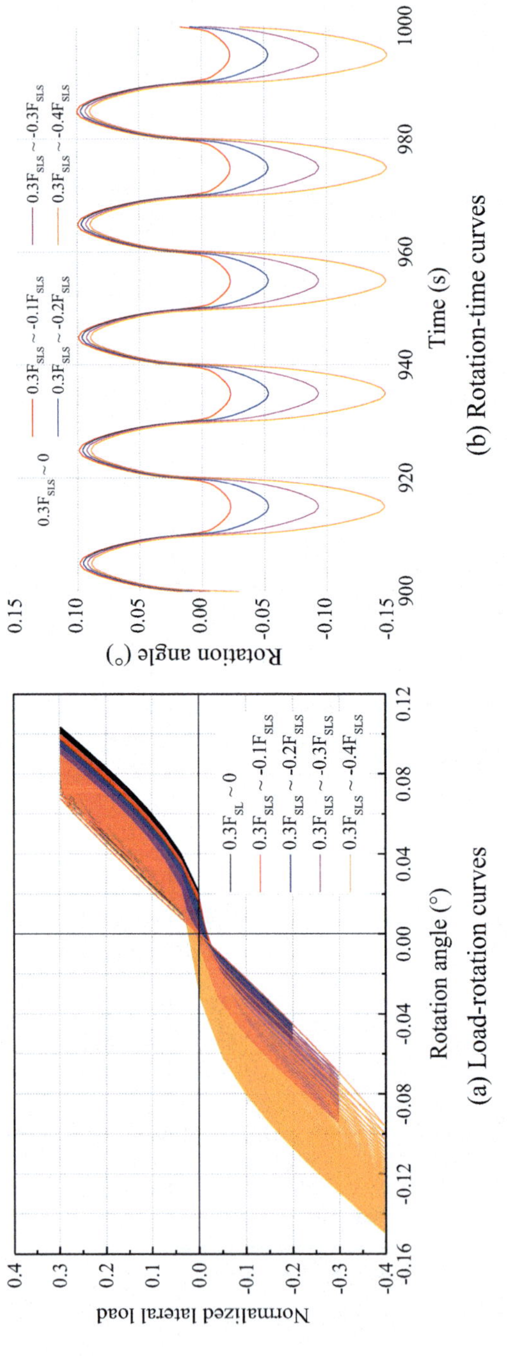

Fig. 4.18 Load-rotation and rotation-time curves at the mud surface under load-time histories shown in Fig. 4.15b

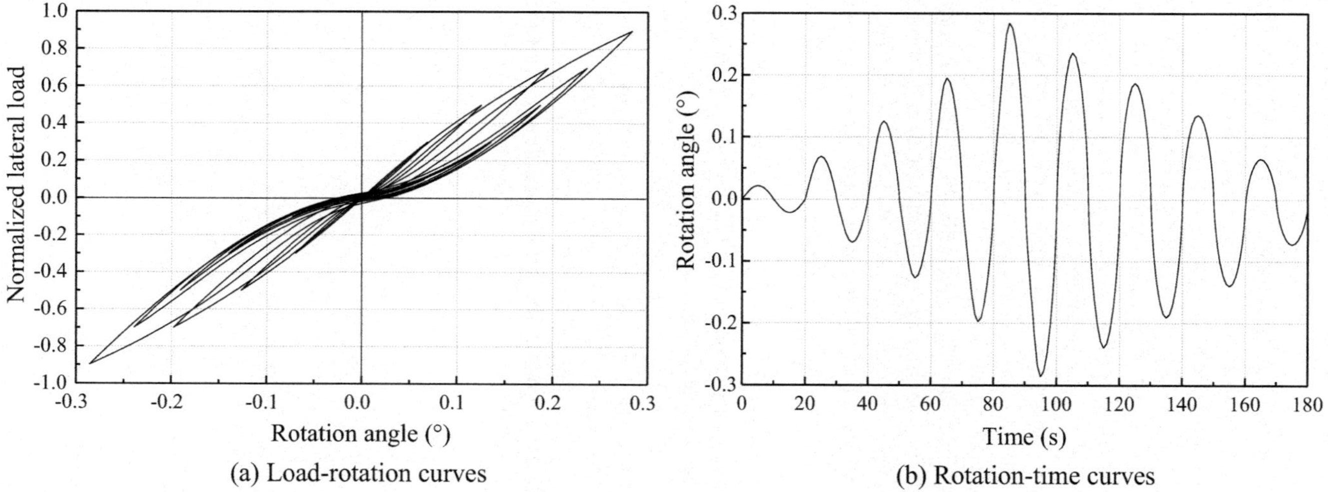

Fig. 4.19 Load-rotation and rotation-time curves at the mud surface under load-time histories shown in Fig. 4.15c

(2) For one-way cyclic loading shown in Fig. 4.8a, the cyclic rotation is significantly smaller than the permanent rotation; hence, the gradual increasing permanent rotation is the main reason for reaching the allowable value of serviceability limit state during the design life of the OWT. The higher the cyclic load level, the less the number of loading cycles for reaching the serviceability limit state.
(3) For symmetric two-way cyclic loading shown in Fig. 4.15a, the rotation accumulates almost symmetric at both the front and rear faces of the pile shaft during cyclic loading. The gradual increasing cyclic rotation is the main reason for reaching the allowable value of serviceability limit state, which can contribute to the degradation of soil strength and stiffness with the cycle number. The nonlinear response of monopiles becomes more and more significant, and the straight line-shape of the cyclic loops gradually changes to the S-shape as the number of loading cycles increases.
(4) For the one-way loading shown in Fig. 4.8b, the rotation accumulation is faster and the peak rotation is larger for the greater load reduction with increasing the number of cycles although the same the peak load. The main reason is that the reloading after each unloading will cause larger soil cumulative plastic deformation and then lead to the greater cyclic rotation for the greater load reduction.
(5) The simulation results for the two-way asymmetric loading indicate that the negative loading seems to have a certain effect on the positive load-rotation response. The larger the negative load amplitude is, the smaller the positive peak rotation is.
(6) The simulation results for variable-amplitude one-way and two-way loading indicate that previous loading history has a certain influence on the rotation caused by subsequent loading. The same load amplitude will cause greater peak rotation after experiencing the maximum load amplitude, which can be attributed to the degradation of soil stiffness and strength caused by previous loading.

References

Achmus, M., Y.S. Kuo, and K. Abdel-Rahman. 2009. Behavior of monopile foundations under cyclic lateral load. *Computers and Geotechnics* 36 (5): 725–735.

American Petroleum Institute (API). 2007. Recommended practice for planning, designing and constructing fixed offshore platforms, RP2A-WSD. Washington.

Arany, L., S. Bhattacharya, J.H.G. Macdonald, et al. 2015. A critical review of serviceability limit state requirements for monopile foundations of offshore wind turbines. In *Offshore Technology Conference Houston.*

Bourgeois, E., M.H.J. Rakotonindriana, A. Le Kouby, et al. 2010. Three-dimensional numerical modelling of the behaviour of a pile subjected to cyclic lateral loading. *Computers and Geotechnics* 37 (7–8): 999–1007.

Chen, L., X. Yang, L. Li, W. Wu, et al. 2020a. Numerical analysis of the deformation performance of monopile under wave and current load. *Energies* 13 (23): 6431.

Chen, D., P. Gao, S. Huang, et al. 2020b. Static and dynamic loading behavior of a hybrid foundation for offshore wind turbines. *Marine Structures* 71: 102727.

Choi, J.I., M.M. Kim, and S.J. Brandenberg. 2015. Cyclic p-y plasticity model applied to pile foundations in sand. *Journal of Geotechnical and Geoenvironmental Engineering* 141 (5): 04015013.

Depina, I., T.M.H. Le, G. Eiksund, et al. 2015. Behavior of cyclically loaded monopile foundations for offshore wind turbines in heterogeneous sands. *Computers and Geotechnics* 65: 266–277.

Det Norske Veritas (DNV). 2013. DNV-OS-J101 Design of Offshore Wind Turbine Structures. Det Norske, Veritas AS.

Dewaikar, D.M., and P.A. Patil. 2006. Analysis of a laterally loaded pile in cohesion-less soil under static and cyclic loading. *Indian Geotechnical Journal* 36: 181.

Esteban, M.D., J.S. Lopez-Gutierrez, V. Negro, et al. 2015. Offshore wind foundation design: Some key issues. *Journal of Energy Resour Technology* 137(5): 051211.1–051211.6.

Fan, C.C., and J.H. Long. 2005. Assessment of existing method for predicting soil response of laterally loaded piles in sand. *Computers and Geotechnics* 32 (4): 274–289.

Gerolymos, N., S. Escoffier, G. Gazetas, et al. 2009. Numerical modeling of centrifuge cyclic lateral pile load experiments. *Earthquake Engineering and Engineering Vibration* 8 (1): 61–76.

Giannakos, S., N. Gerolymos, and G. Gazeta. 2012. Cyclic lateral response of piles in dry sand: Finite element modeling and validation. *Computers and Geotechnics* 44: 116–131.

Hamilton, B., L. Battenberg, M. Bielecki, et al. 2013. *Offshore Wind Market and Economic Analysis: Annual Market Assessment*. Burlington, MA: Navigant Consulting Inc.

Heidari, M., M.H. El Naggar, M. Jahanandish, et al. 2014a. Generalized cyclic p–y curve modeling for analysis of laterally loaded piles. *Soil Dynamics and Earthquake Engineering* 63: 138–149.

Heidari, M., M. Jahanandish, M.H. El Naggar, et al. 2014b. Nonlinear cyclic behavior of laterally loaded pile in cohesive soil. *Canadian Geotechnical Journal* 51 (2): 129–143.

Hong, Y., B. He, L.Z. Wang, et al. 2017. Cyclic lateral response and failure mechanisms of semi-rigid pile in soft clay: Centrifuge tests and numerical modelling. *Canadian Geotechnical Journal* 54 (6): 806–824.

Kuo, Y.S., M. Achmus, and K. Abdel-Rahman. 2011. Minimum embedded length of cyclic horizontally loaded monopiles. *Journal of Geotechnical and Geoenvironmental Engineering* 138 (3): 357–363.

Liang, F., H. Chen, and Y. Jia. 2018. Quasi-static p-y hysteresis loop for cyclic lateral response of pile foundations in offshore platforms. *Ocean Engineering* 148: 62–74.

Liao, W.M., J.J. Zhang, J.B. Wu, et al. 2018. Response of flexible monopile in marine clay under cyclic lateral load. *Ocean Engineering* 147: 89–106.

Ma, H., J. Yang, and L. Chen. 2017. Numerical analysis of the long-term performance of offshore wind turbines supported by monopiles. *Ocean Engineering* 136: 94–105.

Memarpour, M.M., M. Kimiaei, and M. Shayanfar. 2010. Simplified numerical model for analysis of offshore piles under cyclic lateral loading. In *2nd International Symposium on Frontiers in Offshore Geotechnics*, 549–554.

Poulos, H.G., and T.S. Hull. 1989. The role of analytical geomechanics in foundation engineering. *Foundation engineering: Current principles and practices*, 1578–1606. Chicago: ASCE.

Rajashree, S.S., and R. Sundaravadivelu. 1996. Degradation model for one-way cyclic lateral load on piles in soft clay. *Computers and Geotechnics* 19 (4): 289–300.

Shao, W., D.Y. Yang, D.D. Shi, et al. 2019. Degradation of lateral bearing capacity of piles in soft clay subjected to cyclic lateral loading. *Marine Georesources & Geotechnology* 37 (8): 999–1006.

Yang, Q.J., Y.F. Gao, D.Q. Kong, et al. 2019. Centrifuge modelling of lateral loading behaviour of a "semi-rigid" Mono-pile in soft clay. *Marine Georesources & Geotechnology* 37 (10): 1205–1216.

Yang, Q., G.L. Pan, H.J. Liu, et al. 2018. Bearing capacity of offshore umbrella suction anchor foundation in silty soil under varying loading modes. *Marine Georesources & Geotechnology-hnology*: 1–14.

Zhu, B., Z. Zhu, T. Li, et al. 2017. Field tests of offshore driven piles subjected to lateral monotonic and cyclic loads in soft clay. *Journal of Waterway, Port, Coastal, and Ocean Engineering* 143 (5): 05017003.

Chapter 5
Seismic Response of OWT Monopile in Clays

The safe and stable operation of monopile offshore wind turbines (MOWTs) under seismic and environmental loading is a major challenge facing designers. A numerical method was developed for simulating the dynamic interaction of the soil-pile-structure system by employing an elastoplastic bounding surface constitutive model that can capture the cyclic degradation of clay stiffness well. Then, the suitability of the method was validated by comparison with the existing simulation results in the literature. The dynamic responses of MOWTs to seismic and environmental loading were simulated by employing the above numerical method. The effects of soil plasticity and stiffness degradation on the dynamic responses of MOWTs were investigated. The influence of environmental loads on the seismic responses of MOWTs was also analyzed. It was found that the monopile experienced significant rotation and settlement, and the soil plasticity and stiffness degradation significantly increased the peak response of the lateral deflection, rotation angle and bending moment of the MOWT under combined seismic and environmental loads. The input bedrock peak acceleration was amplified to varying degrees from the monopile bottom to the tower top, and the acceleration amplification along the height was the most significant overall when considering the soil stiffness degradation. Environmental loads such as winds and waves significantly increased the absolute values of the seismic response profile, including the deflection, rotation angle and bending moment, as well as the acceleration amplification factor. The research results can support the seismic design of MOWTs.

5.1 Introduction

Offshore wind power is now gaining attention because of its various advantages, including high speed, low wind shear, low turbulence, and high output. The urgent demand for wind power is promoting the development of offshore wind

X. Cheng et al., *Dynamic Analysis of Offshore Wind Turbine Foundations in Soft Clays*,
https://doi.org/10.1007/978-981-97-9454-6_5

turbine (OWT) construction technology (Carter 2007). The design of the foundation supporting an OWT is extremely important because it may account for as much as 30% of the total cost of a typical offshore wind project (Oh et al. 2018; Wu et al. 2019). Nonetheless, monopile foundations remain by far the most popular foundation type, as they were used for more than 75% of the currently installed OWTs (Hamilton et al. 2013; Esteban et al. 2015). In contrast to conventional piles, monopile foundations supporting OWTs usually have large diameters ranging between 4 and 10 m, and even larger diameters are being considered for future design.

In harsh offshore environments, monopile offshore wind turbines (MOWTs) located in seismically active regions are susceptible to earthquakes, in addition to environmental loads such as wind and waves (Zheng et al. 2015). Therefore, the monopile experiences large overturning moments due to these loads. In the current guidelines for OWTs, the requirements for the serviceability limit state (SLS) typically dominate the whole design (Arany et al. 2015; Liao et al. 2018). For instance, the tolerance for permanent accumulated rotation of monopiles at the mudline is specified as 0.25° in terms of the SLS (Det Norske Veritas (DNV) 2013). The ability of monopile foundations to resist environmental and seismic loading is a major challenge facing designers, which has also motivated many researchers to study the dynamic responses of monopile foundations under these loads.

The existing research on large-diameter monopiles mostly focuses on long-term lateral cyclic responses under wind and wave loads that are usually considered quasistatic loads, ignoring the inertia effect due to their low frequency. Long-term lateral cyclic loads lead to the degradation of the soil strength and stiffness, which induces the accumulation of monopile deflections. To consider cyclic degradation, some researchers have proposed a cyclic degradation factor related to the number of cycles and the level of cyclic loads to improve the conventional p-y methods (Rajashree and Sundaravadivelu 1996; Fan and Long 2005; Dewaikar and Patil 2006; Zhu et al. 2017). Some researchers proposed a stiffness degradation model in which the soil modulus around a monopile decreased gradually with the cycle number to reflect the effect of cyclic loading on the soil stiffness (Achmus et al. 2009; Kuo et al. 2011; Depina et al. 2015; Ma et al. 2017; Yang et al. 2018). The lateral cyclic responses of monopiles have also been analyzed by the 3D finite element method based on some advanced soil constitutive models (Bourgeois et al. 2010; Giannakos et al. 2012; Corciulo et al. 2017; Hong et al. 2017; Kementzetzidis et al. 2019; Pisanò 2019; Chen et al. 2020a, b; Cheng et al. 2021; Kementzetzidis et al. 2021; Liu et al. 2022; Ding et al. 2023).

The consideration of seismic load as a part of the required load combination for wind turbines has been highlighted in areas where seismic hazards are high (Prowell and Veers 2009; Kaynia 2019). Different from wind and wave loads, the inertial effect must be considered in the analysis of the seismic load due its high frequency characteristics. Seismic centrifuge tests were performed for traditional and hybrid monopile foundations supporting OWTs (Yu et al. 2015; Wang et al. 2019; Li et al. 2021). It was found that the monopile experienced a large tilt during an earthquake, and the tower head experienced significantly amplified responses. Zheng et al. (2015) conducted scaled shaking table model tests to investigate the dynamic behaviors of

MOWT systems under joint earthquake and wave action and concluded that ignoring the effect of wave action in seismic analysis will lead to underestimation of the structural response. Anastasopoulos and Theofilou 2016 investigated the seismic responses of hybrid monopile OWTs by employing 3D FE modeling and concluded that seismic shaking might lead to a substantial accumulation of rotation and settlement. Vacareanu et al. (2019) analyzed the seismic responses of MOWTs in a non-liquefiable soil deposit via 3D FE modeling incorporating the SANISAND bounding surface plasticity model. Kaynia (2021) investigated the effect of kinematic interaction on the seismic responses of the MOWT by using a rigorous numerical model. The dynamic responses of OWTs subjected to combined wind, wave and earthquake loads were investigated by performing numerical simulations (Wang et al. 2018; Yang et al. 2019; Zhang et al., 2021a, 2021b; Xi et al. 2022).

The soil-structure interaction (SSI) plays a significant role in the seismic behavior of OWTs, as it has a significant influence on the responses of the foundation and superstructure and the natural frequency of the system. It is extremely important to properly consider the interactions among the following three coupled subsystems: the wind turbine structure, the monopile foundation, and the soil ground. In most existing numerical simulation studies on the seismic response of MOWTs, the pile–soil dynamic interaction was usually considered by employing a nonlinear Winkler model with springs (*p–y* spring, *t–z* spring and *Q–z* spring) and dampers (Zhao and Maisser 2006; Zhong et al. 2014; Kim et al. 2014; Sapountzakis et al. 2015; Zuo et al. 2018; Wang et al. 2018; De Risi et al. 2018; Ju and Huang 2019 Yan et al. 2021; Jiang et al. 2021; Xi et al. 2021; Mo et al. 2021 Liang et al. 2022; Kementzetzidis et al. 2022). However, the nonlinear Winkler model is a simplified means of simulating the pile-soil dynamic interaction; it ignores the soil continuity and is inadequate to reveal the soil flow mechanism around piles during earthquakes.

The stiffness degradation of clays under dynamic loads will reduce the overall stiffness of the MOWT system and then affect the soil–pile-structure dynamic interaction. However, most existing analysis methods based on the nonlinear Winkler model do not accurately consider the impact of the stiffness degradation of clays on the seismic dynamic responses of MOWT systems. In this chapter, the authors developed and validated a numerical method for analyzing the dynamic interaction of a soil-pile-structure system based on a simplified bounding surface model that can capture the degradation of the soil stiffness. Then, the dynamic responses of the MOWT system to seismic and environmental loading were simulated by employing the above numerical method. The effects of soil plasticity and stiffness degradation on the dynamic responses of MOWTs were investigated. The influence of environmental loads on the seismic response of wind turbines was also analyzed (Cheng et al. 2023).

5.2 Numerical Method and Verification

Developing a method capable of accurately analyzing the soil-pile-structure dynamic interaction is crucial for assessing the seismic dynamic responses of MOWT systems. In the following, a time-domain numerical method is developed based on the simplified version of single bounding surface model of clay proposed by authors mentioned in Chap. 2, and then its suitability is verified by comparison with the existing simulation results in the literature.

5.2.1 *Numerical Method*

An example of a soil-pile-structure interaction system has been used to validate numerical methods in previous studies (Maheshwari et al. 2004, 2005; Peiris et al. 2014). A 10 m long socketed pile with a square cross-section of 0.5 × 0.5 m was used in these studies. The piles are made of concrete. Young's modulus, mass density and Poisson's ratio are $E_p = 25\,\text{GPa}$, $\rho_p = 2400\,\text{kg/m}^3$, and $v_p = 0.25$, respectively. Young's modulus of clay is $E_s = 11.78\,\text{MPa}$, and its mass density is $\rho_s = 1610\,\text{kg/m}^3$. The superstructure is a 6 m high rectangular column with a cross-section of 0.75 × 0.5 m, which is placed on top of the pile. The total mass of the column is 5400 kg, and its Young's modulus is $E_c = 25\,\text{GPa}$.

A 3D finite element model was established by employing the ABAQUS software package (Dassault Systemes Simulia Corp) to simulate the above soil-pile-structure interaction system. Exploiting the symmetries of the geometry and loading conditions, the model simulates only one half of the system, as shown in Fig. 5.1. The length, width and height of the soil domain are 12 m, 4 m and 10 m, respectively. The elements near the pile are maintained at a size less than one-tenth of the wavelength that corresponds to the highest frequency, which is 20 Hz (Watanabe et al. 2017). The pile, soils and superstructure were all simulated using 8-node linear brick elements with reduced integration (C3D8R).

The pile-soil interaction was modeled by the contact pairs in ABAQUS. In this approach, separation between the interface elements was allowed when they were subjected to tension. When the two surfaces were in contact, the interface behavior was governed by Coulomb's law enforced through the penalty method. A Coulomb frictional coefficient of 0.3 is adopted in this study. It should be emphasized that the combination of friction contact and total stress models is a practical workaround to consider the variation in the interface response with non-uniform soil strength (a linearly increasing strength profile is used in the OWT system in Sect. 5.3) in ABAQUS/Standard. A more appropriate and theoretically sound approach employs an adhesive interface with a strength level proportional to that of the adjacent soils (Huang et al. 2021).

Appropriate boundary conditions should be set to avoid the reflection of seismic waves to simulate an infinite soil medium. Some artificial boundary conditions

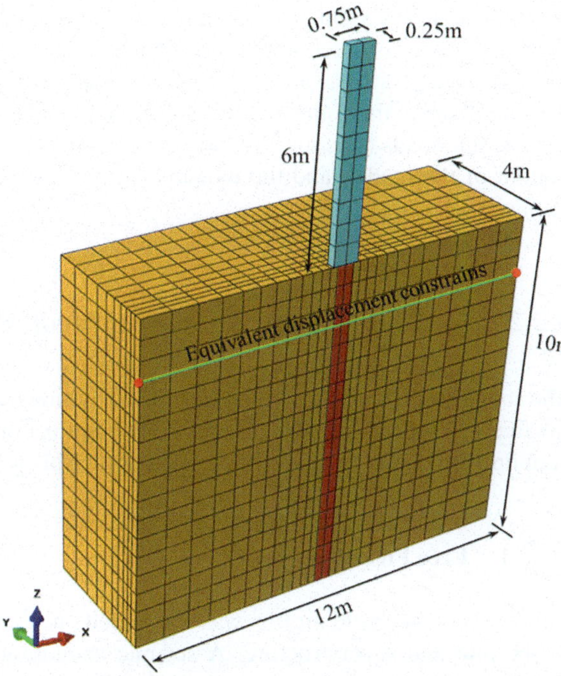

Fig. 5.1 Three-dimensional finite element model for the soil-pile-structure interaction system

have been proposed, including viscous boundaries (Lysmer and Kuhlemeyer 1969) and viscoelastic boundaries (Deeks and Randolph 1994). When using these artificial boundary conditions, it is often necessary to set springs and dashpots on the boundary to characterize stiffness and damping and define relevant constants for them. Although the calculation accuracy of these artificial boundary conditions is relatively high, the numerical implementation is troublesome. In addition, the system is also required to meet certain conditions, such as having a homogeneous and linear elastic soil medium. In this study, a relatively simple and practical method was selected to set boundary conditions. For the vertical boundaries, equivalent displacement constraints were set to simulate seismic responses of infinite horizontal strata. Nodes of the same height on both sides of the analytical region are bound together (the displacement is always consistent), forcing the opposite vertical sides to move simultaneously, thus preventing any rotation (see Fig. 5.1) and eliminating the effect of reflected waves (Tsinidis et al. 2014; Gao et al. 2021). In addition, normal horizontal constraints were applied to the vertical face of symmetry to prevent out-of-plane movement. The base boundary of the model was simulated as rigid bedrock, and the top boundary was fully free.

The dynamic behaviors of clays were simulated by the single bounding surface model above. The model parameters were $A_0 = 2\ S_u = 40$ kPa, $G = 4.15$ MPa, $\mu = 2$, and $\xi = 1$. The Poisson's ratio of the soil was set as 0.49 to simulate the undrained behavior. The soil material damping ratio was assumed to be 5%. The pile behavior was simulated using a linear elastic constitutive model, and the pile damping was considered negligible. Dynamic analysis was performed by following two consecutive steps. Step 1: Gravity was applied to the soil domain to establish the initial stress field; meanwhile, the initial displacement field was set to 0. Step 2: Dynamic loads (seismic excitations or harmonic excitations) were applied to the bottom of the model, and the dynamic responses were calculated by the dynamic implicit method.

5.2.2 Validation of the Numerical Method

The above numerical method can be verified by analyzing the following three aspects: (a) free field response; (b) kinematic interaction for the soil-pile system; and (c) coupling kinematic-inertial interaction for the soil-pile-structure system.

5.2.2.1 Free Field Response

Validation of free field motion was conducted only for the soil profile in the absence of the pile and superstructure. A seismic excitation (El Centro Earthquake 1940 (N–S Component) with a peak ground acceleration of 0.32g, shown in Fig. 5.2a) was applied to the base of the model. The calculated elastic free field response is shown in Fig. 5.2b. For the same soil profile, Peiris et al. (2014) also performed the calculation by using commercial software and general purpose FE software, and the corresponding calculation results are shown in Fig. 5.2c. In their study, the "free horizontal motion and zero vertical motion" boundary was adopted. Figure 5.2b and c show that the current calculated results are in good agreement with those in the literature, which demonstrates the rationality of the boundary conditions used in this study.

5.2.2.2 Kinematic Interaction

Validation for the kinematic interaction was performed only for the soil-pile system in the absence of the superstructure. In accordance with Fan et al. (1991), study a sinusoidal excitation is applied to the base of the soil-pile system, which is described by the following equation:

$$x(t) = A\ \sin(\omega t) \tag{5.1}$$

where $x(t)$ is the displacement, A is the displacement amplitude, ω is the angular frequency and t is the time. Two dimensionless parameters were defined, namely, the kinematic displacement factor I_{u} and dimensionless frequency a_0.

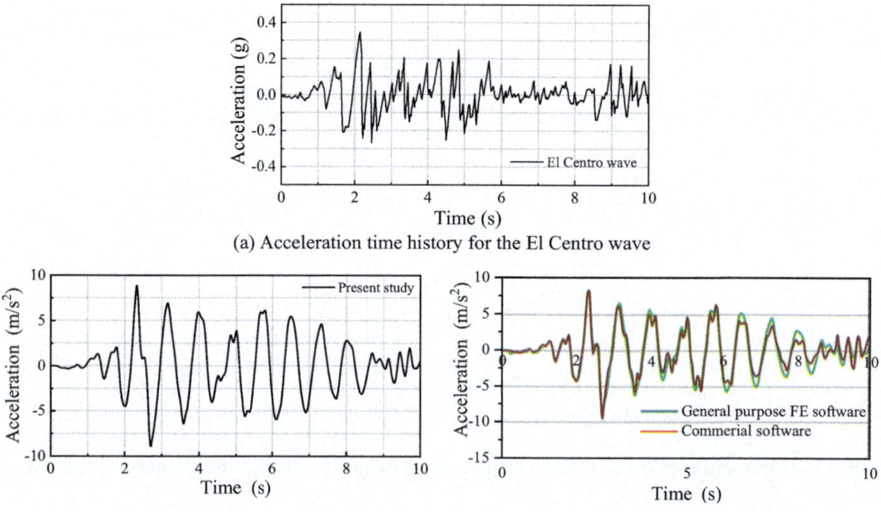

(a) Acceleration time history for the El Centro wave

(b) Current calculation results

(c) Peiris et al.'s calculation results

Fig. 5.2 Comparison of free field response results obtained from the present study and Peiris et al.'s study

$$I_u = |U_p| / U_{ff} \tag{5.2}$$

where U_p is the response of the pile head and U_{ff} is the amplitude of the free field motion.

$$a_0 = \omega \cdot d / v_s \tag{5.3}$$

where ω is the circular frequency of sinusoidal excitation; d is the pile diameter (section width in this study); and v_s is the shear wave velocity of the soil.

The response at the pile head was calculated by employing both the elastic and elastoplastic soil models, and the amplitude of the steady-state response at the pile head was noted. Figure 5.3a shows the results of I_u for different a_0 in this study. Only the calculation results for $a_0 < 0.65$ ($\omega < 10$ Hz) are given in view of the frequency range of greatest interest for earthquake loading. The idealized general shape of I_u versus a_0 for monopiles was proposed by Fan et al. (1991), as shown in Fig. 5.3b. The prediction results agree with the literature results in the overall trend, and both consist of three fairly distinct regions. Figure 5.3b shows a low-frequency region ($0 < a_0 < a_{01}$) in which $I_u \approx 1$, which means that the pile follows the ground deformation in this frequency range. In this study, a_{01} is approximately 0.25, which is within the range of 0.2–0.3 given by Fan et al. (1991). However, for the soil profile in this study, $I_u > 1$, which in the relatively low-frequency region. The main reason is that the amplification of the pile head displacement to the input displacement at the bottom of the model is more significant than the free field amplification because the pile stiffness is significantly higher than the clay stiffness (it can be regarded as

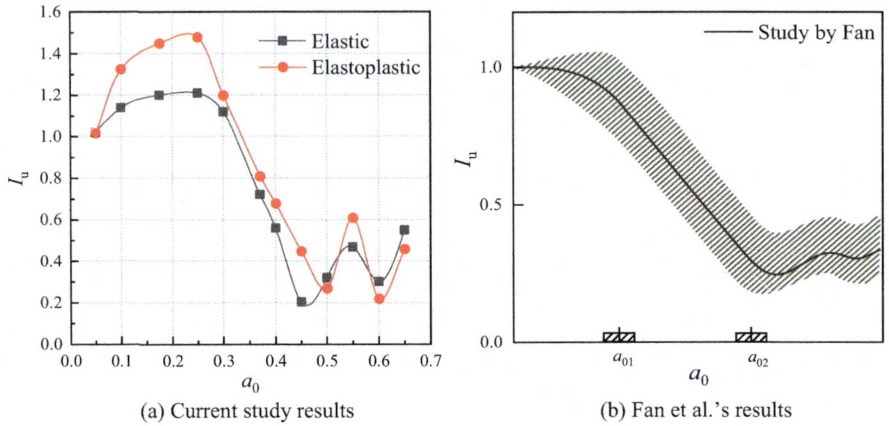

Fig. 5.3 Comparison of the kinematic displacement factor I_u versus dimensionless frequency a_0

soft clay due to the low wave velocity, which is 50 m/s). Therefore, the pile head displacement is slightly greater than the free field displacement for elastic soil, and this phenomenon was also observed in previous studies (Hussein and Naggar 2022). When considering soil plasticity, the wave energy consumption is more significant during the propagation of the sine wave, which leads to a relatively smaller free field displacement. Therefore, compared with elastic soil, I_u is larger for plastic soil. This means that the soil plasticity significantly enhances the pile head response under low frequency. Then, there is an intermediate region ($a_{01} < a_0 < a_{02}$) where I_u declines rapidly with frequency. Finally, I_u fluctuates around the mean value for a relatively high frequency region ($a_0 > a_{02}$). The trend is also true for this study, and the a_{02} values are approximately 0.45 and 0.5 for the elastic and elastoplastic cases, respectively. This comparison also demonstrates that the proposed numerical method can reasonably consider the kinematic interaction effects.

5.2.2.3 Coupling Kinematic-Inertial Interaction

Harmonic seismic excitations with different frequencies were applied to the base of the soil-pile-structure system to validate the coupling kinematic-inertial interaction. The responses at the pile head and at the top of the structure were calculated by employing both the elastic and elastoplastic soil models. The ratio of the steady-state response amplitude at the pile head and at the top of the structure to the input bedrock motion amplitude is defined as the amplification factor; the amplification factor versus the dimensionless frequency a_0 is shown in Fig. 5.4a. Maheshwari et al. (2004) performed finite element analysis for the same soil-pile-structure system using an elastic model and HISS elastoplastic model (Wathugala and Desai 1993). In addition, in their study, the frequency-dependent Kelvin elements proposed by Novak and Mitwally (1988) were used at the boundary to simulate an infinite soil domain.

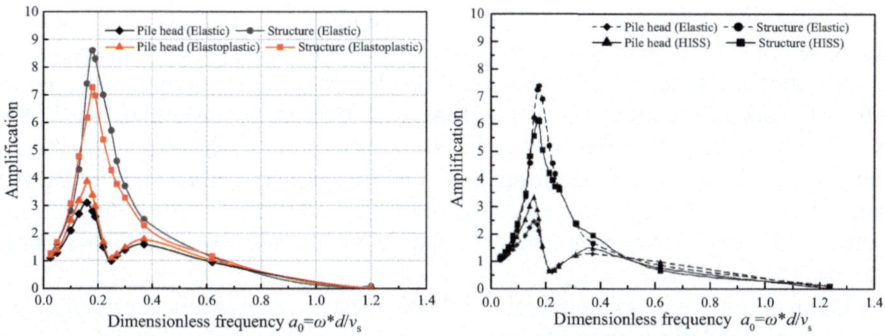

(a) Current study results (b) Maheshwari et al.'s results

Fig. 5.4 Comparison of amplification factor versus dimensionless frequency for elastic and elastoplastic soil models

Their results are shown in Fig. 5.4b. Both studies show that the soil nonlinearity significantly increases the pile head response for low frequencies ($a_0 < 0.2$); however, for higher frequencies, the influence of soil nonlinearity is insignificant. It is also noted that the structural response decreases significantly due to soil nonlinearity for moderate frequencies ($0.15 < a_0 < 0.3$), and for other higher or lower frequencies, the influence of soil nonlinearity is insignificant. In the overall trend, the current calculation results agree with Maheshwari et al.'s (2004) results, although there are some quantitative deviations. The main reason for the deviations is the difference in the numerical methods utilized, such as the boundary conditions and elastic–plastic soil models.

5.3 Numerical Model of a Monopile Offshore Wind Turbine

The dynamic responses of MOWTs subjected to seismic and environmental loads are simulated by employing the above numerical method. The setup of the 3D finite element model for the MOWT system and the generation of loads are introduced.

5.3.1 Finite Element Model

A three-dimensional finite element model of the MOWT system is established according to the specifications of a typical NREL offshore 5-MW baseline wind turbine (see Table 5.1). A schematic of the main environment for the MOWT system is shown in Fig. 5.5. The wind turbine and monopile are constructed from steel with a Young's modulus = 210 GPa, Poisson's ratio = 0.3, and density = 8500 kg/m^3. The assembly of the rotor blades, nacelle, and hub weighs 4.6E5 kg (Jonkman et al. 2009). Exploiting the symmetries of the geometry and loading conditions, the model simulates only one half of the system, as shown in Fig. 5.6 (a), to improve the computational efficiency, while Fig. 5.6b displays the overall outline of the entire MOWT system for clarity. The soil and MOWT structure are discretized by 22,837 8-node linear brick elements with reduced integration (C3D8R), while the blades, hub and nacelle are discretized using 2797 quadratic tetrahedral elements (C3D10). The setup of the 3D numerical model involving the main aspects, including the soil domain size, space discretization and boundary conditions, is described in detail in Appendix.

The material behavior of monopiles is linear elastic. The dynamic behavior of soil was simulated by the simplified version of single bounding surface model proposed by authors mentioned in Chap. 2. The soil model parameters used in the numerical simulation are $A_0 = 2\ S_u$ ($S_u = 10 + 3z$ kPa, and z is the soil depth), $G = 500\ S_u$, and $\mu = 2$. The Poisson's ratio of the soil is set as 0.49 to simulate the undrained behavior. As mentioned in Chap. 2, parameter ξ controls the rate of modulus degradation, and no cyclic degradation occurs when $\xi = 0$; otherwise, cyclic degradation occurs, as shown in Fig. 2.11 of Chap. 2. To investigate the influence of soil plasticity and cyclic degradation of soil stiffness on the dynamic response of MOWTs under combined seismic and environmental loads, three working cases are set in the numerical simulation, as shown in Table 5.2. For Case 1, a linear elastic constitutive model is adopted to simulate soil behavior. For both Case 2 and Case 3, an elastoplastic single bounding surface constitutive model is adopted to simulate the nonlinear behavior of the soil; however, cyclic degradation is not considered by setting $\xi = 0$ for Case 2, while it is considered by setting $\xi = 2$ for Case 3. To investigate the influence of environmental loads on seismic dynamic responses, Case 4 is set as a comparison with Case 3. Blade and the tower structural damping ratios of 2% and a soil damping ratio of 5% are considered in the dynamic analysis.

Table 5.1 Dimensions of the 5 MW OWT model

Property	Symbol	Dimension
Rotor diameter	D_R	126 m
Length of the tower	L_T	90 m
Length of the pile	L_P	75 m
Tower base diameter and thickness	D_T, T_K	6 m, 3 cm
Tower top diameter and thickness	D_T, T_K	3.87 m, 2 cm
Monopile diameter and thickness	D_T, T_K	6 m, 6 cm
Length of embedded pile	L_{EP}	45 m
Mean sea depth	H_w	20 m
Rotor mass	M_R	110,000 kg
Nacelle mass	M_N	240,000 kg
Blade overall mass	M_B	17,740 kg
Hub mass	M_H	56,780 kg

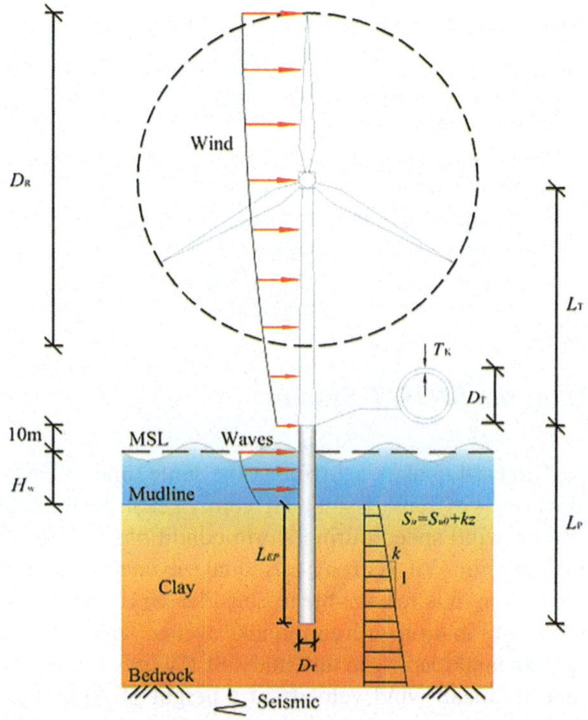

Fig. 5.5 Schematic of the main environment of a 5 MW monopile offshore wind turbine in clay

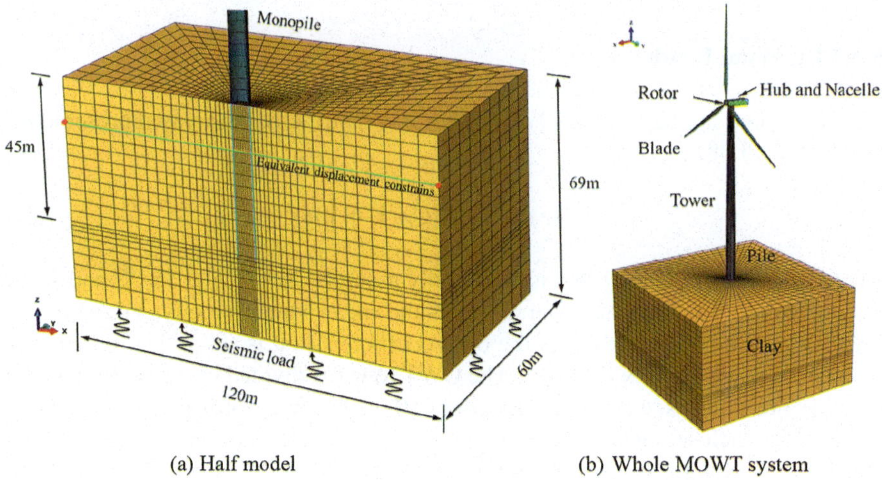

Fig. 5.6 3D finite element model of a monopile offshore wind turbine in clay

Table 5.2 Schedule of working cases in the numerical simulation

Working cases	Loads	Soil behavior	Parameter ξ
Case 1	Combined seismic and environmental loads	Elastic soil	\
Case 2	Combined seismic and environmental loads	Elastoplastic soil without cyclic degradation	$\xi = 0$
Case 3	Combined seismic and environmental loads	Elastoplastic soil with cyclic degradation	$\xi = 2$
Case 4	Only seismic loads	Elastoplastic soil with cyclic degradation	$\xi = 2$

5.3.2 *Loads on the MOWT System*

The probability of an earthquake occurring at the same time as ULS (ultimate limit state) winds and waves is extremely small (Kiyomiya et al. 2002). Earthquake forces can be ignored when wind speed during storm conditions (ULS) is considered for the OWT structural design. When strong earthquake events are adopted for the OWT structural seismic design, it is reasonable to consider serviceability limit state (SLS) winds and waves in combination with earthquake events. Therefore, for the combined seismic and environmental loading in this study, SLS winds and waves were adopted. It is assumed that the mean wind velocity at a height of 10 m $V_{10} = 10$ m/s, the significant wave height $H_{1/3} = 5$ m, and the seismic peak ground acceleration PGA = 0.3 g.

5.3.2.1 Wind Loads

The wind load applied to the OWT tower is a distributed force as represented by (Binh et al. 2008)

$$F_{sh}(z) = 0.5\rho_a C_D (V + v) D^2 \quad (5.4)$$

where ρ_a is the air density that is commonly assumed as 1.225 at 15 °C and 1 atm; D denotes the diameter of tower; $V(z)$ is the mean wind velocity; $v(z, t)$ are the fluctuating wind velocity; C_D is the drag coefficient that is set as 1.2 in this study (Det Norske Veritas (DNV) 2013). The OWT tower is divided into nine segments to simplify the analysis, each with a length of 10 m. The drag force is considered to be uniform for each segment. A reference point is set at the midpoint of each segment that is coupled with the corresponding cross-section in the numerical model. The wind load is applied to these reference points.

The wind load applied to the OWT blades is a thrust force at the tower top that can be represented by Bernoulli's equation (Lee et al. 2010):

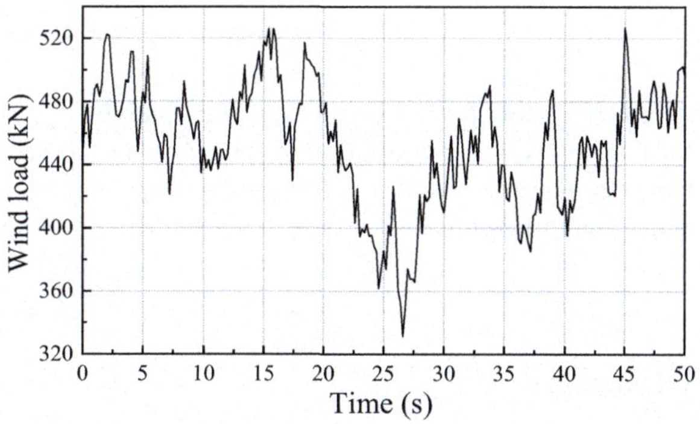

Fig. 5.7 Wind load time history at the hub

$$F_T = 0.5\rho_a \pi R_T^2 V_s^2 (1 + 2v_s/V_s) C_T \tag{5.5}$$

where V_s is the mean wind velocity at the hub height; v_s is the fluctuating wind velocity; R_T is the rotor radius; C_T is the thrust coefficient and is set as 1 in this research (Hansen 2008). F_T is applied to the OWT hub center. The fluctuating wind velocity is simulated using the Davenport wind speed spectrum (Davenport 1961) and the autoregressive (AR) model (Iannuzzi and Spinelli 1987). Figure 5.7 presents simulated typical wind loads time history at the OWT hub.

5.3.2.2 Wave Loads

Sea waves are irregular and random in shape, height, length and speed of propagation. A real sea state is best described by a random wave model. In general, sea waves can be regarded as a stationary random process, which can be formed by the superposition of many (theoretically infinite) cosine waves with different amplitudes, different periods and different random initial phases. The sea surface elevation in the time domain then can then be simulated as (Ren and Wang 2004):

$$\eta(t) = \sum_{i=1}^{m} \sqrt{2S_{\eta\eta}(\overline{\omega}_i)\Delta\omega_i} \cos(\omega_i t + \varepsilon_i) \tag{5.6}$$

where η(t) is the instantaneous height of fluctuating water surface relative to static water surface; m is the number of cosine waves; ω_i represents angular frequency for the ith component wave respectively; t is time and ε_i is an independent random variable that represents the phase of each component wave uniformly distributed in the range of [0, 2π].

Sea waves can be simulated by wave spectrum. The Pierson-Moskowitz spectrum is used to simulate the stochastic wave elevation (Pierson and Moskowitz 1964), as represented by

$$S_{\eta\eta}(\omega) = 0.78\omega^{-5} \exp\left(-3.11\omega^{-4} H_{1/3}^{-2}\right) \tag{5.7}$$

where $H_{1/3}$ denotes the significant wave height.

The sea wave load applied to the tower can be calculated by using the Morison formula. Based on that, the longitudinal sea wave load per unit length of the tower can be calculated as:

$$f(z,t) = f_D(z,t) + f_I(z,t) \tag{5.8a}$$

velocity force

$$f_D(z,t) = \phi_D \cdot u(z,t)|u(z,t)| \tag{5.8b}$$

inertia force

$$f_I(z,t) = \phi_M \cdot \partial u(z,t)/\partial t \tag{5.8c}$$

where $\phi_D = \rho C_D D/2$ and $\phi_M = \pi\rho C_M D^2/4$; ρ is water density and equals to 1030 kg/m^3; C_D and C_M are the drag and inertia coefficients, which equal 1.2 and 2.0 respectively; D is the tower diameter; $u(z,t)$ and $\partial u(z,t)/\partial t$ are the horizontal velocity and acceleration of the fluctuating water quality point at z position. They can be simulated by, respectively:

$$u(z,t) = \sum_{i=1}^{m} \sqrt{2S_{\eta\eta}(\overline{\omega}_i)\Delta\omega_i}\omega_i \frac{\cosh k_i z}{\sinh k_i d} \cos(\omega_i t + \varepsilon_i) \tag{5.9a}$$

$$\frac{\partial}{\partial t} u(z,t) = -\sum_{i=1}^{m} \sqrt{2S_{\eta\eta}(\overline{\omega}_i)\Delta\omega_i}\omega_i^2 \frac{\cosh k_i z}{\sinh k_i d} \sin(\omega_i t + \varepsilon_i) \tag{5.9b}$$

where d is the water depth, k_i is the sea wave number.

The tower part in the water is divided into two segments; each is 10 m. In the finite element model, a reference point is defined in each segment and coupled with the cross-section of the corresponding segment, and then sea wave loads are applied on these reference points. Figure 5.8 presents the simulated typical wave load time history.

5.3.2.3 Hydrodynamic Pressure Caused by Earthquake

A vibrating tower in sea water will bear additional force due to hydrodynamic pressure associated with its motion. The interaction between water and tower can be simulated by the equivalent mass of water, that is, the additional mass method (Liaw and Chopra 1974). The added mass of tower can be expressed by

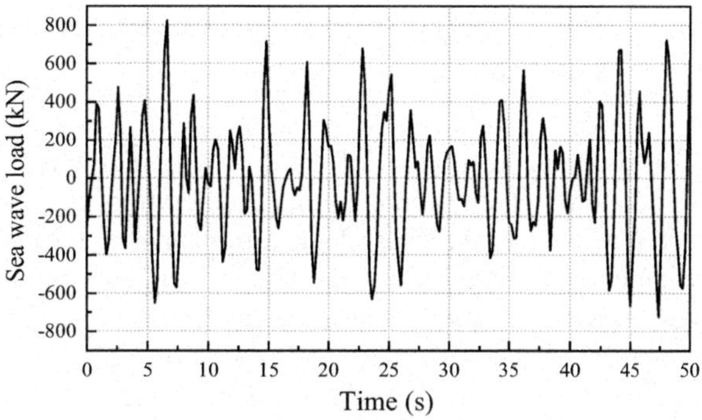

Fig. 5.8 Typical sea wave load time history

$$m_a = C_a A_p \rho_w \tag{5.10}$$

where C_a is the added mass coefficient and $C_a = 1.0$ in the simulation (Cao et al. 2020); ρ_w is the sea water density; A_p is the section area of the tower. In dynamic analysis, the calculated mass of the tower part in water should be the sum of the added mass and the actual mass of the tower.

5.3.2.4 Seismic Loads

The seismic record of Northern California is selected. The acceleration time history and response spectra for this ground motion record are shown in Fig. 5.9. The peak ground acceleration (PGA) is scaled to be 0.3 g. The bottom of the finite element model is considered to be rigid bedrock, and the input ground motion is applied as vertically propagating shear wave (SV-wave) represented by the acceleration time history at the bottom of the model.

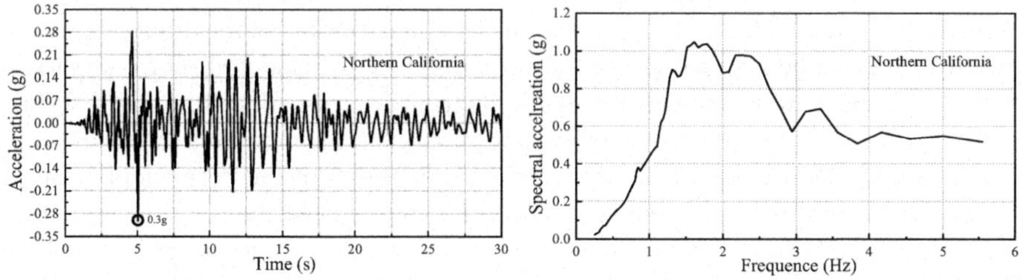

(a) Acceleration time history (b) Acceleration response spectra

Fig. 5.9 Acceleration time history and response spectra for the earthquake record

5.4 Seismic Dynamic Response of Monopile Offshore Wind Turbine

The dynamic responses of the MOWT, including the rotation angle, settlement, deflection, bending moment and acceleration, etc., are analyzed. The effects of the soil plasticity and stiffness degradation on dynamic responses under seismic and environmental loads are clarified. The influence of environmental loads on the seismic responses of MOWTs is also discussed.

5.4.1 Cumulative Rotation and Settlement

For the design of OWTs, the deformation of the foundation is of primary significance rather than the ultimate capacity (Byrne et al. 2010). For the monopile, the serviceability criterion of the deflection at the mudline should be less than 1/500 of the embedment length (Germanischer 2010), and the permanent accumulated rotation angle at the mudline should not exceed 0.25 degrees in the SLS of OWTs (Det Norske Veritas (DNV) 2013). For the turbine tower, the maximum deflection at the tower top should be limited to 1.25% of the tower height to avoid excessive motion (Nicholson 2011).

Figure 5.10 shows the rotation angle time history of the monopile section at the mudline under different working cases. It can be observed that the monopile experienced significant rotation under all working cases. Herein, we define the rotation that can be recovered after unloading as the vibration rotation (or temporary rotation), the unrecoverable rotation as the permanent rotation, and the sum of the two as the peak rotation, as shown in Fig. 5.10b. Figure 5.10a shows the influence of soil plasticity on the rotation angle time history under combined seismic and environmental loading. The rotation angle corresponding to Case 1 is significantly smaller than those of Cases 2 and 3, and the maximum peak rotation angle in the latter two cases exceeds the limit of 0.25°. This demonstrates that the rotation of the monopile will be underestimated when the soil plastic deformation is not considered. The comparison of Cases 2 and 3 indicates that the cyclic degradation of soil stiffness will significantly increase the rotation of the monopile; moreover, the increases in the vibration rotation and peak rotation are more significant than that of the permanent rotation. Figure 5.10b shows the influence of environmental loads on the rotation angle time history. Although instantaneous reverse rotation (positive rotation angle) occurs, the rotation angle of the monopile will gradually accumulate in the negative direction under only seismic loading (Case 4). The rotation angle corresponding to Case 3 is significantly larger than that of Case 4, which demonstrates that the environmental loads can aggravate the rotation caused by earthquakes.

Figure 5.11 shows the settlement time history of the monopile under different working cases. The settlement of the monopile is almost 0 for elastic soil (Case 1),

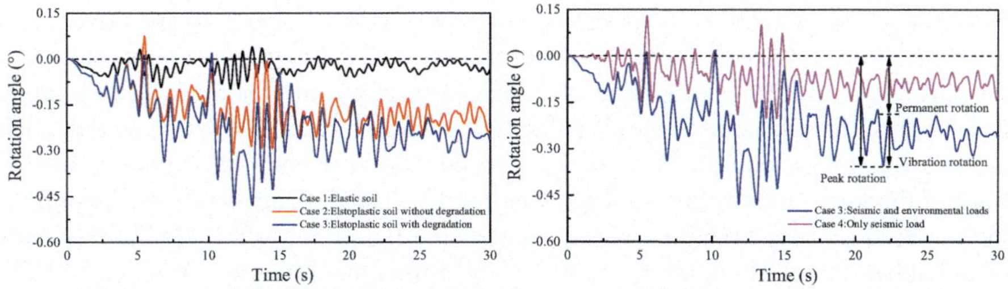

(a) Influence of soil plasticity and stiffness degradation (b) Influence of environmental loads

Fig. 5.10 Rotation angle time history of the monopile section at the mudline

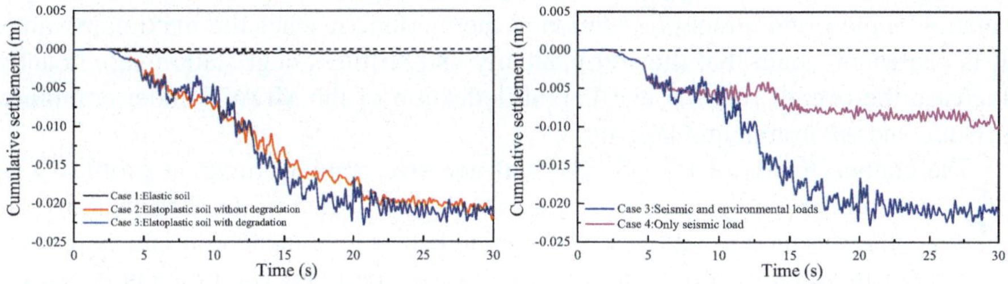

(a) Influence of soil plasticity and stiffness degradation (b) Influence of environmental loads

Fig. 5.11 Settlement time history of the monopile

as shown in Fig. 5.11a. Significant settlement occurs when considering soil plasticity (Cases 2 and 3). The permanent settlement is significantly greater than the vibration settlement, and the permanent settlement increases continuously during the earthquake and eventually tends to be stable. It is noted that the settlement is relatively larger when considering the cyclic degradation of soil stiffness (Case 3). Figure 5.11b shows the influence of environmental loads on the settlement time history. As expected, the environmental loads can also aggravate the settlement caused by earthquakes, and hence, the settlement of the monopile for Case 3 is larger than that for Case 4.

5.4.2 Peak Response Profile of the Monopile and Tower

Figure 5.12 presents the maximum deflection, rotation angle and bending moment values at different elevations from the monopile bottom to the tower top under the three working cases. It can be observed from Fig. 5.12a that lateral deflection mainly occurs in the negative direction, while the value in the positive direction is very small. The maximum deflection increases monotonically from the bottom to the maximum value at the top. Compared with Cases 2 and 3, the lateral deflection profile for Case

1 is relatively smaller in the negative direction but slightly larger in the positive direction (the values for Cases 2 and 3 almost reach 0). The comparison demonstrates that the plasticity and stiffness degradation of the clay significantly increase the maximum deflection. When considering soil stiffness degradation (Case 3), the peak deflection at the tower top reaches 1.63 m, which exceeds the maximum allowable deflection limit of the tower top (1.25% of the tower height). The deflection of the monopile at the mudline reaches 20.6 cm, which exceeds the maximum serviceability criterion of the deflection at the mudline (1/500 of the embedment length). As expected, the rotation angle profile is similar to the deflection profile under three different working cases, as shown in Fig. 5.12b, because the two are positively correlated, i.e., the greater the deflection, the greater the rotation angle. Figure 5.13 shows the displacement field of the MOWT system under three working cases at the moment when the rotation angle of the monopile section at the mudline reaches the maximum value. It is confirmed again that the soil plasticity and stiffness degradation significantly increase the overall rotation and lateral deflection of the MOWT under combined seismic and environmental loading.

The change trends of the positive and negative bending moment profiles with height are similar but asymmetric, as shown in Fig. 5.12c. For the section above the mudline, the variation trends of the maximum bending moment along the height are similar for the three working cases, i.e., the maximum bending moment decreases non-monotonically to almost 0 with height in the overall trend. However, there are significant differences for the section below the mudline. For elastic soil (Case 1), the maximum bending moment occurs at the mudline; an inflection point (zero bending moment point) appears at a certain position of the pile section, in which the bending moment direction reverses. When considering soil plasticity (Cases 2 and 3), the value

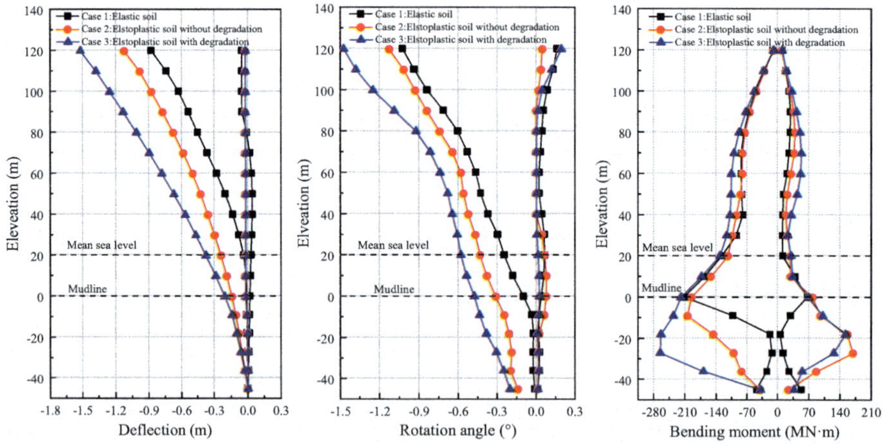

(a) Maximum deflection (b) Maximum rotation angle (c) Maximum bending moment

Fig. 5.12 Influence of soil plasticity and stiffness degradation on the peak response profile of the MOWT under combined seismic and environmental loading

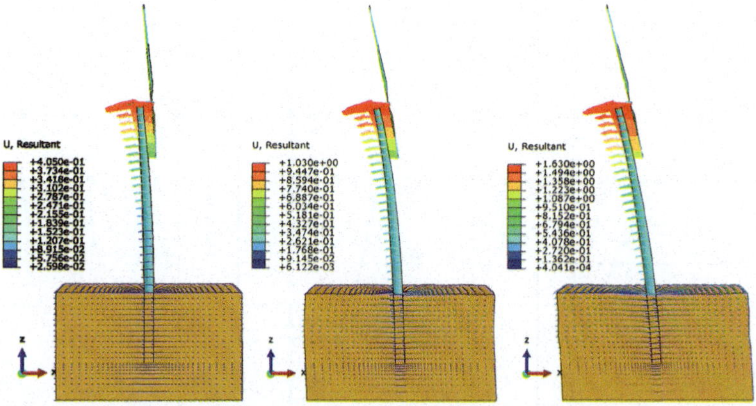

(a) Case 1: Elastic soil (b) Case 2: Elastoplastic soil without degradation (c) Case 3: Elastoplastic soil with degradation

Fig. 5.13 Displacement field of the MOWT system under combined seismic and environmental loading (amplification factor of 16)

of the maximum bending moment is larger and occurs at a certain position below the mudline. As expected, the maximum bending moment is the largest when the soil stiffness degradation is considered. Compared with elastoplastic soil, the stiffness of elastic soil is greater, so the pile-soil relative stiffness is relatively smaller, i.e., the pile is more flexible, and hence, an inflection point appears in Case 1. On the other hand, the soil plasticity is more significant when considering the soil stiffness degradation, so the pile is relatively more rigid, and hence, the maximum bending moment is also greater.

Figure 5.14 shows the influence of environmental loads on the seismic peak response profile of the MOWT. The peak response profiles for Case 3 (combined seismic and environmental loading) are significantly larger than those for Case 4 (only seismic loading) in the negative direction but are relatively smaller in the positive direction. In general, environmental loads, such as wind and wave loads, significantly increase the absolute values of the seismic maximum response profile, including the deflection, rotation angle and bending moment, although they weaken the response values in the opposite direction.

5.4.3 Acceleration Amplification

Figure 5.15 shows the acceleration time history at several different positions from the pile bottom to the tower top under the three working cases. The input bedrock peak acceleration is amplified to varying degrees. The amplified peak acceleration time lags significantly behind the input bedrock peak acceleration time (see Fig. 5.9a). The frequency of structural vibration decreases gradually as the height increases. Compared with the elastic soil (Case 1), when soil plasticity is considered (Cases 2

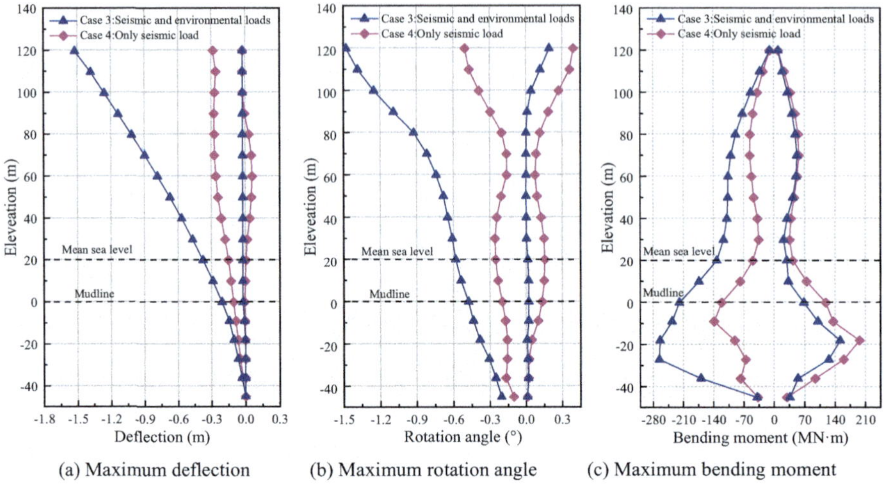

Fig. 5.14 Influence of environmental loads on the seismic peak response profile of the MOWT

and 3), the attenuation of peak acceleration with time slows down, and the duration of the earthquake is prolonged.

Figure 5.16 presents the acceleration amplification factor profile along elevations from the pile bottom to the tower top. The acceleration amplification factor is defined as the ratio of the peak acceleration for a section of the MOWT to the input bedrock

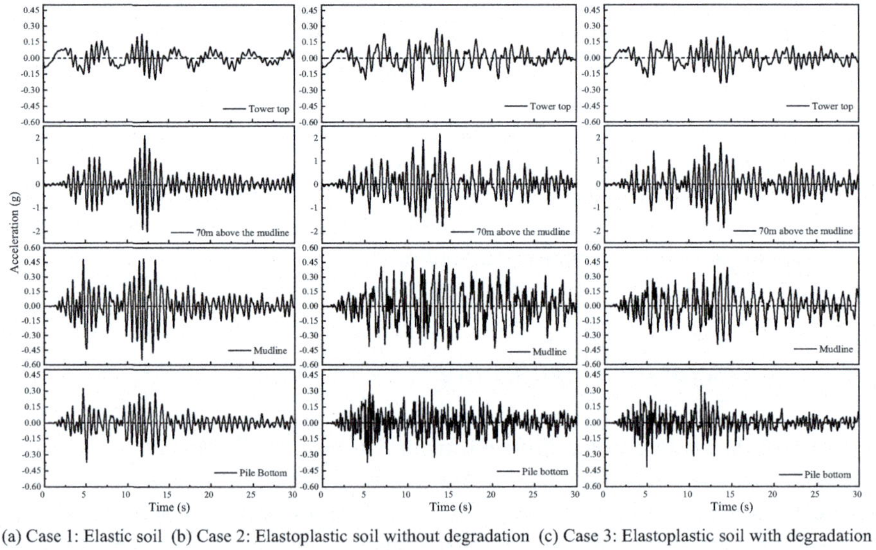

Fig. 5.15 Acceleration time history at several different positions from the pile bottom to the tower top under the three working cases

peak acceleration. For the section below the mudline, the acceleration amplification factor increases monotonically for elastic soil, while it increases non-monotonically for elastoplastic soil, as shown in Fig. 5.16a. The difference in the variation trend may be attributed to the pile-soil relative stiffness, i.e., as mentioned above, the pile for elastic soil is relatively more flexible than that for elastoplastic soil. For the section above the mudline, the variation trend of the acceleration amplification factor along the height is similar for the three working cases, i.e., it increases gradually with the height, reaches the maximum value at approximately 70 m above the mudline, and then decreases gradually to the minimum value at the tower top. In general, when considering the cyclic degradation of soil stiffness, the acceleration amplification effect is most significant, and the maximum amplification factor reaches 7.2. The small acceleration amplification at the tower top may be attributed to the large, concentrated mass, including the nacelle and blade. Figure 5.16b compares the acceleration amplification factor profiles for Cases 3 and 4. This comparison shows that the environmental loads increase the amplification of acceleration along the pile and tower overall during an earthquake, and the increasing effect is more significant in the middle section of the tower (50–80 m above the mudline).

To investigate the variation law of acceleration along the vertical and horizontal directions of the soil layer, four column nodes from A1 to A4 and four row nodes from B1 to B4 are selected, as shown in Fig. 5.17a. Figure 5.17b and c shows the acceleration amplification factors for these nodes, i.e., the ratio of the peak acceleration for any node to the input bedrock peak acceleration. For these four column nodes, the amplification factor increases non-monotonically as the soil depth decreases in

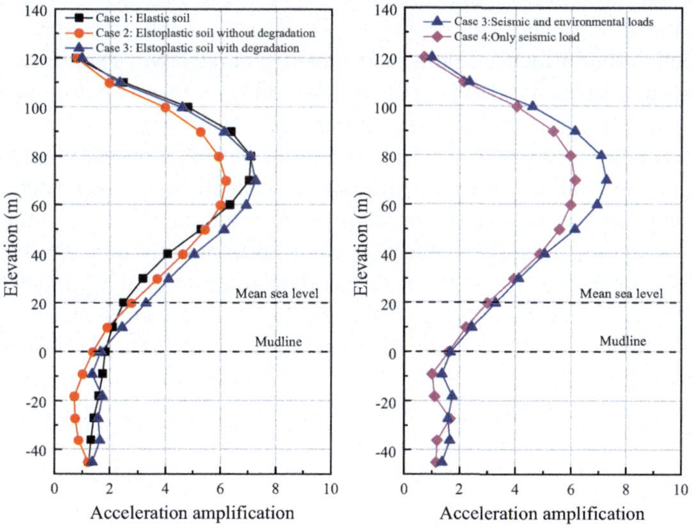

(a) Influence of soil plasticity and stiffness degradation

(b) Influence of environmental loads

Fig. 5.16 Acceleration amplification factor along elevations from the pile bottom to the tower top the overall trend, as shown in Fig. 5.17b. When the depth of the soil layer is within 10 m below the mudline, the amplification factors for Columns A2 and A3 near the pile side are greater than those for Columns A1 and A4 far from the pile side. This phenomenon can be further confirmed by the amplification factor for the Row B1 nodes shown in Fig. 5.17c. For Row B1 at the mudline, the acceleration amplification is more significant near the pile side, while for Rows B2 and B3 far away from the mudline, this trend seems to be opposite. For Row B4 below the pile tip, the acceleration amplification in the horizontal direction seems to be the same. This phenomenon can be attributed to the influence of the pile-soil dynamic interaction near the pile side. This dynamic interaction is most significant near the mudline, which causes the lateral flow of soil to be more significant and enhances the acceleration amplification. However, the existence of piles actually hinders the horizontal movement of soil, and the soil flow away from the mudline is not as significant as that close to the mudline, so the acceleration amplification usually decreases near the pile side. Obviously, there is almost no pile-soil dynamic interaction below the pile tip, so the acceleration amplification is basically consistent in the horizontal direction, such as in Row B4.

5.4.4 Soil Flow Mechanism

Figure 5.18 shows the typical displacement field of the soil domain under the three working cases. For elastic soil (Case 1), the surface acceleration direction is opposite to the soil layer movement direction, as shown in Fig. 5.18a. The movement of the monopile and that of the soil layer are almost synchronous during the earthquake, whether the surface acceleration is positive or negative. Therefore, separation between the pile and soil did not occur, and the soil layer mainly exhibited horizontal displacement, while the vertical displacement was almost negligible. The movement direction from the bottom of the soil layer to the mudline is completely consistent, and the displacement of the soil layer increases monotonically as the depth decreases. The displacement reaches the maximum value at the mudline. Obviously, the displacement of the pile top is greater than that of the pile bottom, and the displacement difference between the two will cause a certain rotation of the pile section at the mudline. The pile-soil dynamic interaction effect is relatively insignificant overall for Case 1, which is mainly because the relatively high stiffness of elastic soil causes the pile to be more flexible.

Compared with the elastic soil (Case 1), the displacement of the elastoplastic soil layer (Cases 2 and 3) increases significantly. The movement of the pile and that of the soil layer will no longer be synchronized when considering soil plasticity. For the elastoplastic soil without consideration of the cyclic degradation of stiffness (Case 2), the pile and the surrounding soil demonstrate a certain dynamic interaction effect. The ground surface in the passive earth pressure area near the pile swells due to the lateral pressure of the pile, while the ground surface in the active earth pressure area settles due to lateral unloading, and a gap occurs between the pile and

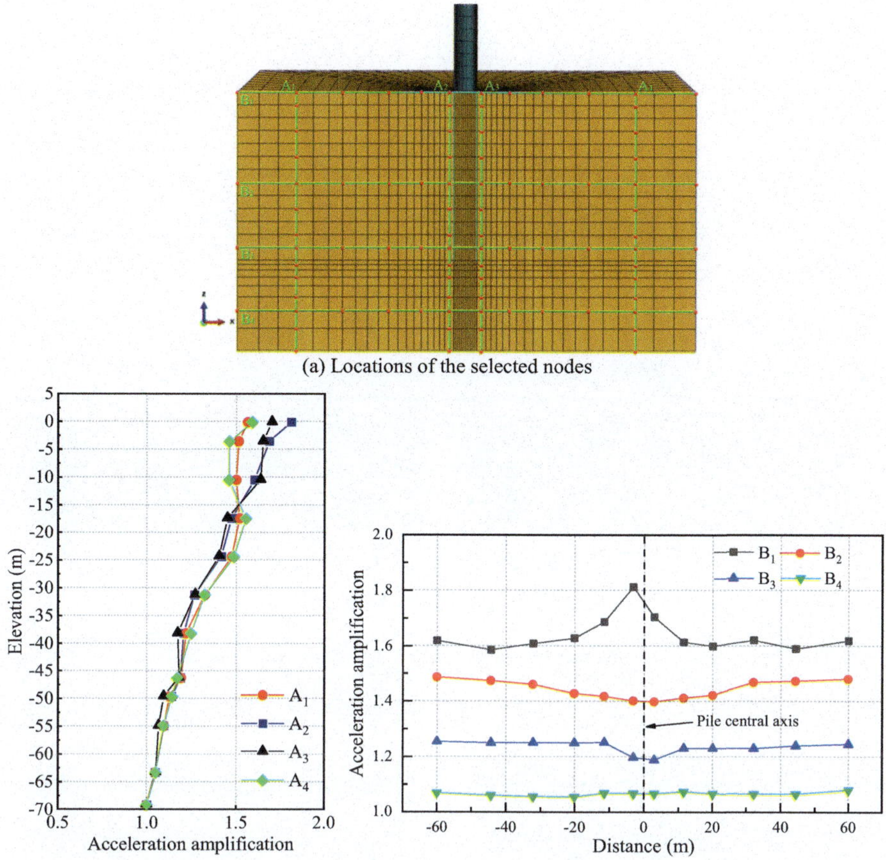

(a) Locations of the selected nodes

(b) Vertical direction

(c) Horizontal direction

Fig. 5.17 Acceleration amplification along the vertical and horizontal directions of the soil layer

soil, as shown in Fig. 5.18b. When the surface acceleration is negative (the left one of Fig. 5.18b), the direction of movement of the soil layer along the depth is not completely consistent. The soil layer near the surface in the passive earth pressure area (left side of pile) moves in the positive x-axis direction, while the lower soil layer moves in the negative x-axis direction. The maximum displacement of the soil layer does not occur on the surface but at a certain depth below the surface. However, the maximum displacement along the pile still occurs at the mudline. When the surface acceleration is positive (the right image of Fig. 5.18b), the pile still moves in the negative x-axis direction, which is consistent with the movement direction of the soil layer, and the lateral displacement of the pile increases significantly.

For the elastoplastic soil considering the cyclic degradation of stiffness (Case 3), the pile-soil dynamic interaction effect is the most significant, as shown in Fig. 5.18c. The heave deformation of the ground surface in the passive earth pressure area

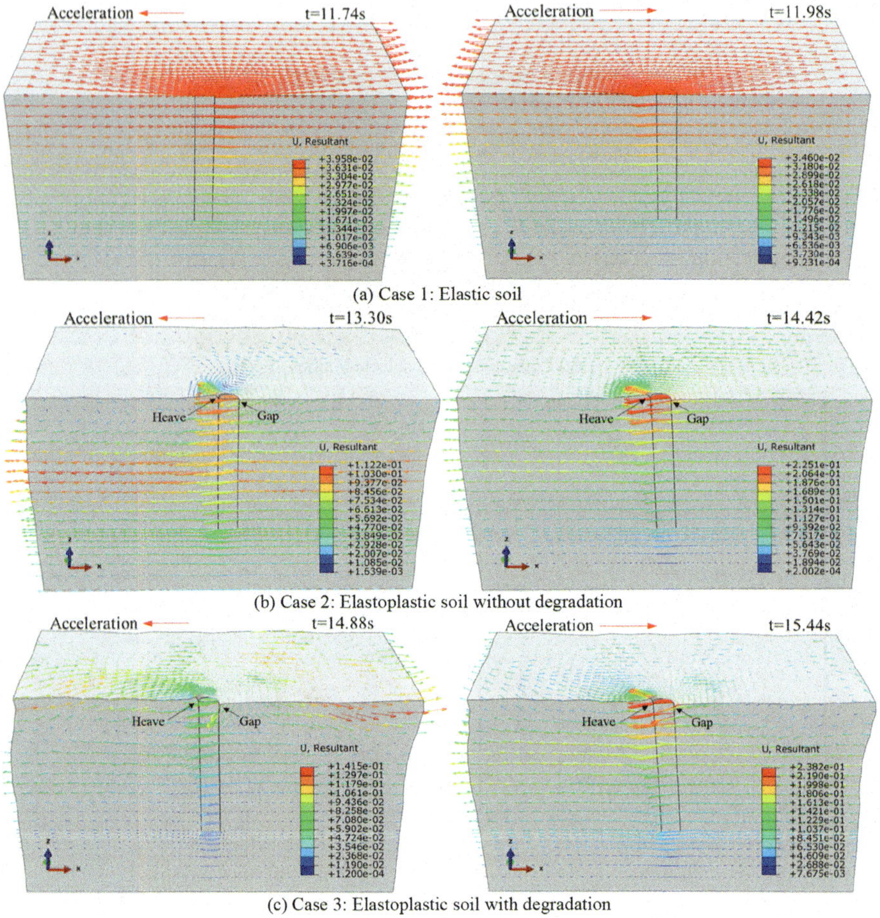

Fig. 5.18 Typical displacement field of the soil domain (amplification factor of 16)

of the pile side and the settlement deformation in the active earth pressure area become more significant, and the gap between the pile and the surrounding soil after separation becomes larger. Compared with the elastoplastic soil without stiffness degradation (Case 2), the influence range of the pile-soil dynamic interaction becomes larger. When the surface acceleration is negative (the left image of Fig. 5.18c), the movement of the soil layer near the mudline is opposite to that of the pile; the significant pile-soil interaction complicates the soil flow by forming a "vortex" flow area on the left side of the pile. When the surface acceleration is positive (the right one of Fig. 5.18c), similar to Case 2, the movement direction of the pile becomes consistent with that of the soil layer. However, the ground surface in the passive earth pressure area still has a significant upwards displacement component due to the pile-soil interaction. In fact, the pile displacement always occurs in the negative x-axis direction during an earthquake, and the rotation angle of the pile section at the mudline always accumulates in the counterclockwise direction, which can also be confirmed by Fig. 5.12a and b. The soil plasticity is more significant when the cyclic degradation of soil stiffness is considered, which will significantly increase the dynamic interaction effect of the soil-pile-superstructure and then increase the complexity of the soil flow.

5.5 Summary

The dynamic responses of MOWTs to seismic and environmental loading considering the stiffness degradation of clay were investigated by performing 3D finite element analysis. The key conclusions can be summarized as follows:

(1) A numerical method was developed for the interaction of the soil-pile-structure system based on a simplified bounding surface constitutive model of clay that can capture the stiffness degradation well. The numerical method can simulate the (a) free field response, (b) kinematic interaction for the soil-pile system and (c) coupling kinematic-inertial interaction for the soil-pile-structure system well.
(2) The monopile experienced significant rotation and settlement under combined seismic and environmental loading. The rotation and settlement will be underestimated if the soil plasticity is not considered. The cyclic degradation of soil stiffness significantly increases the rotation and settlement. The environmental loads can also aggravate the rotation and settlement caused by earthquakes.
(3) The maximum deflection and rotation angle increase monotonically from the pile bottom to the maximum value at the tower top. The change trends of the positive and negative bending moment profiles are similar but asymmetric. The soil plasticity and stiffness degradation significantly increase the peak response of the lateral deflection, rotation angle and bending moment of the MOWT under combined seismic and environmental loading.
(4) The input bedrock peak acceleration is amplified to varying degrees from the pile bottom to the tower top. The amplified peak acceleration time lags significantly behind the input bedrock peak acceleration time. The frequency of structural vibration decreases gradually as the height increases. The acceleration amplification along the height is the most significant overall when the degradation of soil stiffness is considered. The acceleration amplification of the soil layer increases non-monotonically as the soil depth decreases, and it is most significant near the pile side at the mudline.
(5) Environmental loads. such as wind and wave loads, significantly increase the absolute values of the seismic peak response profile, including the deflection, rotation angle and bending moment. The environmental loads also increase the acceleration amplification along the pile and tower during an earthquake, and the increasing effect is more significant in the middle section of the tower.

(6) For the elastic soil, the movement of the pile and that of the soil layer are almost synchronous during the earthquake, and the effect of the pile-soil dynamic interaction is relatively insignificant because the relatively high stiffness of elastic soil causes the pile to be more flexible. The movement of the pile and soil will no longer be synchronized when the soil plasticity is considered. The soil plasticity is more significant when considering the soil stiffness degradation, which will significantly increase the dynamic interaction effect of the soil-pile-superstructure and then increase the complexity of the soil flow.

Appendix

Setup of the Numerical Model for the MOWT System

(1) Soil domain size

A sensitivity study was conducted to establish the length, width and height of the soil domain to eliminate the effects of the boundary on the results. Soil domains with different sizes were meshed to calculate the dynamic responses of monopiles under earthquakes. The size is considered to be appropriate when it has little effect on the calculation results. Figure 5.19 shows various finite element models with different soil domain sizes. Figure 5.20 shows the calculated rotation angle time histories of the monopile sections at the mudline for different soil domain sizes. It can be observed from Fig. 5.20 that the rotation angle (absolute value) increases slightly as the size increases. When the size exceeds 120 m × 60 m × 69 m (Soil domain 3), it has little effect on the calculation results. Hence, a mesh size of 120 m × 60 m × 69 m (length × width × height) is considered sufficient to avoid boundary effects on the obtained results.

(2) Space discretization

To meet the requirements of computational accuracy, a suitable grid spacing should be determined by considering the minimum relevant wavelength (or highest frequency) of the input signal. The maximum grid size is usually less than one-tenth of the minimum wavelength (Watanabe et al. 2017), as follows:

$$\Delta x \leq \frac{\lambda_{\min}}{10} = \frac{V_s}{10 f_{\max}} = 1.42\,\text{m} \tag{5.11}$$

where Δx is the maximum grid size; $\lambda_{\min}$ is the minimum relevant wavelength; $f_{\max}$ is the highest frequency, and $f_{\max} = 20$ Hz in this study; V_s is the shear wave velocity of soil, and $V_s = \sqrt{G/\rho} = 284$ m/s, where the soil shear modulus $G = 46.3$ MPa (average value of the soil layer along the pile shaft), and the soil density $\rho = 1628$ kg/m^3.

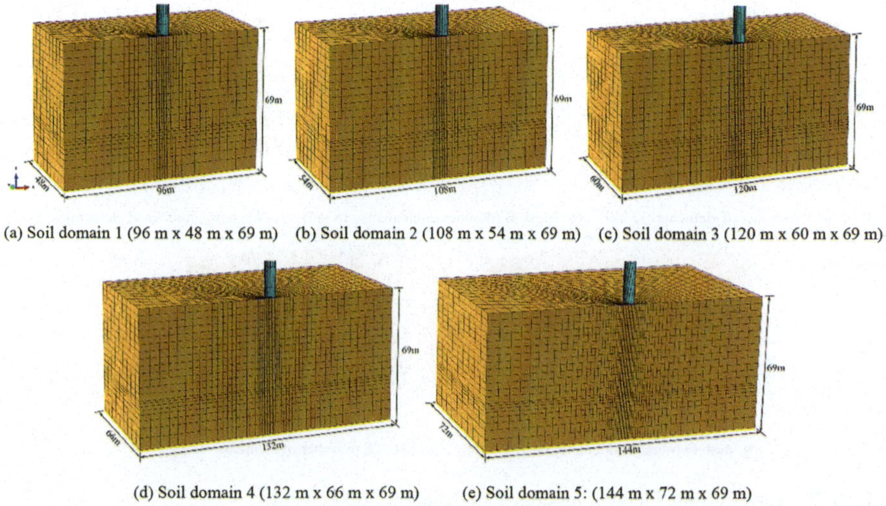

(a) Soil domain 1 (96 m x 48 m x 69 m) (b) Soil domain 2 (108 m x 54 m x 69 m) (c) Soil domain 3 (120 m x 60 m x 69 m)

(d) Soil domain 4 (132 m x 66 m x 69 m) (e) Soil domain 5: (144 m x 72 m x 69 m)

Fig. 5.19 Various finite element models with different soil domain sizes

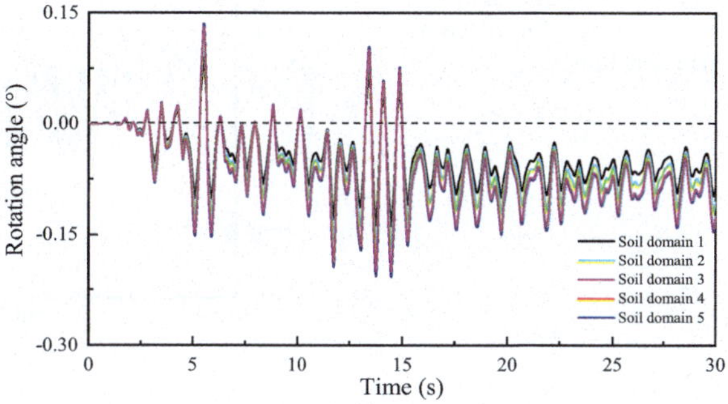

Fig. 5.20 Rotation angle time histories of monopiles at the mudline for different soil domain sizes

A sensitivity study was conducted by establishing 3D finite element models with different grid densities (see Fig. 5.21) in view of the above requirement for grid space. To improve the computational efficiency, a finer mesh is applied near the monopile, where nonlinearity is expected to be more prominent, while a coarser mesh is applied away from the monopile. The rotation angle time history of the monopile section at the mudline computed by employing different models is shown in Fig. 5.22. It can be observed that the rotation angle (absolute value) increases as the grid density increases. Mesh C is found to be appropriate because the calculation results do not change much when using a denser grid. Therefore, the finite element model shown in Fig. 5.21c was selected as the final calculation model considering computational accuracy and efficiency. In addition, 8-node linear brick elements with reduced integration (C3D8R) are used in the finite element calculation, which can effectively avoid volumetric locking when the material behavior is (almost) incompressible.

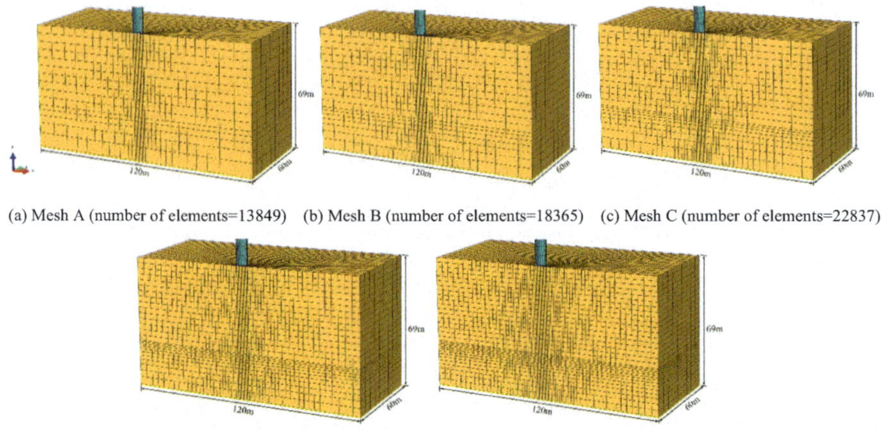

(a) Mesh A (number of elements=13849) (b) Mesh B (number of elements=18365) (c) Mesh C (number of elements=22837)

(d) Mesh D (number of elements=31875) (e) Mesh E (number of elements=43923)

Fig. 5.21 Various finite element models with different grid densities

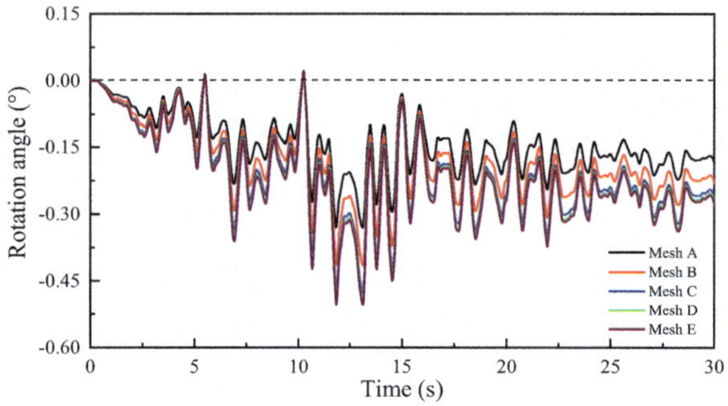

Fig. 5.22 Rotation angle time histories of monopiles at the mudline for different grid densities

(3) Boundary conditions

The specific settings of the boundary conditions for the finite element model are shown in Fig. 5.23. For the two vertical boundaries parallel to the YOZ plane, equivalent displacement constraints were set to simulate the seismic responses of infinite horizontal strata. The nodes on both boundaries with the same height are bound together (multipoint constraint (MPC) method), forcing the opposite vertical sides to move simultaneously, thus preventing any rotation and eliminating the effect of reflected waves. Normal horizontal constraints ($Y = 0$) were applied to the two vertical boundaries parallel to the XOZ plane. The base boundary of the model is simulated as rigid bedrock, and the seismic acceleration in the X-axis direction is applied to the bedrock ($Y = Z = 0$) and propagates upwards.

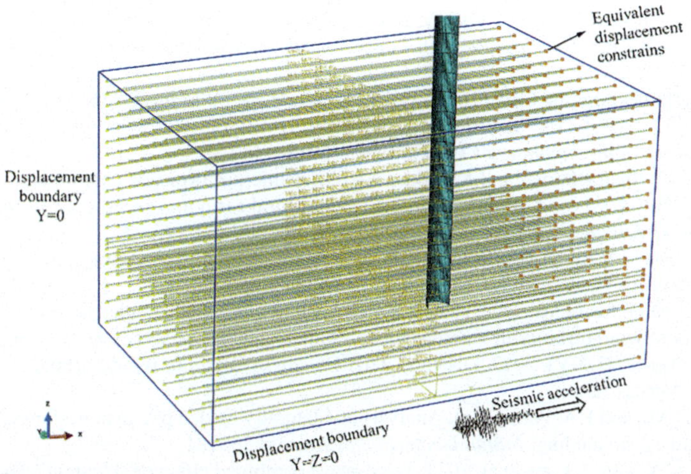

Fig. 5.23 Boundary conditions for the finite element model

References

Achmus, M., Y.S. Kuo, and K. Abdel-Rahman. 2009. Behavior of monopile foundations under cyclic lateral load. *Computers and Geotechnics* 36 (5): 725–735.

Anastasopoulos, I., and M. Theofilou. 2016. Hybrid foundation for offshore wind turbines: Environmental and seismic loading. *Soil Dynamics and Earthquake Engineering* 80: 192–209.

Arany, L., S. Bhattacharya, J.H.G. Macdonald, et al. 2015. A critical review of serviceability limit state requirements for monopile foundations of offshore wind turbines. In *Offshore Technology Conference Houston.*

Binh, L., T. Ishihara, P. Phuc, et al. 2008. A peak factor for non-Gaussian response analysis of wind turbine tower. *Journal of Wind Engineering and Industrial Aerodynamics* 96: 2217–2227.

Bourgeois, E., M.H.J. Rakotonindriana, A. Le Kouby, et al. 2010. Three-dimensional numerical modelling of the behaviour of a pile subjected to cyclic lateral loading. *Computers and Geotechnics* 37 (7–8): 999–1007.

Byrne, B.W., C. Leblanc, and G.T. Houlsby. 2010. Response of stiff piles to long term cyclic loading. *Géotechnique* 60 (2): 79–90.

Cao, G., Z. Chen, C. Wang, et al. 2020. Dynamic responses of offshore wind turbine considering soil nonlinearity and wind-wave load combinations. *Ocean Engineering* 217 (2): 108155.

Carter, J.M.F. 2007. North Hoyle offshore wind farm: Design and build. *Proceedings of the Institution of Civil Engineers-Energy* 160 (1): 21–29.

Chen, D., P. Gao, and S. Huang, et al. 2020a. Static and dynamic loading behavior of a hybrid foundation for offshore wind turbines. *Marine Structures* 71: 102727.

Chen, L., X. Yang, L. Li, W. Wu, et al. 2020b. Numerical analysis of the deformation performance of monopile under wave and current load. *Energies* 13(23): 6431.

Cheng, X.L., T.J. Wang, J.X. Zhang, et al. 2021. Finite element analysis of cyclic lateral responses for large diameter monopiles in clays under different loading patterns. *Computers and Geotechnics* 134: 104104.

Cheng, X.L., T.J. Wang, J.X. Zhang, et al. 2023. Dynamic response analysis of monopile offshore wind turbines to seismic and environmental loading considering the stiffness degradation of clay. *Computers and Geotechnics* 155: 105210.

Corciulo, S., O. Zanoli, and F. Pisanò. 2017. Transient response of offshore wind turbines on monopiles in sand: Role of cyclic hydro–mechanical soil behaviour. *Computers and Geotechnics* 83: 221–238.

Dassault Systemes Simulia Corp. 2014. *Abaqus Analysis User's Manual Version 6.14*. Providence.

Davenport, A.G. 1961. The application of statistical concepts to the wind loading of structures. *Proceedings of the Institution Civil Engineers* 19 (4): 449–472.

De Risi, R., S. Bhattacharya, and K. Goda. 2018. Seismic performance assessment of monopile-supported offshore wind turbines using unscaled natural earthquake records. *Soil Dynamics and Earthquake Engineering* 109: 154–172.

Deeks, A.J., and M.F. Randolph. 1994. Axisymmetric time-domain transmitting boundaries. *Journal of Engineering Mechanics* 120 (1): 25–42.

Depina, I., T.M.H. Le, G. Eiksund, et al. 2015. Behavior of cyclically loaded monopile foundations for offshore wind turbines in heterogeneous sands. *Computers and Geotechnics* 65: 266–277.

Det Norske Veritas (DNV). 2013. DNV-OS-J101 Design of Offshore Wind Turbine Structures. Det Norske, Veritas AS.

Dewaikar, D.M., and P.A. Patil. 2006. Analysis of a laterally loaded pile in cohesion-less soil under static and cyclic loading. *Indian Geotechnical Journal* 36: 181.

Ding, X., S.C. Chian, J. Lian, et al. 2023. Wind-wave combined effect on dynamic response of soil-monopile-OWT system considering cyclic hydro-mechanical clay behavior. *Computers and Geotechnics* 154: 105124.

Esteban, M.D., J.S. Lopez-Gutierrez, V. Negro, et al. 2015. Offshore wind foundation design: Some key issues. *Journal of Energy Resource Technology* 137(5): 051211.1–051211.6.

Fan, C.C., and J.H. Long. 2005. Assessment of existing method for predicting soil response of laterally loaded piles in sand. *Computers and Geotechnics* 32 (4): 274–289.

Fan, K., G. Gazetas, A. Kaynia, et al. 1991. Kinematic seismic response of single piles and pile groups. *Journal Geotechnical Engineering* 117 (12): 1860–1879.

Gao, B., G. Ye, Q. Zhang, et al. 2021. Numerical simulation of suction bucket foundation response located in liquefiable sand under earthquakes. *Ocean Engineering* 235: 109394.

Germanischer, L. 2010. *Guideline for the Certification of Offshore Wind Turbines*. Hamburg: Germanischer Lloyd.

Giannakos, S., N. Gerolymos, and G. Gazeta. 2012. Cyclic lateral response of piles in dry sand: Finite element modeling and validation. *Computers and Geotechnics* 44: 116–131.

Hamilton, B., L. Battenberg, M. Bielecki, et al. 2013. *Offshore Wind Market and Economic Analysis: Annual Market Assessment*. Burlington, MA: Navigant Consulting Inc.

Hansen, M.O.L. 2008. *Aerodynamics of Wind Turbines*, 2nd ed. London, UK: Earthscan.

Hong, Y., B. He, L.Z. Wang, et al. 2017. Cyclic lateral response and failure mechanisms of semi-rigid pile in soft clay: Centrifuge tests and numerical modelling. *Canadian Geotechnical Journal* 54 (6): 806–824.

Huang, M.S., L. Liu, Z.H. Shi, et al. 2021. Modeling of laterally cyclic loaded monopile foundation by anisotropic undrained clay model. *Ocean Engineering* 228: 108915.

Hussein, A.F., and M.H. El Naggar. 2022. Seismic behaviour of piles in non-liquefiable and liquefiable soil. *Bulletin of Earthquake Engineering* 20 (1): 77–111.

Iannuzzi, A., and P. Spinelli. 1987. Artificial wind generation and structural response. *Journal of Structural Engineering* 113 (12): 2382–2398.

Jiang, W., C. Lin, and M. Sun. 2021. Seismic responses of monopile-supported offshore wind turbines in soft clays under scoured conditions. *Soil Dynamics and Earthquake Engineering* 142: 106549.

Jonkman, J.M., S. Butterfield, W. Musial, et al. 2009. *Definition of a 5MW Reference Wind Turbine for Offshore System Development*. J. Office of Scientific & Technical Information Technical Reports.

Ju, S., and Y. Huang. 2019. Analyses of offshore wind turbine structures with soil-structure interaction under earthquakes. *Ocean Engineering* 187: 106190.

Kaynia, A.M. 2019. Seismic considerations in design of offshore wind turbines. *Soil Dynamics and Earthquake Engineering* 124: 399–407.

Kaynia, A.M. 2021. Effect of kinematic interaction on seismic response of offshore wind turbines on monopiles. *Earthquake Engineering & Structural Dynamics* 50 (3): 777–790.

Kementzetzidis, E., S. Corciulo, W.G. Versteijlen, et al. 2019. Geotechnical aspects of offshore wind turbine dynamics from 3D non-linear soil-structure simulations. *Soil Dynamics and Earthquake Engineering* 120: 181–199.

Kementzetzidis, E., A.V. Metrikine, W.G. Versteijlen, et al. 2021. Frequency effects in the dynamic lateral stiffness of monopiles in sand: Insight from field tests and 3D FE modelling. *Géotechnique* 71 (9): 812–825.

Kementzetzidis, E., F. Pisanò, and A.V. Metrikine. 2022. A memory-enhanced p-y model for piles in sand accounting for cyclic ratcheting and gapping effects. *Computers and Geotechnics* 148: 104810.

Kim, D.H., S.G. Lee, and I.K. Lee. 2014. Seismic fragility analysis of 5Mw offshore wind turbine. *Renewable Energy* 65: 250–256.

Kiyomiya, O., T. Rikiji, and P.H. Van Gelder. 2002. Dynamic response analysis of onshore wind energy power units during earthquakes and wind. In *Proceedings 12th International Offshore and Polar Engineering Conference*, 520–526.

Kuo, Y.S., M. Achmus, and K. Abdel-Rahman. 2011. Minimum embedded length of cyclic horizontally loaded monopiles. *Journal of Geotechnical and Geoenvironmental Engineering* 138 (3): 357–363.

Lee, S., H. Kim, and S. Lee. 2010. Analysis of aerodynamic characteristics on a counter-rotating wind turbine. *Current Applied Physics* 10: S339–S342.

Li, X., X. Zeng, X. Yu, et al. 2021. Seismic response of a novel hybrid foundation for offshore wind turbine by geotechnical centrifuge modeling. *Renewable Energy* 172: 1404–1416.

Liang, F., Z. Yuan, X. Liang, et al. 2022. Seismic response of monopile-supported offshore wind turbines under combined wind, wave and hydrodynamic loads at scoured sites. *Computers and Geotechnics* 144: 104640.

Liao, W.M., J.J. Zhang, J.B. Wu, et al. 2018. Response of flexible monopile in marine clay under cyclic lateral load. *Ocean Engineering* 147: 89–106.

Liaw, C.Y., and A.K. Chopra. 1974. Dynamics of towers surrounded by water. *Earthquake Engineering & Structural Dynamics* 3 (1): 33–49.

Liu, H., F. Pisanò, H.P. Jostad, et al. 2022. Impact of cyclic strain accumulation on the tilting behaviour of monopiles in sand: An assessment of the Miner's rule based on SANISAND-MS 3D FE modelling. *Ocean Engineering* 250: 110579.

Lysmer, J., and R.L. Kuhlemeyer. 1969. Finite dynamic model for infinite media. *Journal of the Engineering Mechanics Division* 95 (4): 859–878.

Ma, H., J. Yang, and L. Chen. 2017. Numerical analysis of the long-term performance of offshore wind turbines supported by monopiles. *Ocean Engineering* 136: 94–105.

Maheshwari, B.K., K.Z. Truman, M.H. El Naggar, et al. 2004. Three-dimensional nonlinear analysis for seismic soil-pile-structure interaction. *Soil Dynamics and Earthquake Engineering* 24 (4): 343–356.

Maheshwari, B.K., K.Z. Truman, P.L. Gould, et al. 2005. Three-dimensional nonlinear seismic analysis of single piles using finite element model: effects of plasticity of soil. *International Journal of Geomechanics* 5 (1): 35–44.

Mo, R., R. Cao, M. Liu, et al. 2021. Seismic fragility analysis of monopile offshore wind turbines considering ground motion directionality. *Ocean Engineering* 235: 109414.

Nicholson, J.C. 2011. *Design of Wind Turbine Tower and Foundation Systems: Optimization Approach*. J. Dissertations & Theses Gradworks.

Novak, M., and H. Mitwally. 1988. Transmitting boundary for axisymmetrical dilation problems. *Journal of Engineering Mechanics* 114 (1): 181–187.

Oh, K.Y., W. Nam, M.S. Ryu, et al. 2018. A review of foundations of offshore wind energy convertors: Current status and future perspectives. *Renewable and Sustainable Energy Reviews* 88: 16–36.
Peiris, T., D. Thambiratnam, N. Perera, et al. 2014. Soil-pile interaction of pile embedded in deep-layered marine sediment under seismic excitation. *Structural Engineering International* 24 (4): 521–531.
Pierson, W.J., and L. Moskowitz. 1964. A proposed spectral form for fully developed wind seas based on the similarity theory of SA Kitaigorodskii. *Journal of Geophysical Research Atmospheres* 69 (24): 5181–5190.
Pisanò, F. 2019. Input of advanced geotechnical modelling to the design of offshore wind turbine foundations. In *Proceedings of the XVII ECSMGE-2019: Geotechnical Engineering Foundation of the Future*, 1–26.
Prowell, I., and P. Veers. 2009. *Assessment of wind Turbine Seismic Risk: Existing Literature and Simple Study of Tower Moment Demand*. Sandia National Laboratories Technical Report SAND2009-1100 (Albuquerque, New Mexico).
Rajashree, S.S., and R. Sundaravadivelu. 1996. Degradation model for one-way cyclic lateral load on piles in soft clay. *Computers and Geotechnics* 19 (4): 289–300.
Ren, B., and Y.X. Wang. 2004. Numerical simulation of random wave slamming on structures in the splash zone. *Ocean Engineering* 31: 547–560.
Sapountzakis, E.J., I.C. Dikaros, A.E. Kampitsis, et al. 2015. Nonlinear response of wind turbines under wind and seismic excitations with soil-structure interaction. *Journal of Computational and Nonlinear Dynamics* 10 (4): 41001–41007.
Tsinidis, G., K. Pitilakis, and A.D. Trikalioti. 2014. Numerical simulation of round robin numerical test on tunnels using a simplified kinematic hardening model. *Acta Geotechnica* 9: 641–659.
Vacareanu, V., E. Kementzetzidis, and F. Pisano. 2019. 3D FE seismic analysis of a monopile-supported offshore wind turbine in a non-liquefiable soil deposit. In *Proceedings of the 2nd International Conference on Natural Hazards & Infrastructure (ICONHIC2019)*, 23–26.
Wang, P., M. Zhao, and X. Du. 2018. Wind, wave and earthquake responses of offshore wind turbine on monopile foundation in clay. *Soil Dynamics and Earthquake Engineering* 113: 47–57.
Wang, X., X. Zeng, X. Yang, et al. 2019. Seismic response of offshore wind turbine with hybrid monopile foundation based on centrifuge modelling. *Applied Energy* 235: 1335–1350.
Watanabe, K., F. Pisanò, and B. Jeremić. 2017. Discretization effects in the finite element simulation of seismic waves in elastic and elastic-plastic media. *Engineering with Computers* 33 (3): 519–545.
Wathugala, G.W., and C.S. Desai. 1993. Constitutive model for cyclic behavior of clays. I: Theory. *Journal of Geotechnical Engineering* 119(4): 714–729.
Wu, X., Y. Hu, Y. Li, et al. 2019. Foundations of offshore wind turbines: A review. *Renewable & Sustainable Energy Reviews* 104: 379–393.
Xi, R., P. Wang, X. Du, et al. 2021. Dynamic behaviors of wind turbines under wind and earthquake excitations. *Journal of Renewable and Sustainable Energy* 13 (4): 043306.
Xi, R., C. Xu, X. Du, et al. 2022. Framework for dynamic response analysis of monopile supported offshore wind turbine excited by combined wind-wave-earthquake loading. *Ocean Engineering* 247: 110743.
Yan, Y., C. Li, and Z. Li. 2021. Buckling analysis of a 10 MW offshore wind turbine subjected to wind-wave-earthquake loadings. *Ocean Engineering* 236: 109452.
Yang, M., R. Luo, and W. Li. 2018. Numerical study on accumulated deformation of laterally loaded monopiles used by offshore wind turbine. *Bulletin of Engineering Geology and the Environment* 77: 911–921.
Yang, Y., Bashir, M., Li, C., et al., 2019. Analysis of seismic behaviour of an offshore wind turbine with a flexible foundation. *Ocean Engineering* 178: 215–228.
Yu, H., X. Zeng, B. Li, et al. 2015. Centrifuge modeling of offshore wind foundations under earthquake loading. *Soil Dynamics and Earthquake Engineering* 77: 402–415.

Zhang, J.X., W.L. Cheng, X.L. Cheng, et al. 2021a. Seismic responses analysis of suction bucket foundation for offshore wind turbine in clays. *Ocean Engineering* 232: 109159.

Zhang, Q.Y., Y. Zhang, H.D. Lin, et al. 2021b. Numerical investigation on bearing capacity of OWT foundation with large diameter monopile under seismic load. *Applied Ocean Research* 108 (7): 102518.

Zhao, X., and P. Maisser. 2006. Seismic response analysis of wind turbine towers including soil-structure interaction. *Proceedings of the Institution of Mechanical Engineers, Part k: Journal of Multi-Body Dynamics* 220 (1): 53–61.

Zheng, X.Y., H. Li, W. Rong, et al. 2015. Joint earthquake and wave action on the monopole wind turbine foundation: An experimental study. *Marine Structures* 44: 125–141.

Zhu, B., Z. Zhu, T. Li, et al. 2017. Field tests of offshore driven piles subjected to lateral monotonic and cyclic loads in soft clay. *Journal of Waterway, Port, Coastal, and Ocean Engineering* 143 (5): 05017003.

Zuo, H., K. Bi, and H. Hao. 2018. Dynamic analyses of operating offshore wind turbines including soil-structure interaction. *Engineering Structures* 157: 42–62.

Chapter 6
Dynamic Responses of OWT Tripod Pile in Clays

Tripod pile foundation is pursued as an efficient foundation system for offshore wind turbines (OWTs) installed in deeper waters (20–50 m). In this application, the foundations would be subjected to dynamic loads including due to wind, waves, and earthquakes. This chapter presents a numerical method for analyzing the dynamic responses of tripod pile foundation installed in clay based on a simplified bounding surface model to capture the clay stiffness degradation. Their behaviors under lateral monotonic, cyclic and seismic loads were investigated, and their bearing mechanism was analyzed. The results revealed that the loading direction significantly affects the ultimate bearing capacity of tripod pile foundation as the foundation capacity results from either a Pile A in tension and a Pile B in compression or vice versa. The evolution of axial force, bending moment and lateral displacement profiles of the Pile A and Pile B with the number of cycles exhibit different characteristics under lateral one-way and two-way cyclic loading. The foundation experiences cumulative rotation angles toward the Pile A side under seismic load due to the lower vertical bearing capacity of the Pile A compared to the Pile B. The tower top experiences the maximum lateral displacement and rotation angle, and the top of tripod support experiences the maximum bending moment. These findings should be considered in the design of OWT tripod pile foundations.

6.1 Introduction

Numerous offshore wind farms are being planned worldwide to satisfy the growing energy demand. Offshore wind turbines (OWTs) construction still encounters many challenges due to the harsh offshore environment, which imposes complex loading conditions due to wind, waves, current, and earthquakes. Therefore, the design of OWT foundations is both challenging and economically crucial as it may account for as much as 30% of the total cost of a typical offshore wind project (Oh et al. 2018;

X. Cheng et al., *Dynamic Analysis of Offshore Wind Turbine Foundations in Soft Clays*,
https://doi.org/10.1007/978-981-97-9454-6_6

Wu et al. 2019). Nonetheless, monopile foundations remain by far the most popular foundation type, as they were used for more than 65% of the currently installed OWTs (Esteban et al. 2015). Compared with large diameter monopile foundation, tripod pile foundation has great potential advantages for deep water (20–50 m) OWT installation. It comprises the tripod support structure and three steel pipe piles of equal diameter arranged in equilateral triangle (see Fig. 6.1). It has relatively large structural stiffness, good overall stability, and a relatively simple construction process; and has been used in major wind farm projects such as Alpha Ventus and Trianel Windpark Borkum in Germany (Koh and Ng 2016).

In harsh offshore environment, OWTs located in seismically active regions are susceptible to earthquakes, in addition to wind and waves loads. These loads may cause the OWT foundation experience a large overturning moment, and therefore induces excessive rotation angular displacement, which will affect the normal operation of the wind turbine, and may even cause overturning damage. In the current guidelines for OWTs, the requirements for the serviceability limit state (SLS) typically dominate the whole design of foundations (Arany et al. 2015). For instance, the tolerance for permanent accumulated rotation of monopiles at the mudline is specified as 0.25° in terms of the SLS (Det Norske Veritas (DNV) 2013). The ability of foundations to resist environmental and seismic loading is a major challenge facing

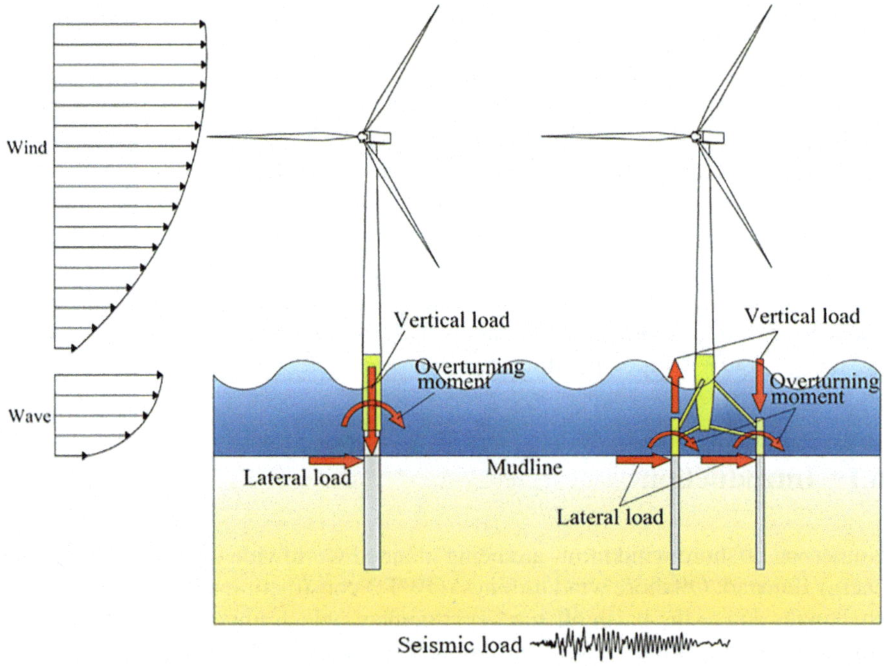

Fig. 6.1 Schematic diagram of monopile and tripod pile foundation for OWTs

designers, which has also motivated many researchers to study dynamic responses of foundations under these loads.

Most existing research has focused primarily on large diameter monopiles. For example, the long-term lateral cyclic responses of monopiles under wind and wave loads have been studied widely. Field tests (Zhu et al. 2017; Xu et al. 2020), centrifuge model tests (Jeanjean 2009; Zhang et al. 2011; Hong et al. 2017; Lai et al. 2020) and scaled model tests (Liao et al. 2018; Li et al. 2021a) have been conducted to study lateral cyclic responses of monopiles in clay. The lateral cyclic responses of monopiles have also been analyzed employing three-dimensional (3D) finite element simulations that incorporated some stiffness degradation models (Achmus et al. 2009; Kuo et al. 2011; Depina et al. 2015; Yang et al. 2018) and advanced soil constitutive models (Bourgeois et al. 2010; Giannakos et al. 2012; Ma et al. 2017; Corciulo et al. 2017; Chen et al. 2020a, b; Cheng et al. 2021; Wu et al. 2022; Ding et al. 2023). In general, important achievements have been made on the bearing and deformation characteristics of large-diameter monopile foundation under horizontal cyclic loads; and good understanding was attained with regard to the factors that affect the cumulative rotation angle, lateral deflection and bending moment of monopile with the number of cycles, amplitude, frequency, and direction of the cyclic load.

For OWTs located in seismically active regions, the seismic loads should be considered in their design (Kaynia 2019). Different from wind and wave loads, the inertial effect of the seismic load must be considered in the analysis due its high frequency characteristics. Seismic centrifuge tests of monopiles supporting OWTs (Yu et al. 2015; Wang et al. 2019; Li et al. 2021b) indicate that they experience large tilt during earthquakes, and the tower head experiences significantly amplified response. Similarly, 3D numerical simulation results (Vacareanu et al. 2019; Cheng et al. 2023a) indicate that earthquake loading leads to accumulation of monopile rotation and settlement. Some studies have evaluated seismic vulnerability of monopile OWTs employing the nonlinear Winkler foundation models to simulate pile-soil interaction (Kim et al. 2014; De Risi et al. 2018). Mo et al. (2021) and investigated the influence of seismic motion direction on the monopile seismic response while Kaynia (2021) studied the influence of kinematic interaction on the monopile seismic response. The dynamic responses of monopile-supported OWTs under the combined action of wind, wave, and earthquake were investigated, and the effects of factors such as wind, wave and seismic load intensity on the dynamic response of wind turbines were evaluated (Zheng et al. 2015; Wang et al. 2018b; Yang et al. 2019; Zhang et al. 2021; Xi et al. 2022). Collectively, the above studies provided good understanding of the seismic dynamic response characteristics of monopiles, including their bearing capacity and deformation as well the influence of pile-soil interaction on the seismic response of supported OWTs.

Different from the resistance mode of monopile foundation, multi-pile foundation will withstand large overturning moment through axial resistance, i.e., axial push–pull forces of individual piles as demonstrated in Fig. 6.1, in addition to horizontal resistance (Lu et al. 2017). Some studies investigated the axial cyclic behavior of multi-pile foundations through model tests (e.g., Jardine and Standing 2012; Tsuha et al. 2012) and numerical simulations (e.g., D'Aguiar et al. 2011; Huang and Liu

2015). These studies revealed that the pile shaft friction significantly affects the pile axial dynamic response. Therefore, the pile-soil interface behavior was investigated through interface shear tests, and the evolution mechanism of soil-pile interface strength was evaluated (Liu et al. 2019; Zhou et al. 2020a, b). However, each pile of multi-pile foundation experiences axial and horizontal loads simultaneously, so it is prudent to study the overall dynamic response of the multi-pile foundation.

The overall dynamic response of four-pile foundation of jacket OWTs under horizontal cyclic loads were studied through centrifugal model tests (Kong et al. 2019) and numerical simulations (Wei et al. 2013). In addition, Wang et al. (2020) studied the evolution characteristics of natural frequency and cumulative tilt of four-pile foundation of jacket OWTs under long-term horizontal cyclic loads. Abhinav and Saha (2018) investigated the response of a jacket supporting a 5-MW OWT under the hydrodynamic and aerodynamic loading by incorporating soil-structure interaction using a Winkler-spring model. Ma et al. (2018) studied the impact of scouring on the dynamic response of tripod pile foundations of OWTs by conducting 3D finite element simulation. Asumadu et al. (2022) investigated the wave-induced oscillatory seabed response around the tripod pile foundation of OWTs by using an integrated 3D numerical model.

In conclusion, the dynamic responses of OWT monopile foundations to wind, wave and earthquake loads is well investigated. However, studies on the dynamic response of the OWT tripod pile foundations in clays are scarce, especially research on their seismic response is rarely reported. In this chapter, a numerical method is developed to analyze the dynamic response of OWT tripod pile foundations based on a simplified bounding surface model to capture the clay stiffness degradation (Cheng et al. 2023b). The developed model is then employed to investigate their behavior under horizontal monotonic, cyclic and seismic loads, and establish their bearing mechanism.

6.2 3D Numerical Model of Tripod Pile OWT System

6.2.1 3D Finite Element Model

Figure 6.2 presents the 3D finite element model of the tripod pile OWT system, which is established according to the specifications of the NREL offshore 5-MW baseline wind turbine described in Table 6.1. The OWT is supported by a cylindrical tower that connects with the three-pile foundation by a tripod; and a flange connects the tower and tripod. The tower has a variable cross-section with diameter varying from 6 m at the base to 3.87 m at the top and wall thickness varying from 0.03 m to 0.02 m, respectively. The tower has a total height of 77.5 m, and its bottom is located 10 m above mean sea level (MSL). The wind turbine, tripod support and pile foundation are constructed from steel with Young's modulus, $E_p = 210$ GPa and Poisson's ratio of 0.3. The steel density for the wind turbine is 7850 kg/m^3, and the steel density of tripod support and pile foundation is 8500 kg/m^3 (Jonkman et al. 2009). The rotor blades, nacelle and hub are simulated as a lumped mass of 350,000 kg at the top of tower, as shown in Fig. 6.2. Based on the results of a sensitivity study, the finite element mesh size is set to 80 m × 80 m × 60 m (length × width × height) to avoid boundary effects on the obtained results. The piles-tripod-OWT system is discretized by 55,704 8-node linear brick elements with reduced integration (C3D8R). The soil element size is chosen based on the soil shear wave velocity (v_s) and the dominant frequency ($f_{\max}$) of the input ground motion. The minimum size (Δl) of the elements should be at least one-eighths to one-tenth of the seismic wave length (λ) (Kuhlemeyer and Lysmer 1973), i.e.,

$$\Delta l = \left(\frac{1}{8} - \frac{1}{10}\right)\lambda = \left(\frac{1}{8} - \frac{1}{10}\right)\frac{v_s}{f_{\max}} \tag{6.1}$$

The pile-soil interaction is simulated utilizing the surface-to-surface contact method. In the normal direction of contact surface, hard contact is set to allow separation between the interface elements when they are subjected to tension. The penalty contact method is used in the tangential direction; when the two surfaces are in contact, the interface behavior is governed by Coulomb's friction theory. The critical friction shear stress at the contact surface can be expressed by $\tau_{\text{crit}} = \mu \cdot p_c$ in terms of frictional coefficient μ (0.4 is set in this study) and contact pressure p_c. When the shear stress at the contact surface exceeds τ_{crit}, tangential slip occurred.

The model base boundary is simulated as rigid bedrock, and the top boundary is free surface. In the static analysis, horizontal constraints are applied to the vertical boundaries, while equivalent displacement constraints are set in the seismic dynamic

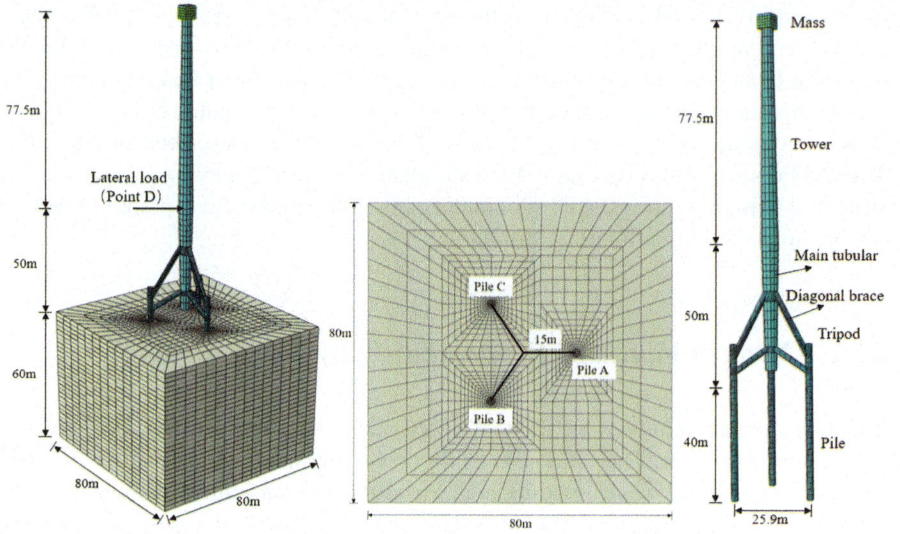

Fig. 6.2 3D finite element model of a tripod pile OWT system in clay

Table 6.1 Dimensions of the 5 MW OWT model

Property	Symbol	Dimension
Rotor diameter	D_{R}	126 m
Length of the tower	L_{T}	77.5 m
Length of the pile	L_{P}	40 m
Tower base diameter and thickness	D_{T}, T_{K}	6 m, 3 cm
Tower top diameter and thickness	D_{T}, T_{K}	3.87 m, 2 cm
Pile diameter and thickness	D_{T}, T_{K}	2.4 m, 3 cm
Main tubular diameter and thickness	D_{T}, T_{K}	3–6 m, 6 cm
Diagonal brace diameter and thickness	D_{T}, T_{K}	2 m, 3 cm
Mean sea depth	H_{W}	40 m
Mass	M	350,000 kg

Table 6.2 Material parameters of soil layers

Effective density (kg/m^3)	Poisson's ratio	S_u (kPa)	G (kPa)	A_0 (kPa)	μ	ξ
800	0.49	20 + 2z	167 S_u	2 S_u	2	1

analysis to simulate the seismic responses of infinite horizontal strata. The nodes on both boundaries with the same height are bound together (multipoint constraint (MPC) method), forcing the opposite vertical sides to move simultaneously, thus preventing any rotation and eliminating the effect of reflected waves (Tsinidis et al. 2014; Cheng et al. 2022).

The simplified version of single bounding surface model proposed by authors mentioned in Chap. 2 is adopted in the present simulation. The model can capture complicated nonlinear response of soil under dynamic loads by accounting for the nonlinear hysteresis, plastic strain accumulation, and cyclic degradation characteristics. In this study, S_u varied linearly with soil depth z, as represented by $S_u = 20 + 2z$ (z is soil depth). The material properties of soil layers are described in Table 6.2. The soil Poisson's ratio is set as 0.49 to simulate undrained behavior. In addition, the structural damping ratios 0.43% (Jonkman et al. 2009) and the soil damping ratio of 5% are considered in numerical simulation.

6.2.2 *Natural Vibration Frequency of OWT*

The first natural frequency of the OWT system including the foundation governs its dynamic behavior. The closeness of the tower frequency to excitation frequencies of the wind and waves loading could lead to large-amplitude responses of the structure. This must be avoided to prevent accelerated fatigue damage. In particular, there are two important frequencies associated with the OWT operation: the rotor

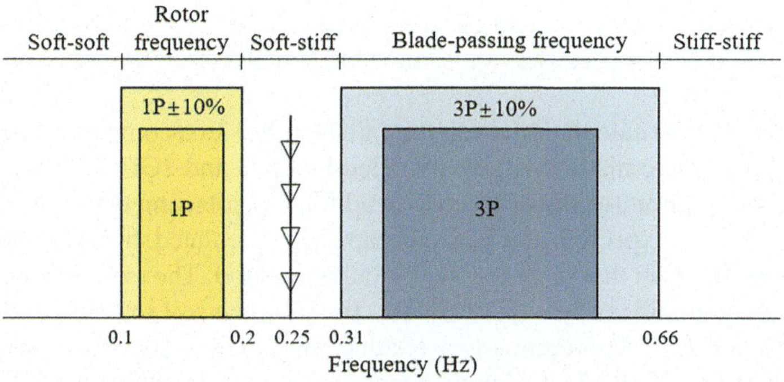

Fig. 6.3 Natural vibration frequency of NREL 5 MW OWT

rotation frequency, known as 1P frequency; and the frequency of three blades passing through the tower, known as the 3P frequency (see Fig. 6.3). These two frequencies correspond to the specific model of the OWT. For conventional OWT foundations, resonance avoidance is generally achieved by using a soft-stiff design, i.e., the first natural frequency of the structure is set between the 1P and 3P by adjusting the foundation stiffness. The size of tripod pile foundation in the finite element model is determined by multiple trial calculations to ensure that its natural frequency meets the requirements of engineering design. In the trial calculation, modal analysis step in ABAQUS was used to determine the vibration frequency of the OWT system, which involved binding constraints to avoid contact nonlinearity in pile-soil contact. The first natural frequency of the selected tripod pile OWT is 0.25 Hz, which is between 1 and 3P, and thus meeting the engineering design requirements.

6.3 Response of Tripod Pile Foundation Under Monotonic Loading

Offshore wind turbines may be subjected to wind and wave loads in any horizontal direction. Due to the asymmetric layout of the tripod pile foundation, the loading direction will affect its bearing characteristics. The angle between the horizontal load direction and the x-axis positive direction is defined as the loading angle, as shown in Fig. 6.4. The direction of 0° and 180° are selected as two extreme loading cases to discuss the impact of loading directions on the bearing capacity of tripod pile foundation. Given the foundation equilateral triangular plan arrangement, the forces applied on the two piles on the left side (Pile B and Pile C) are identical. Therefore, only Piles A and B are discussed in the following text. In the finite element simulation, the geostatic stress equilibrium analysis is applied first to establish soil initial stress field and ensuring initial displacement field of nearly zero. The horizontal monotonic load is then applied to Point D (see Fig. 6.2). The rotation angle of Point D represents the rotation angle of the overall tripod pile foundation. To ensure normal operation of the wind turbine, the permanent accumulated rotation angle at Point D should not exceed 0.25° in terms (i.e., Service Limit State (SLS)) (Det Norske Veritas (DNV) 2013).

3D finite element calculation result for a single pile is first compared with the traditional API *p–y* curve method (Recommended Practice and 2GEO 2014). Figure 6.5 shows the calculation results of the pile head load–displacement curve obtained by two methods. As expected, the bearing capacity calculated by 3D finite element method is higher than that obtained by *p–y* curve method. The main reason is that the API *p–y* curve significantly underestimates the ultimate soil reaction and the initial stiffness of soft clays. This comparison result is supported by previous research (Jeanjean 2009; Zhu et al. 2017), which demonstrates the rationality of the 3D numerical method in this study.

Figure 6.6a, b, and c, respectively, displays the load-rotation, load-vertical displacement and load-lateral displacement curves for the Pile A and Pile B at the mud surface and for Point D. Figure 6.6 demonstrates that the maximum rotation angle reaches 6° which far exceeds the rotation angle limit of 0.25°. This means the

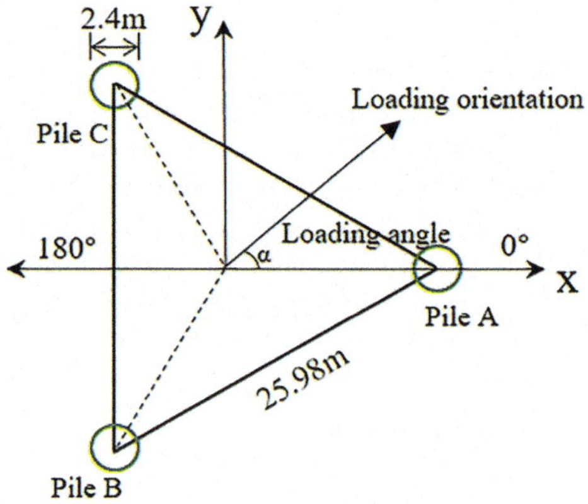

Fig. 6.4 Plan view of tripod pile foundation

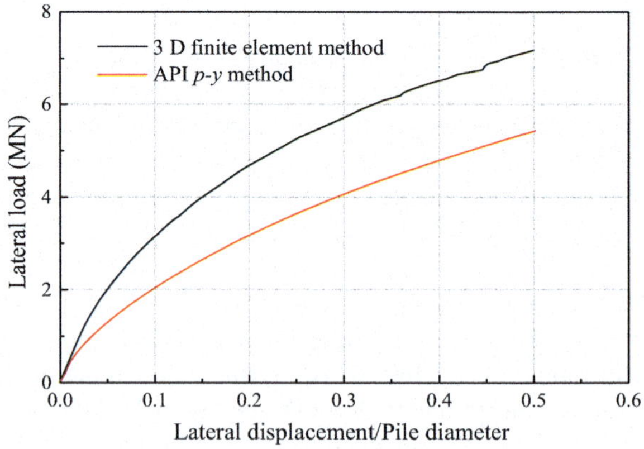

Fig. 6.5 Comparison of calculation results between 3D finite element method and API *p–y* curve method

OWT would experience very large rotation, and the structure would fail. Figure 6.6a shows that the increase in rotation becomes nonlinearly proportional to the load as the horizontal load increases, indicating gradual degradation of soil stiffness. Regardless of loading direction (0° or 180°), the cumulative rotation angles of Pile A and Pile B are generally consistent. However, in the initial stage of loading, rotation angle of Pile B under the same load is slightly smaller, which indicates the stiffness of Pile B is slightly greater than that of Pile A, so Pile A is more likely to rotate at lower load. In addition, under the same lateral load, for both Pile A and Pile B, the corresponding rotation angle for 0° is larger than 180° loading, which indicates that the bearing capacity of tripod pile foundation under 0° loading is smaller than for 180° loading. It is also observed that the rotation angle of Point D is larger than that of Pile A and Pile B under the same horizontal load, which indicates the rotation angle of the overall tripod pile foundation is larger than that of each pile.

Figure 6.6b shows that when loading is in 0° direction, Pile A experiences large settlement, while the vertical displacement of the Pile B is insignificant; When loading is along 180° direction, both Pile A and Pile B experience significant vertical displacement, but Pile A experiences upward displacement, while Pile B experiences downward displacement. This is because the Pile B has larger compressive and uplift bearing capacities than Pile A. However, the rotation of the overall tripod

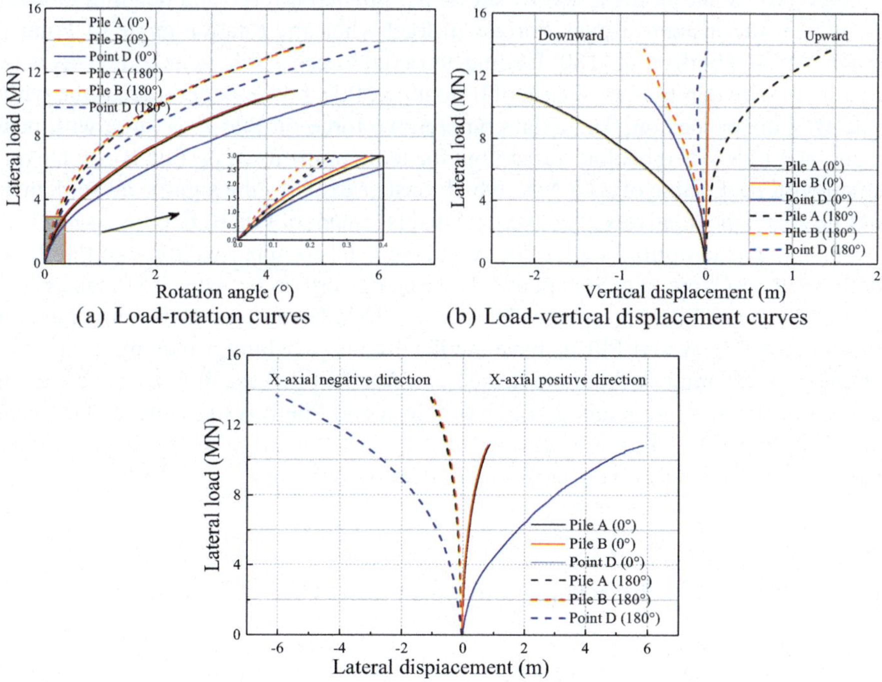

(a) Load-rotation curves (b) Load-vertical displacement curves

(c) Load-lateral displacement curves

Fig. 6.6 Responses of tripod pile foundation under lateral monotonic loading pile foundation is mainly controlled by the pile compressive capacity. Obviously, the compression bearing capacity of one pile (0° loading) is smaller than that of two piles (180° loading), so the bearing capacity of tripod pile foundation under 0° loading is smaller and hence the settlement difference is larger. In addition, regardless of loading direction, the vertical displacement difference between Pile A and Pile B increases gradually as the horizontal load increases. This increase in differential vertical displacement between the two is the main reason for the overall rotation of the tripod pile foundation. It is noted that when loading is along the 180° direction, Point D initially experiences a slight downward displacement, and then gradual upward trends because the Pile A experiences significant upward displacement. Meanwhile, for loading along the 0° direction, the Pile A experiences a significant downward displacement, which causes the whole tripod pile foundation to rotate with an obvious downward displacement. Figure 6.6c demonstrates that for the same lateral load level, the horizontal displacements for Pile A and Pile B at the mud surface and Point D are greater for loading along the 0° direction than for loading in 180° direction. No matter loading in 0° or 180°, the horizontal displacement at the mud surface of Pile A and Pile B is basically the same, and both are much less than the horizontal displacement at Point D. This is because Point D is 50 m above the mud surface, where the rotation of the tripod support structure causes its horizontal displacement to be more pronounced.

Figure 6.7a and b, respectively, shows the pile axial force and bending moment profiles at the instant of peak horizontal load when the rotation angle of Point D reaches 6°. For both 0° and 180° loading, the axial forces of Pile A and Pile B decrease gradually with depth. For the case of 0° loading, Pile A is under compression, while Pile B is under tension. The axial compressive force on Pile A is large with peak value of 14.58 MN (absolute value), but the axial uplift force on Pile B is relatively small with peak value of 2.07 MN, which further explains the significant settlement of Pile A and the relatively small vertical displacement of Pile B. For the case of 180° loading, the axial uplift force of Pile A and the axial compressive force of Pile B are both relatively large, which explains the large upward and downward displacement of the Pile A and Pile B, respectively. Figure 6.7b indicates that the bending moment profiles for Pile A and Pile B have similar trends, reaching a maximum at about 10 m below the mud surface, and decrease to 0 at the pile base. For the same loading angles, the bending moment direction of Pile A and Pile B is the same, and the peak bending moment of Pile B is greater than that of Pile A; however, the difference between the two is relatively small under loading in 0°.

6.4 Lateral Cyclic Responses of Tripod Pile Foundation

The structure of OWT is usually subjected to winds and sea wave loads that vary along the height of the tower and the support structure. This article focuses on discussing the bearing behavior of tripod pile foundations under lateral cyclic loads. For the

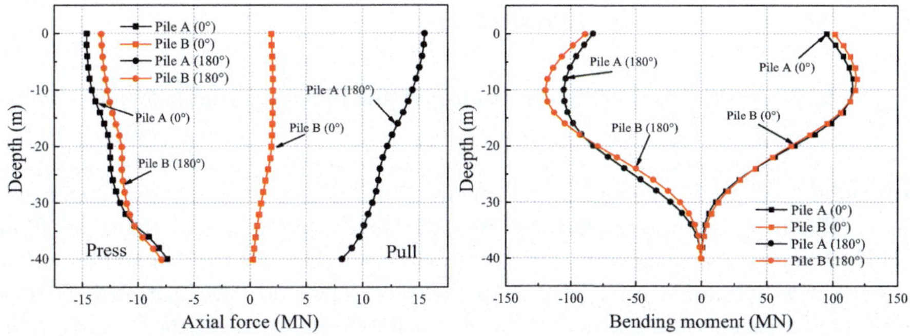

(a) Axial force of pile (b) Bending moment of pile

Fig. 6.7 Axial force and bending moment profile of pile at peak horizontal load

convenience of calculation, the distributed load that varies along the height is simplified as a point load applied to Point D at the bottom of the tower. This simplification pattern has been adopted by many scholars (e.g., Bourgeois et al. 2010; Giannakos et al. 2012; Hong et al. 2017). F_{SLS} is defined as the horizontal load when the rotation angle of Point D reaches 0.25° (i.e., the limit rotation angle of SLS), which is found to be $F_{SLS} = 1.9$ MN for the considered piles-soil system. F_{SLS} is used as a reference to determine the cyclic load level. Horizontal one-way and two-way cyclic loads of 0.4 F_{SLS} are applied to Point D in two separate analyses. The loading angle of one-way cyclic load is 0° (the most unfavorable direction, with the lowest ultimate bearing capacity); the two-way cyclic load is alternately applied in 0° and 180° directions. The schematic diagrams of one-way and two-way cyclic load time histories are shown in Fig. 6.8a and b respectively. The cyclic response of tripod pile foundation under the two different loading patterns is discussed below.

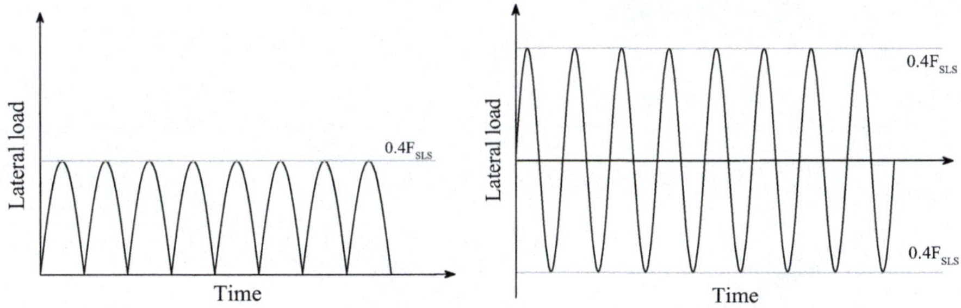

(a) One-way cyclic loading (b) Two-way cyclic loading

Fig. 6.8 Schematic diagram of load time histories for two different cyclic loading patterns

6.4.1 One-Way Cyclic Loading

Figure 6.9 shows the response curves of tripod pile foundation subjected to one-way cyclic loading. In Fig. 6.9a, the load-rotation angle curves exhibit some nonlinear hysteresis characteristics under one-way cyclic loading. In the first loading cycle, when the load reaches its peak, rotation angle of Pile A is significantly greater than that of Pile B. This is because the bending stiffness of Pile A is smaller than that of Pile B. The load-rotation angle curve of Point D indicates that the tripod pile foundation experiences cumulative rotation under the lateral cyclic load. The rotation that can be recovered after unloading is denoted herein as temporary rotation, the unrecoverable rotation as permanent rotation, and the maximum value of rotation as peak rotation. The peak rotation is equal to the sum of the temporary rotation and permanent rotation as shown in Fig. 6.9b. The rotation angles of Pile A, Pile B and Point D accumulate as the number of cycles increases. The peak rotation angle at Point D is the highest, followed by Pile A, and the minimum for Pile B.

Figure 6.9c shows that both Pile A and Pile B experience cumulative settlement as the number of cycles increases, and the cumulative settlement of Pile A is much greater than that of Pile B. The gradually increasing cumulative differential settlement between the two, which is mainly caused by the cumulative settlement of Pile A rather than Pile B, causes the rotation of tripod pile foundation under lateral cyclic

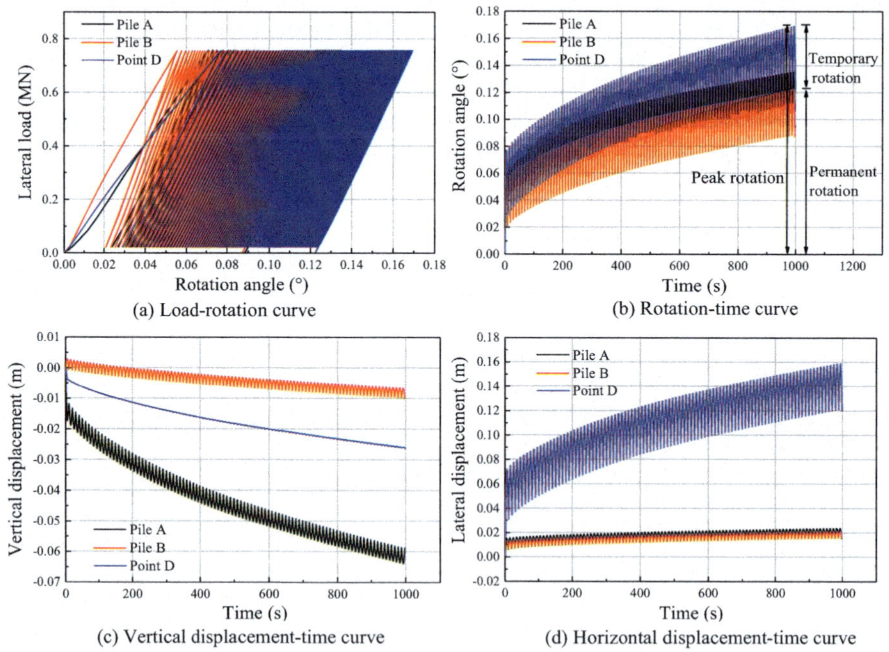

Fig. 6.9 Response of tripod pile foundation subjected to one-way cyclic loading

loads. This is because the compressive capacity of Pile A is relatively small. It should be noted that at the initial stage of cyclic loading, Pile B experiences slight upward displacement; however, as the number of cycles increases, the downward displacement of Pile B gradually increases, and eventually accumulative settlement occurs. This is because the overturning moment generated by the horizontal load causes an upward load on Pile B, which is smaller than the weight of the wind turbine structure, so the resultant vertical force on the Pile B is still downward. In addition, the vertical temporary displacement of Pile A is obviously greater than that of Pile B, because the vertical bearing capacity of Pile A is smaller than that of Pile B, and the upward and downward displacement is more likely to occur under reciprocating loads. The vertical temporary displacement of Point D is insignificant, and mainly experiences vertical cumulative settlement. As shown in Fig. 6.9d, both Pile A and Pile B experience small horizontal displacement as the number of cycles increases, and their horizontal displacement is almost the same (Pile A is slightly larger). The temporary and cumulative horizontal displacements of Point D are significantly larger than those of Pile A and Pile B, which indicates that the pile foundation and tripod support structure rotate significantly during loading. It is noted that for individual pile, vertical displacement is more significant than horizontal displacement, while for Point D, horizontal displacement is more significant than vertical displacement.

Figure 6.10a–f shows the evolution of axial force, bending moment and lateral displacement profiles of the piles at peak horizontal load as the number of cycles increases. Figure 6.10a demonstrates that the axial force of Pile A gradually decreases, while axial force of Pile B gradually increases. The pile axial force is related to the weight of OWT and the pile-soil friction resistance. When the pile is pulled, pile axial force is equal to the weight of OWT plus the pile-soil friction resistance; when the pile is under compression, the pile axial force is equal to the weight of OWT minus the pile-soil friction resistance. The Pile A is under compression while the Pile B is pulled (despite the total displacement is downward) at peak horizontal load. The soil strength and stiffness degrade gradually as the cycle number increases, and then the permanent rotations of Pile A and Pile B gradually increase. Correspondingly, the soil around the pile is subjected to greater passive pressure, and hence the pile-soil frictional resistance increases. Figure 6.10b reveals that at depths greater than 35 m, axial force of Pile B gradually decreases as the number of cycles increases. This is due to the reversal of the pile-soil frictional resistance below this depth. Figure 6.10c and d shows the bending moments of Pile A and Pile B gradually increase as the number of cycles increases. Figure 6.10e and f demonstrates that the maximum horizontal displacement of piles occurs at the mud surface, and gradually increases as the number of cycles increases and the piles curvature becomes more significant. In addition, the horizontal displacement at the mud surface of Pile A is slightly larger than that of Pile B, which can be attributed to the relatively small bending stiffness of Pile A. Figure 6.11 shows the displacement field of tripod pile foundation at peak horizontal load for different cycle number. Figure 6.11 presents the evolution of displacement for the Pile A and Pile B and the relative motion between piles and soil. The observed increase of peak bending moment and horizontal displacement of piles with the number of cycles can be attributed to soil stiffness degradation around pile gradually.

6.4.2 Two-Way Cyclic Loading

Figure 6.12 displays the response curves of tripod pile foundation subjected to two-way cyclic loading. Figure 6.12a clearly indicates that the Pile A, Pile B and Point D all experience positive permanent rotation angle (positive direction of x-axis in Fig. 6.4), i.e., the tripod pile foundation rotates toward the Pile A side. In the first loading cycle, rotation angle of Pile A is significantly greater than that of Pile B under positive and negative peak loads, which can be attributed to the relative weak bending stiffness of Pile A. When lateral load changes from positive to negative (approach to 0), the growth rate of rotation angle suddenly accelerates, and load-rotation angle curve shows a certain degree of deflection, which is due to the separation between the pile and soil.

Figure 6.12b shows more clearly that Pile A, Pile B, and Point D all experience significant temporary rotation angle and some permanent rotation angle. Peak rotation angle and cumulative rotation angle at Point D are the largest, and peak rotation angle and temporary rotation angle of Pile A are greater than those of Pile B. It is worth noting that as the cycle number increases, the peak rotation angle and cumulative rotation angle reach the maximum value when the load cycles are approximately 40 times, and then gradually decrease; the tripod pile foundation exhibits a favorable feature of "self-healing", i.e., the reduction of accumulated rotation, which agrees with the results on cyclic behavior of tripod suction bucket foundation obtained by Wang et al. (2018a). Figure 6.12c shows that cumulative settlement of Pile A, Pile B and Point D gradually increases with the cycle number increases, and cumulative vertical displacement and temporary vertical displacement of Pile A are larger than those of Pile B. In addition, at the beginning of cyclic loading, settlement rate of Pile A is faster than that of Pile B, and the settlement difference between the two gradually increases; As the number of cycles increases, the settlement rate of Pile A gradually decreases, and hence the differential settlement between the two gradually decreases, which leads to gradual decrease in cumulative rotation angle of Point D after reaching its peak, as shown in Fig. 6.12b. The vertical displacement of Point D is between that of the Pile A and the Pile B; however the vertical temporary displacement of Point D is very weak. This is because the projection of Point D on the horizontal plane is located at the center of the tripod pile foundation, and hence it experiences very weak alternating trend of uplift and compression during the rotation process of the foundation.

Figure 6.12d shows that Pile A and Pile B experience insignificant horizontal cumulative displacement under two-way cyclic loading; they mainly experience some temporary displacement that gradually decreases as the cycle number increases. In addition, Point D experiences significant temporary displacement and permanent

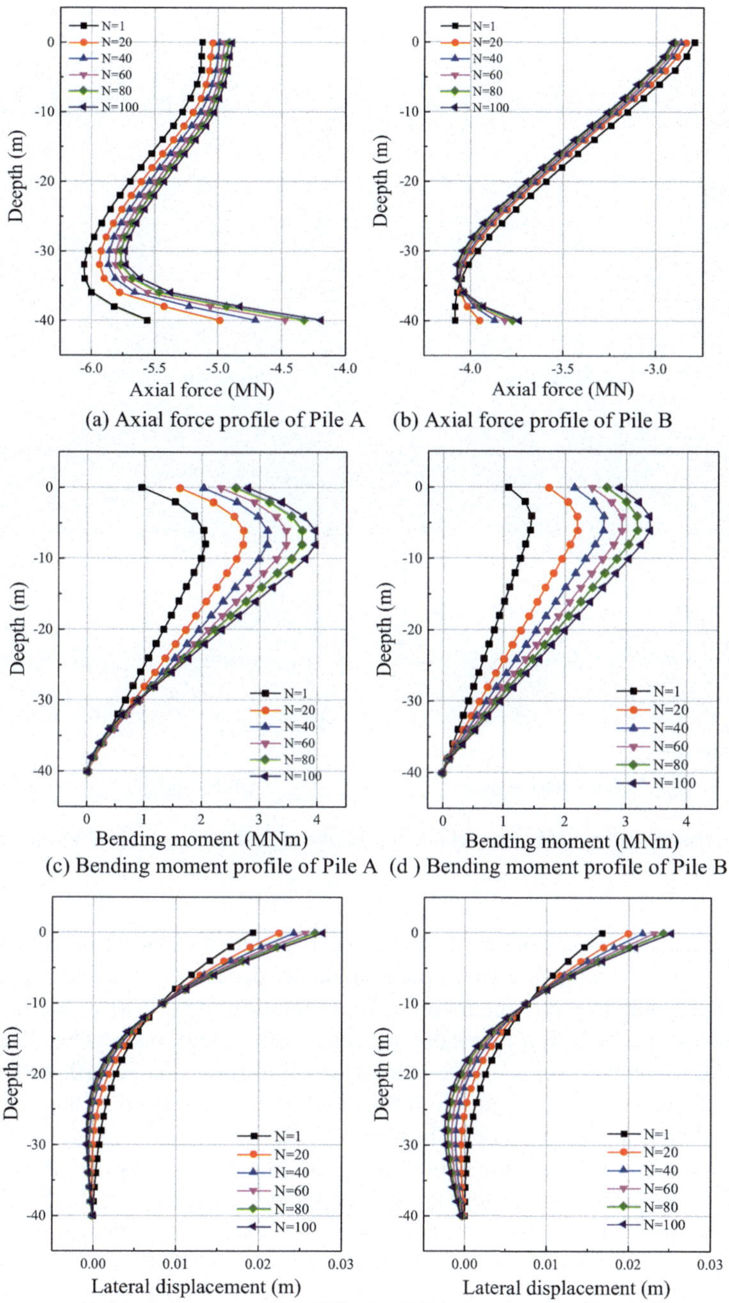

(a) Axial force profile of Pile A (b) Axial force profile of Pile B

(c) Bending moment profile of Pile A (d) Bending moment profile of Pile B

(e) Displacement profile of Pile A (f) Displacement profile of Pile B

Fig. 6.10 Evolution of axial force, bending moment and lateral displacement profiles of pile at peak horizontal load as the number of cycles increases.

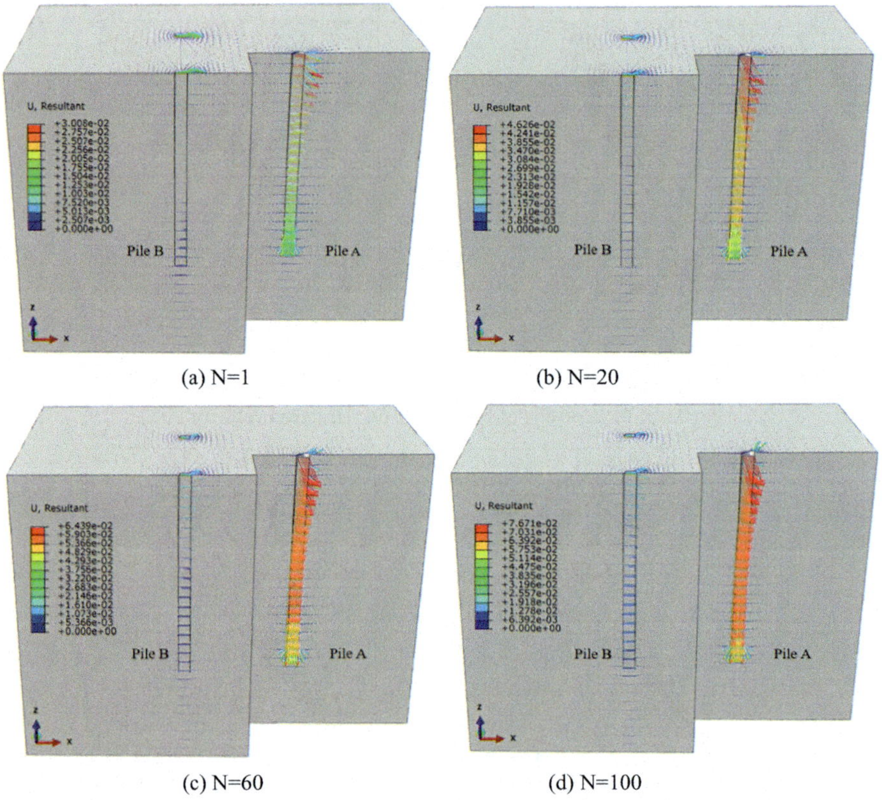

Fig. 6.11 Displacement field of tripod pile foundation at peak horizontal load for different cycle number

displacement, and the temporary displacement is greater than the permanent displacement. However, horizontal displacement gradually decreases after reaching the peak value, which indicates that the reciprocating vibration of the tripod pile foundation in the horizontal direction gradually weakens after a few load cycles. That is the tripod pile foundation seems to be more difficult to rotate. This further verifies the self-healing behavior of tripod pile foundation, which is related to the degradation rate of soil around Pile A and Pile B during cyclic loading.

Figure 6.13a–f presents the evolution of the piles axial force, bending moment and lateral displacement profiles at positive and negative peak horizontal load as the number of cycles increases. Figure 6.13a and b indicates that as the number of cycles increases, axial force of Pile A gradually decreases, while that of Pile B slightly increases (below 35 m depth), which is consistent with the results of one-way cyclic loading. However, the axial force (absolute value) of Pile A at the positive peak load (Pile A under compression) is significantly greater than that at the negative peak load (Pile A under uplift) and decreases more significantly as the number of cycles increases. Compressive axial force of Pile B at the negative peak load is significantly

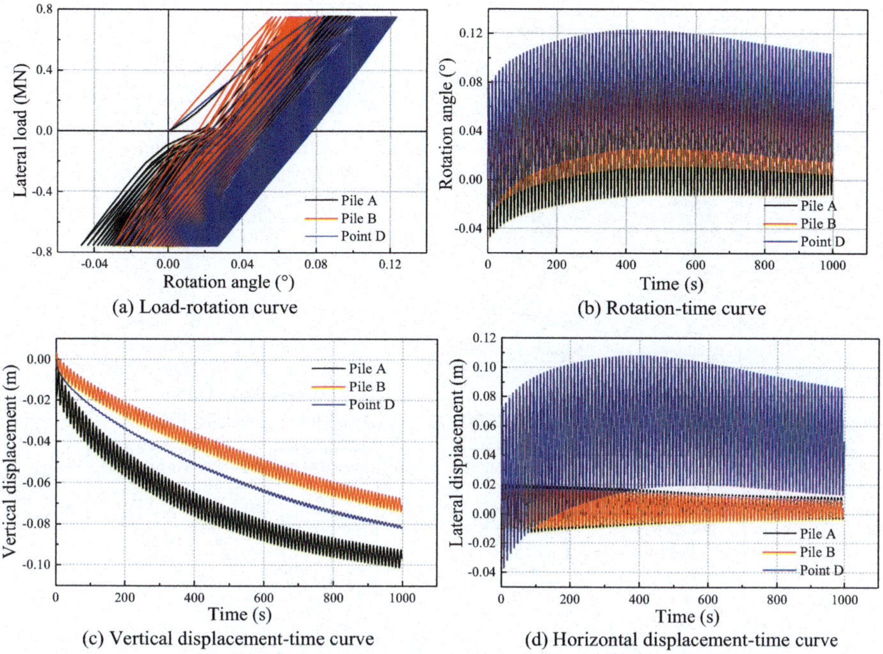

Fig. 6.12 Responses of tripod pile foundation subjected to two-way cyclic loading

greater than that the uplift load it experiences at the positive peak load. The pile axial force is more significantly affected by the number of cycles at depth greater than 35 m.

Figure 6.13c and d indicates that for both Pile A and Pile B, the corresponding bending moment profile at the positive peak load is greater than the corresponding value at the negative peak load. It is noted that the bending moment for Pile A and Pile B does not increase monotonically with the number of cycles; it reaches the peak value at a certain number of cycles then decreases gradually. This complex trend is related to the non-monotonic increase of the pile rotation angle observed in Fig. 6.12b. Figure 6.13e and f demonstrates that the maximum horizontal displacements of Pile A and Pile B occur at the mud surface. As the number of load cycles increases, the horizontal displacements gradually decrease, consistent with the lateral displacement–time history presented in Fig. 6.12d. This behavior is opposite to the variation trend of displacement profile under one-way cyclic load noted in Fig. 6.10e and f.

Figure 6.14 displays the displacement field of tripod pile foundation at positive and negative peak horizontal load for different cycle number. Figure 6.14 clearly demonstrates that the horizontal displacements for both Pile A and Pile B decrease gradually, but their vertical displacements increase significantly. Furthermore, the vertical cumulative displacement of Pile A is greater than that of Pile B for both

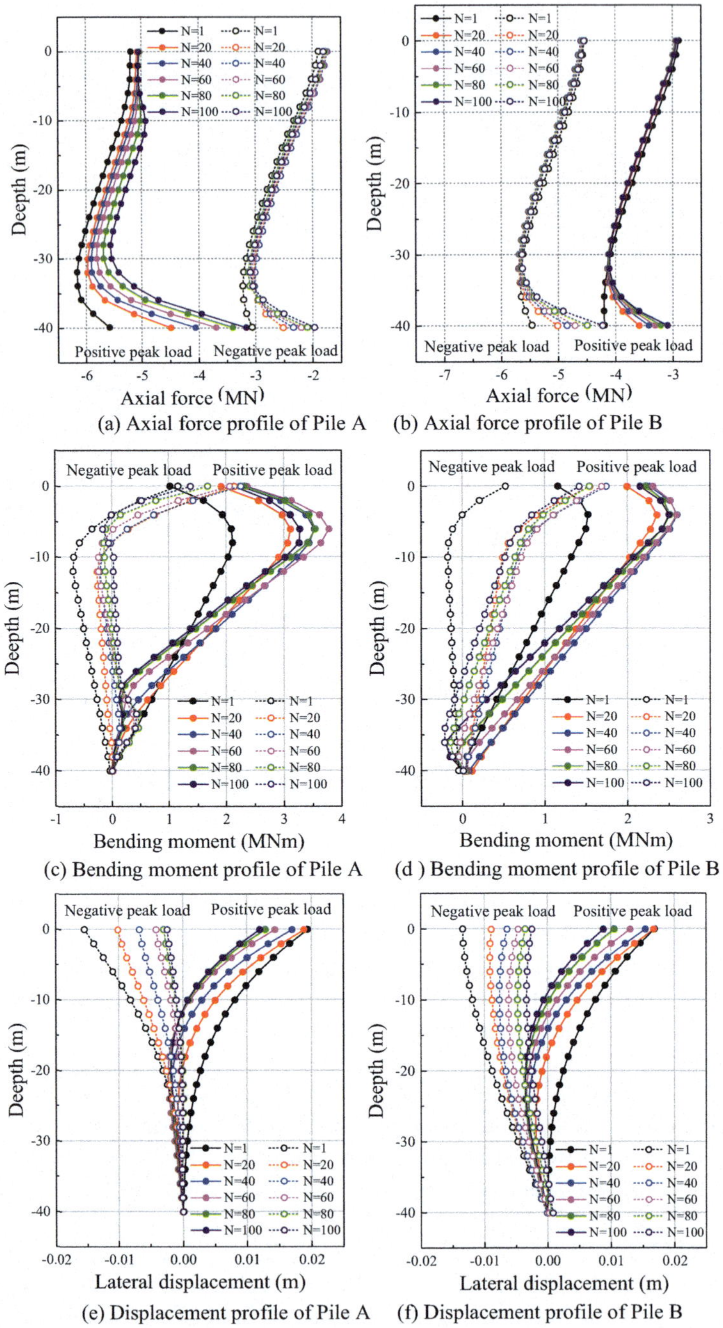

Fig. 6.13 Evolution of axial force, bending moment and lateral displacement profiles of pile at positive and negative peak horizontal load as the number of cycles increases

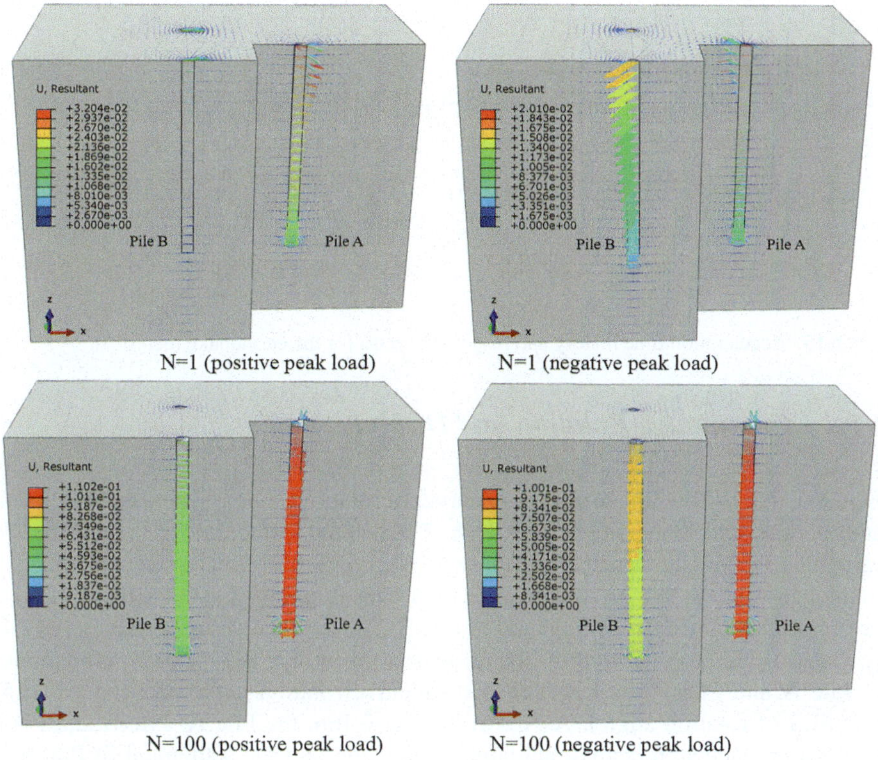

Fig. 6.14 Displacement field of tripod pile foundation at the instant of positive and negative peak load for different cycle number

positive and negative peak horizontal loads, which means that the overall tripod pile foundation gradually rotates toward the Pile A side.

6.5 Seismic Responses of Tripod Pile Foundation

The Kobe (KAKOGAWA (CUE90), 1995) and Northern California (Ferndale City Hall, 1967) earthquake records are applied as the input ground motion. The acceleration time history and response spectra for this ground motion record with the peak ground acceleration (PGA) scaled to 0.3*g* are shown in Fig. 6.15. In this analysis, the bottom of the finite element model is considered to be rigid bedrock, and the input ground motion is applied as vertically propagating shear wave (SV-wave) represented by the acceleration time history at the bottom of the model. The seismic response of the tripod pile OWT system is calculated, and the results are discussed herein.

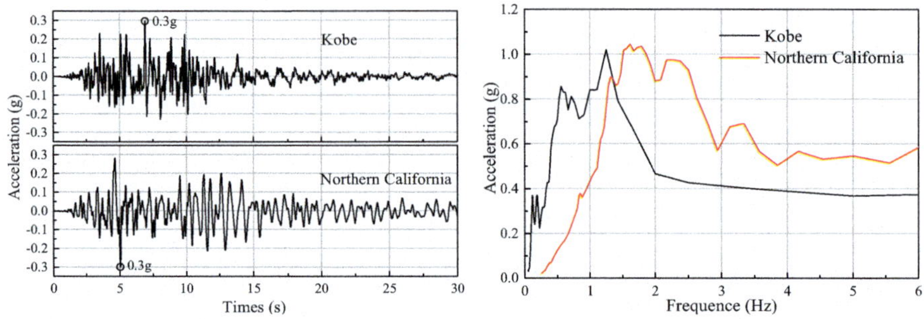

Fig. 6.15 Acceleration time history and response spectra for the earthquake record

6.5.1 Cumulative Rotation and Displacement

Figure 6.16a–c shows the time history curves of rotation angle, vertical displacement, and lateral displacement of pile A and pile B at the mud surface and of Point D for two earthquake records, respectively. Figure 6.16a shows that all experience cumulative rotation toward the pile A under seismic load. This is mainly due to the lower vertical bearing capacity of Pile A compared to Pile B. The cumulative rotation angle of Pile A is slightly larger than that of Pile B. It is worth noting that peak rotation angles of Pile A and Pile B are larger than cumulative rotation angle and exceed 0.25° (the SLS of rotation angle). As shown in Fig. 6.16b, the Pile A, Pile B and Point D all experience significant cumulative settlement, and the settlement of Pile A is greater than that of Pile B, which led to the rotation of the tripod pile foundation. Fig. 6.16c shows that Pile A, Pile B and Point D all experience cumulative horizontal displacement in the x-axis positive direction, and the cumulative displacement of Pile A and Pile B are almost the same, while that of Point D is greater. The variation trend of time history curves is basically consistent for two earthquake records; however, the rotation angle and vertical displacement for Northern California earthquake is larger than that for Kobe earthquake although both PGA are the same. The main reason is that the duration of Northern California earthquake is longer than that of Kobe earthquake.

6.5.2 Peak Response Profile of the Tripod Pile and Tower

Figure 6.17 presents the maximum lateral displacement, rotation angle and bending moment at different elevations from the pile bottom to the tower top for Kobe earthquake. As shown in Fig. 6.17a, the displacement in positive direction of *x*-axis is significantly greater than that in negative direction of *x*-axis. The lateral peak displacement of Pile A and Pile B are almost identical, and the lateral peak displacement changes monotonically along the pile with a maximum value at the

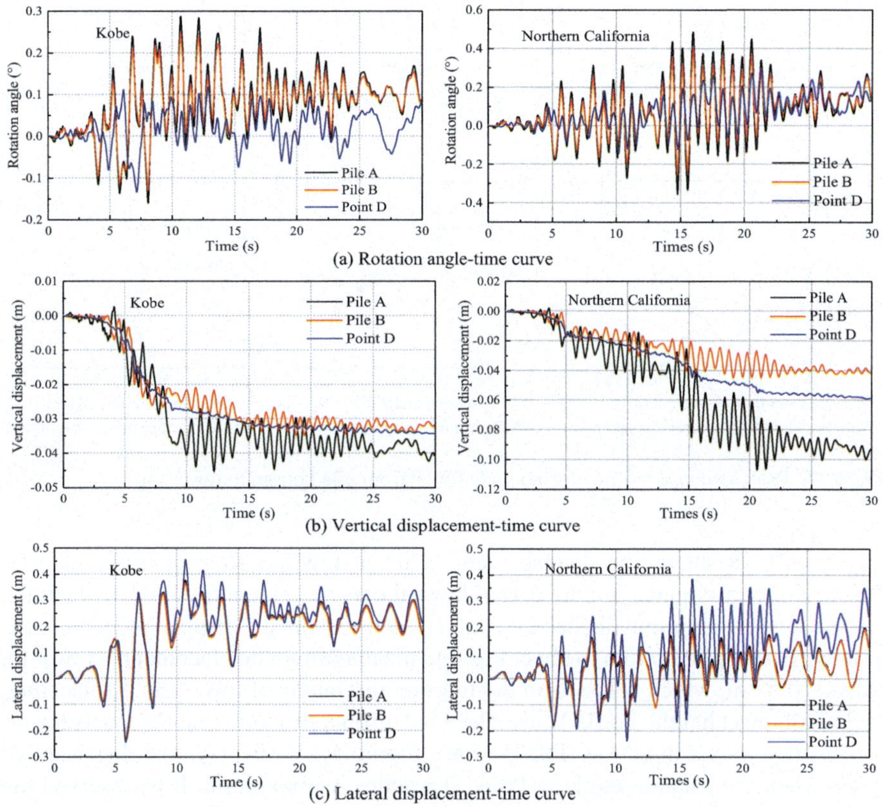

Fig. 6.16 Time history of rotation angle, vertical and lateral displacements of 'Pile A', 'Pile B' and Point D for two earthquake records

mud surface. The lateral displacement along the tripod support and tower remains almost unchanged with height, and the peak displacement at the tower top is the largest. Figure 6.17b shows that rotation angle of Pile A is slightly larger than that of the Pile B near the mud surface, and positive rotation angle is greater than negative rotation angle. The maximum peak rotation angles of both Pile A and Pile B occur at about 25 m below the mud surface. The peak rotation angle of tripod support and tower gradually increases with height, and the rotation angle of tower top is maximum. Figure 6.17c shows that the profiles of positive and negative peak bending moment are very similar. The pile peak bending moment increases non-monotonically along the height, and the maximum peak bending moment occurs at about 10 m below the mud surface. As the height increases, the tower bending moment decreases gradually, and the value at tower top is almost 0. The bending moment of tripod support increases significantly with height between 15 and 35 m, and the peak bending moment of OWT reaches the maximum value at the top of tripod support.

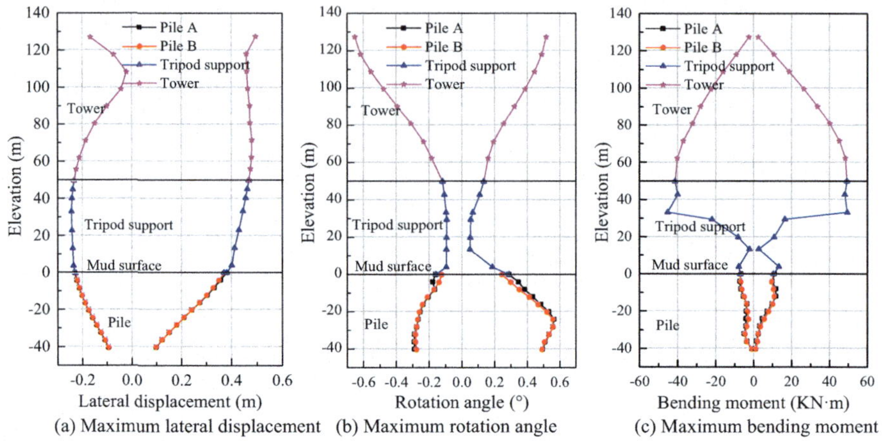

Fig. 6.17 Peak response profile of tripod pile OWT from pile bottom to tower top

Figure 6.18 shows the displacement fields of OWT system at different times during Kobe earthquake. Figure 6.18a shows the displacement fields when the tower top reaches its maximum displacement. It is observed that the displacement increases gradually from pile bottom to tower top, and the maximum displacement at tower top reaches 0.52 m, which is less than the limit requirements of tower top displacement (1.25% of tower height, i.e., 0.97 m). The wind turbine structure tilts obviously toward the positive *x*-axis direction (Pile A side). Figure 6.18b displays the displacement fields when the rotation angle at Point D reaches its maximum. It is observed that the tower top displacement is less than its maximum, and the displacement changes direction along the tower height. Also, the tripod pile foundation tilts toward to the positive *x*-axis direction, while the tower top tilts toward to the negative *x*-axis direction. Figure 6.18c shows the displacement fields when the tower top acceleration reaches its maximum; the displacement field is similar with that in Fig. 6.18b but with larger tripod pile foundation rotation.

6.5.3 Acceleration Amplification

Figure 6.19a shows the acceleration time history at several positions from the pile bottom to tower top for Kobe earthquake. The results indicate that the input bedrock peak acceleration is amplified to varying degrees. The acceleration amplification factor is defined as the ratio of the peak acceleration for a section of tripod pile OWT to the input bedrock peak acceleration. Figure 6.19b presents the profile of the acceleration amplification factor along elevations from the pile bottom to the tower top. It should be noted that within the height range of the tripod support, the position points on the main tubular were selected to determine the acceleration. The

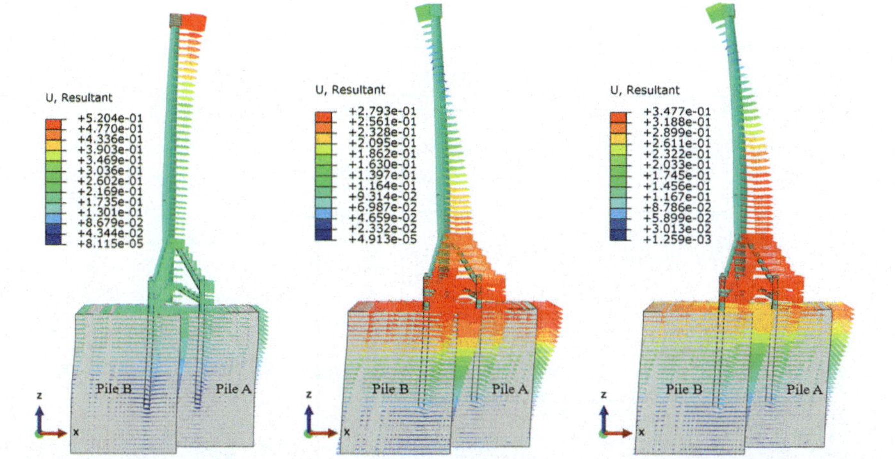

(a) Maximum displacement at tower top (b) Maximum rotation at Point D (c) Maximum acceleration at tower top

Fig. 6.18 Displacement fields of tripod pile OWT system at different times during earthquake (amplified 10 times)

acceleration amplification factor of Pile B below the mud surface is significantly greater than that of Pile A. However, the amplification factors at the mud surface and pile bottom are almost equal. In addition, the variation of acceleration amplification factor along the pile height is non-monotonic. Overall, the acceleration amplification factor initially increases then decreases along the height above the mud surface. The maximum amplification factor occurs at about 80 m. However, the acceleration amplification factor at tower top is relatively small, which is attributed to the presence of a large, concentrated mass at the tower top. Cheng et al. (2023a) reported similar observation for monopile OWT.

6.6 Summary

This chapter investigated the response of tripod pile foundation for offshore wind turbine in clays under monotonic loads, cyclic loads and seismic loads by performing three-dimensional finite element analysis. A simplified bounding surface constitutive model of clay has been incorporated in the ABAQUS software package to simulate the degradation in clay stiffness and strength due to cyclic loading. The key conclusions can be summarized as follows:

(1) The loading direction greatly affects the ultimate bearing capacity of tripod pile foundation. The ultimate bearing capacity in the 0° loading direction (Pile A under compression) is smaller than that in the 180° loading direction (Pile B under compression) because the bearing capacity of tripod pile foundation is determined by the pile compressive capacity.

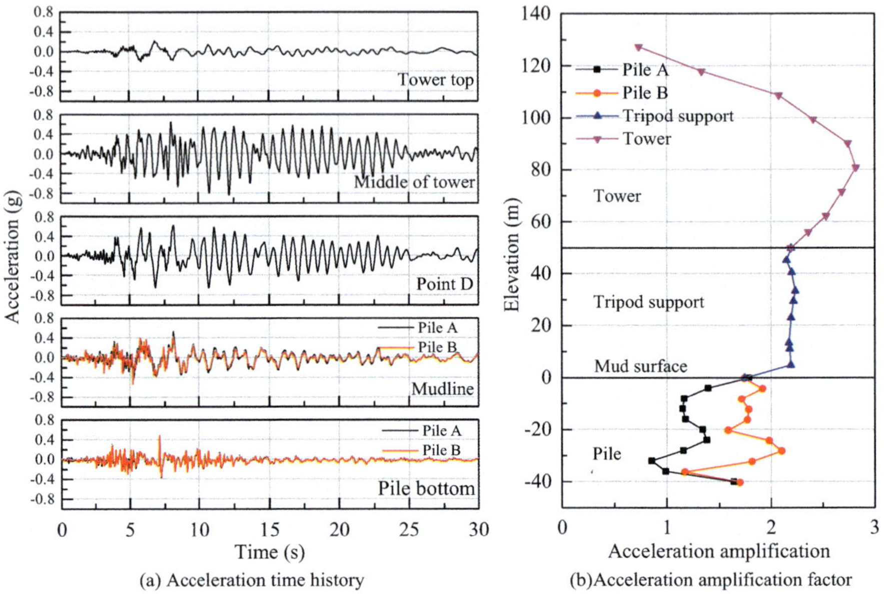

Fig. 6.19 Acceleration amplification effect at different positions from the pile bottom to the tower top

(2) For one-way cyclic loading in the 0° direction, the cumulative settlement of Pile A is much greater than that of Pile B, and the cumulative differential settlement between the two increases gradually, which leads to the rotation of tripod pile foundation. The axial force of Pile A decreases and the axial force of Pile B increases. The bending moments of Pile A and Pile B gradually increase and the piles curvatures become increase as the number of cycles increases.
(3) For two-way cyclic loading along 0° and 180° directions, the tripod pile foundation exhibits a favorable feature of "self-healing" in accumulated rotation. The Pile A and Pile B experience insignificant horizontal cumulative displacement but experience a certain of horizontal temporary displacement that gradually decreases as the number of cycles increases. The bending moments of Pile A and Pile B do not increase monotonically with the number of cycles.
(4) Under seismic load, the Pile A, Pile B, and Point D all experience cumulative rotation angles toward the Pile A side due to the lower vertical bearing capacity of the Pile A compared to the Pile B. The tower top experiences maximum lateral displacement and rotation angle, and the top of tripod support experiences maximum bending moment. The acceleration amplification factor of Pile B below the mud surface is significantly greater than that of Pile A. However, amplification factor at the mud surface and pile bottom are almost the same. The maximum amplification factor occurs at about 80 m above the mud surface. The acceleration amplification factor at tower top is relatively small.

There are still some limitations that need to be emphasized in this study. It should be noted that the total stress-based bounding surface constitutive model was used to describe the dynamic behavior of clay. The soil is assumed to be undrained during the numerical calculation process. This assumption is relatively reasonable for short-term dynamic loads such as earthquake, however, for long-term wind and wave cyclic loads, the undrained assumption will result in conservative calculation results. In addition, the influence of soil pore pressure on pile-soil dynamic interaction has not been properly considered. It is worthwhile to conduct more in-depth research employing effective stress-based constitutive models in subsequent work.

References

Abhinav, K.A., and N. Saha. 2018. Nonlinear dynamical behaviour of jacket supported offshore wind turbines in loose sand. *Marine Structures* 57: 133–151.

Achmus, M., Y.S. Kuo, and K. Abdel-Rahman. 2009. Behavior of monopile foundations under cyclic lateral load. *Computers and Geotechnics* 36 (5): 725–735.

API Recommended Practice 2GEO. 2014. *Geotechnical and Foundation Design Considerations*, 1st ed. Addendum 1, Published by the American Petroleum Institute.

Arany, L., S. Bhattacharya, J.H.G. Macdonald, et al. 2015. A Critical review of serviceability limit state requirements for monopile foundations of offshore wind turbines. In *Offshore Technology Conference, Houston.*

Asumadu, R., J. Zhang, and H. Osei-Wusuansa. 2022. 3-D numerical study of offshore tripod wind turbine pile foundation on wave-induced seabed response. *Ocean Engineering* 255: 111421.

Bourgeois, E., M.H.J. Rakotonindriana, A. Le Kouby, et al. 2010. Three-dimensional numerical modelling of the behaviour of a pile subjected to cyclic lateral loading. *Computers and Geotechnics* 37 (7–8): 999–1007.

Chen, D., P. Gao, S. Huang, et al. 2020a. Static and dynamic loading behavior of a hybrid foundation for offshore wind turbines. *Marine Structures* 71: 102727.

Chen, L., X. Yang, L. Li, W. Wu, et al. 2020b. Numerical analysis of the deformation performance of monopile under wave and current load. *Energies* 13(23): 6431.

Cheng, X.L., T.J. Wang, J.X. Zhang, et al. 2021. Finite element analysis of cyclic lateral responses for large diameter monopiles in clays under different loading patterns. *Computers and Geotechnics* 134: 104104.

Cheng, X.L., W.L. Cheng, P.G. Wang, et al. 2022. Response of offshore wind turbine tripod suction bucket foundation to seismic and environmental loading. *Ocean Engineering* 257: 111708.

Cheng, X.L., T.J. Wang, J.X. Zhang, et al. 2023a. Dynamic response analysis of monopile offshore wind turbines to seismic and environmental loading considering the stiffness degradation of clay. *Computers and Geotechnics* 155: 105210.

Cheng, X.L., M. Li, C. Ma, et al. 2023b. Dynamic analysis of tripod pile foundation in clays for offshore wind turbines. *Ocean Engineering* 287: 115832.

Corciulo, S., O. Zanoli, and F. Pisanò. 2017. Transient response of offshore wind turbines on monopiles in sand: Role of cyclic hydro-mechanical soil behaviour. *Computers and Geotechnics* 83: 221–238.

D'Aguiar, S.C., A. Modaressi, A. Jaime, et al. 2011. Piles under cyclic axial loading: Study of the friction fatigue and its importance in pile behavior. *Canadian Geotechnical Journal* 48 (10): 1537–1550.

De Risi, R., S. Bhattacharya, and K. Goda. 2018. Seismic performance assessment of monopile-supported offshore wind turbines using unscaled natural earthquake records. *Soil Dynamics and Earthquake Engineering* 109: 154–172.

Depina, I., T.M.H. Le, G. Eiksund, et al. 2015. Behavior of cyclically loaded monopile foundations for offshore wind turbines in heterogeneous sands. *Computers and Geotechnics* 65: 266–277.

Det Norske Veritas (DNV). 2013. DNV-OS-J101 Design of Offshore Wind Turbine Structures. Det Norske, Veritas AS.

Ding, X., S.C. Chian, J. Lian, et al. 2023. Wind-wave combined effect on dynamic response of soil-monopile-OWT system considering cyclic hydro-mechanical clay behavior. *Computers and Geotechnics* 154: 105124.

Esteban, M.D., J.S. Lopez-Gutierrez, V. Negro, et al. 2015. Offshore wind foundation design: Some key issues. *Journal of Energy Resource Technology* 137(5): 051211.1–051211.6.

Giannakos, S., N. Gerolymos, and G. Gazeta. 2012. Cyclic lateral response of piles in dry sand: Finite element modeling and validation. *Computers and Geotechnics* 44: 116–131.

Hong, Y., B. He, L.Z. Wang, et al. 2017. Cyclic lateral response and failure mechanisms of semi-rigid pile in soft clay: Centrifuge tests and numerical modelling. *Canadian Geotechnical Journal* 54 (6): 806–824.

Huang, M.S., and Y. Liu. 2015. Axial capacity degradation of single piles in soft clay under cyclic loading. *Soils and Foundations* 55 (2): 315–328.

Jardine, R.J., and J.R. Standing. 2012. Field axial cyclic loading experiments on piles driven in sand. *Soils and Foundations* 52 (4): 723–736.

Jeanjean, P. 2009. Re-assessment of P-Y curves for soft clays from centrifuge testing and finite element modelling. In *Proceedings of the Offshore Technology Conference*, Paper number: OTC-20158-MS.

Jonkman, J.M., S. Butterfield, W. Musial, et al. 2009. *Definition of a 5MW Reference Wind Turbine for Offshore System Development*. J. office of scientific & technical information technical reports.

Kaynia, A.M. 2019. Seismic considerations in design of offshore wind turbines. *Soil Dynamics and Earthquake Engineering* 124: 399–407.

Kaynia, A.M. 2021. Effect of kinematic interaction on seismic response of offshore wind turbines on monopiles. *Earthquake Engineering & Structural Dynamics* 50 (3): 777–790.

Kim, D.H., S.G. Lee, and I.K. Lee. 2014. Seismic fragility analysis of 5MW offshore wind turbine. *Renewable Energy* 65: 250–256.

Koh, J.H., and E.Y.K. Ng. 2016. Downwind offshore wind turbines: Opportunities, trends and technical challenges. *Renewable & Sustainable Energy Reviews* 54: 797–808.

Kong, D., K. Wen, and B. Zhu. 2019. Centrifuge modeling of cyclic lateral behaviors of a tetrapod piled jacket foundation for offshore wind turbines in sand. *Journal of Geotechnical and Geoenvironmental Engineering* 145 (11): 04019099.

Kuhlemeyer, R., and J. Lysmer. 1973. Finite element method accuracy for wave propagation problems. *ASCE Soil Mechanics and Foundation Division Journal* 99 (5): 421–427.

Kuo, Y.S., M. Achmus, and K. Abdel-Rahman. 2011. Minimum embedded length of cyclic horizontally loaded monopiles. *Journal of Geotechnical and Geoenvironmental Engineering* 138 (3): 357–363.

Lai, Y., L. Wang, Y. Hong, et al. 2020. Centrifuge modeling of the cyclic lateral behavior of large-diameter monopiles in soft clay: Effects of episodic cycling and reconsolidation. *Ocean Engineering* 200: 107048.

Li, L., H. Liu, and W. Wu. 2021a. Investigation on the behavior of hybrid pile foundation and its surrounding soil during cyclic lateral loading. *Ocean Engineering* 240: 110006.

Li, X., X. Zeng, X. Yu, et al. 2021b. Seismic response of a novel hybrid foundation for offshore wind turbine by geotechnical centrifuge modeling. *Renewable Energy* 172: 1404–1416.

Liao, W.M., J.J. Zhang, J.B. Wu, et al. 2018. Response of flexible monopile in marine clay under cyclic lateral load. *Ocean Engineering* 147: 89–106.

Liu, J.W., L. Cui, and N. Zhu. 2019. Investigation of cyclic pile-sand interface weakening mechanism based on large-scale CNS cyclic direct shear tests. *Ocean Engineering* 194 (6): 106650.

Lu, W., G. Zhang, and A. Wang. 2017. Bearing behavior of multiple piles for offshore wind driven generator. *Ocean Engineering* 129: 538–548.

Ma, H., J. Yang, and L. Chen. 2017. Numerical analysis of the long-term performance of offshore wind turbines supported by monopiles. *Ocean Engineering* 136: 94–105.
Ma, H., J. Yang, and L. Chen. 2018. Effect of scour on the structural response of an offshore wind turbine supported on tripod foundation. *Applied Ocean Research* 73: 179–189.
Mo, R., R. Cao, and M. Liu. 2021. Effect of ground motion directionality on seismic dynamic responses of monopile offshore wind turbines. *Renewable Energy* 175: 179–199.
Oh, K.Y., W. Nam, M.S. Ryu, et al. 2018. A review of foundations of offshore wind energy convertors: Current status and future perspectives. *Renewable and Sustainable Energy Reviews* 88: 16–36.
Tsinidis, G., K. Pitilakis, and A.D. Trikalioti. 2014. Numerical simulation of round robin numerical test on tunnels using a simplified kinematic hardening model. *Acta Geotechnica* 9: 641–659.
Tsuha, C., P.Y. Foray, and R.J. Jardine. 2012. Behaviour of displacement piles in sand under cyclic axial loading. *Soils and Foundations* 52 (3): 393–410.
Vacareanu, V., E. Kementzetzidis, and F. Pisano. 2019. 3D FE seismic analysis of a monopile-supported offshore wind turbine in a non-liquefiable soil deposit. In *Proceedings of the 2nd International Conference on Natural Hazards & Infrastructure (ICONHIC2019)*, 23–26.
Wang, X., X. Zeng, X. Yang, et al. 2019. Seismic response of offshore wind turbine with hybrid monopile foundation based on centrifuge modelling. *Applied Energy* 235: 1335–1350.
Wang, L., W. Zhou, and Z. Guo. 2020. Frequency change and accumulated inclination of offshore wind turbine jacket structure with piles in sand under cyclic loadings. *Ocean Engineering* 217: 108045.
Wang, P., M. Zhao, and X. Du, 2018b. Wind, wave and earthquake responses of offshore wind turbine on monopile foundation in clay. *Soil Dynamics and Earthquake Engineering* 113: 47–57.
Wang, L.Z., H. Wang, B. Zhu, et al. 2018a. Comparison of monotonic and cyclic lateral response between monopod and tripod bucket foundations in medium dense sand. *Ocean Engineering* 155: 88–105.
Wei, S., H. Park, and J. Han. 2013. A study on the effect of different modeling parameters on the dynamic response of a jacket-type offshore wind turbine in the Korean Southwest Sea. *Renewable Energy* 58: 50–59.
Wu, X., Y. Hu, Y. Li, et al. 2019. Foundations of offshore wind turbines: A review. *Renewable & Sustainable Energy Reviews* 104: 379–393.
Xi, R., C. Xu, X. Du, et al. 2022. Framework for dynamic response analysis of monopile supported offshore wind turbine excited by combined wind-wave-earthquake loading. *Ocean Engineering* 247: 110743.
Xu, D.S., X.Y. Xu, W. Li, et al. 2020. Field experiments on laterally loaded piles for an offshore wind farm. *Marine Structures* 69: 102684.
Yang, M., R. Luo, and W. Li. 2018. Numerical study on accumulated deformation of laterally loaded monopiles used by offshore wind turbine. *Bulletin of Engineering Geology and the Environment* 77: 911–921.
Yang, Y., Bashir, M., Li, C., et al., 2019. Analysis of seismic behaviour of an offshore wind turbine with a flexible foundation. *Ocean Engineering* 178: 215-228.
Yu, H., X. Zeng, B. Li, et al. 2015. Centrifuge modeling of offshore wind foundations under earthquake loading. *Soil Dynamics and Earthquake Engineering* 77: 402–415.
Zhang, C., D. White, and M. Randolph. 2011. Centrifuge modeling of the cyclic lateral response of a rigid pile in soft clay. *Journal of Geotechnical and Geoenvironmental Engineering* 137 (7): 717–729.
Zhang, J.X., W.L. Cheng, X.L. Cheng, et al. 2021. Seismic responses analysis of suction bucket foundation for offshore wind turbine in clays. *Ocean Engineering* 232: 109159.
Zheng, X.Y., H. Li, W. Rong, et al. 2015. Joint earthquake and wave action on the monopole wind turbine foundation: An experimental study. *Marine Structures* 44: 125–141.
Zhou, X., D.C. Lu, X.L. Du, G.S. Wang, and F.P. Meng. 2020b. A 3d non-orthogonal plastic damage model for concrete. *Computer Methods in Applied Mechanics and Engineering*. 360: 112716.

Zhou, W., Z. Guo, and L. Wang. 2020. Sand-steel interface behaviour under large-displacement and cyclic shear. *Soil Dynamics and Earthquake Engineering* 138: 106352.
Zhu, B., Z. Zhu, T. Li, et al. 2017. Field tests of offshore driven piles subjected to lateral monotonic and cyclic loads in soft clay. *Journal of Waterway, Port, Coastal, and Ocean Engineering* 143 (5): 05017003.

Chapter 7
Seismic Responses of OWT Monopod Suction Bucket in Clays

Suction buckets are widely used as foundations for offshore wind turbines because of its many advantages such as simple structure, convenient installation, low construction cost and reusable. It is very essential to perform the seismic response analysis of suction bucket foundations for the design of offshore wind turbines. However, the related research is still relatively rare. In this chapter, three-dimensional finite element analysis is performed to predict seismic responses of suction buckets for offshore wind turbines in clays by using a simplified kinematic hardening constitutive model available in the Abaqus library. The seismic bearing mechanism of bucket foundation in clays is investigated, and various factors affecting seismic responses of suction bucket are analyzed, including the soil strength, seismic intensity and frequency, and bucket geometry. The research results can provide a reference for the seismic design of suction bucket foundations for offshore wind turbines.

7.1 Introduction

Offshore wind energy has developed rapidly in recent years due to its abundant, clean and renewable advantages. Related infrastructure construction is also rising on a large scale along with the development of offshore wind energy resources (Perveen et al. 2014). In the harsh marine environment, the offshore wind turbine not only suffers from the long-term cyclic load such as wind, wave and current, but also the accidental load such as earthquake and impact. The coupling effect of these loads makes the offshore wind turbine bear huge overturning moment, and then causes excessive rotation angle and settlement deformation, which will affect the normal operation of the wind turbine, and even lead to the overall instability of the wind turbine (Byrne and Houlsby 2003).

The selection of foundation is very important for the safe service and engineering cost of offshore wind turbines. The depth of sea water and the engineering properties

X. Cheng et al., *Dynamic Analysis of Offshore Wind Turbine Foundations in Soft Clays*,
https://doi.org/10.1007/978-981-97-9454-6_7

of seabed soil should be fully considered. At present, the foundation types for offshore wind turbines include gravity foundations, monopiles, tripod foundations and suction buckets. Among them, suction buckets are often installed in offshore soft clays seabed with shallow water depth. Because of its simple structure, convenient installation, low construction cost, reusable and strong anti-overturning moment, it is more and more widely used in construction of offshore wind turbines (Houlsby and Byrne 2000).

So far, most researches have focused on analyzing the responses of suction bucket foundation under monotonic and cyclic loads such as wind, wave and current. Numerical simulations or model tests were performed to study the ultimate bearing capacity of suction bucket foundation under vertical loads, horizontal loads, bending moment loads, and the combination of these loads (Sukumaran et al. 1999; Cao et al. 2003; Maniar 2004; Monajemi and Razak 2009; Li et al. 2014; Ahn et al. 2014; Jin et al. 2019; Faizi et al. 2019). The bearing capacity, failure mechanism and load–displacement responses of suction bucket foundation under cyclic loads induced by wind, wave and current were also investigated by some researchers (Chen and Randolph 2007; Kim et al. 2014a; Cheng et al. 2018; Zhu et al. 2017; Chen et al. 2018; Gelagoti et al. 2018; Guo et al. 2018; Zhang et al. 2019; Yin et al. 2020).

Kourkoulis et al. (2014) studied the influence of different soil-foundation interface on seismic response of suction bucket foundation by using nonlinear three-dimensional finite element analyses. Kim et al. (2014b) performed seismic fragility analysis of 5 MW offshore wind turbine considering nonlinear soil-pile interaction using nonlinear spring elements. Wang et al. (2017) investigated the seismic response of suction bucket with different geometric design in dry sand and saturated sand. Li et al (2017) performed a comprehensive analysis on the dynamic characteristics of the umbrella suction anchor foundation based on the finite element program ANSYS. Yang et al. (2017) performed a three-dimensional dynamic finite element analysis to predict the seismic responses of suction caisson foundation for an offshore wind turbine located in a sandy seabed by using a unified plasticity model for the large post-liquefaction shear deformation of the sand.

It is very essential to perform the seismic response analysis of suction bucket foundation for the design of offshore wind turbine. However, the research on seismic responses of suction buckets located in a clayey seabed is still relatively rare. In this study, three-dimensional finite element analysis was performed to predict the seismic responses of suction bucket by using a simplified kinematic hardening constitutive model available in the Abaqus library. The seismic bearing mechanism of bucket foundation in clays was investigated. Various factors affecting seismic responses of suction bucket were analyzed, including the soil strength, seismic intensity and frequency, and bucket geometry.

7.2 Dynamic Kinematic Hardening Constitutive Model of Undrained Clay

It is assumed that the behaviors of saturated clay foundations occur with no volume change under undrained conditions because of low permeability when they are subjected to rapid seismic loads. For this case, from the perspective of total stress, the yielding of clays is independent of normal stress and only depends on deviatoric stress (Prevost 1977; Anastasopoulos et al. 2011; Huang and Liu 2015). Hence, the undrained dynamic nonlinear behavior of clays is modeled by a total stress-based simple kinematic hardening model. The model is available in the Abaqus library and has been used to the numerical simulation of soil-structure interaction under cyclic or seismic loading (Anastasopoulos et al. 2011; Giannakos et al. 2012; Kourkoulis et al. 2014). In the following sections, the basic relationship of the model is briefly introduced, and then the method of determining model parameters and the predictive ability of the model are discussed.

7.2.1 Basic Relationship of the Model

The Von Mises yield criterion is adopted in this model, as represented by Eq. (7.1) (Boldface letters in bold denote tensors unless otherwise specified below).

$$F = f(\boldsymbol{\sigma} - \boldsymbol{\alpha}) - \sigma^0 = 0 \tag{7.1}$$

$\boldsymbol{\sigma}$ is the stress tensor; $\boldsymbol{\alpha}$ is the backstress tensor; σ^0 is the yield stress, indicating the size of the yield surface; $f(\boldsymbol{\sigma} - \boldsymbol{\alpha})$ is the equivalent Mises stress with respect to $\boldsymbol{\alpha}$, which is defined by Eq. (7.2).

$$f(\boldsymbol{\sigma} - \boldsymbol{\alpha}) = \sqrt{\frac{3}{2}\left(\boldsymbol{s} - \boldsymbol{\alpha}^{\text{dev}}\right)\left(\boldsymbol{s} - \boldsymbol{\alpha}^{\text{dev}}\right)} \tag{7.2}$$

where $\boldsymbol{s}$ is deviatoric stress tensor, and $\boldsymbol{\alpha}^{\text{dev}}$ is the deviatoric tensor of $\boldsymbol{\alpha}$.

Given the associated plastic flow, the plastic flow rate is defined as follows

$$\dot{\boldsymbol{\varepsilon}}^{pl} = \dot{\bar{\varepsilon}}^{pl}\frac{\partial f(\boldsymbol{\sigma} - \boldsymbol{\alpha})}{\partial \boldsymbol{\sigma}} \tag{7.3}$$

where $\dot{\boldsymbol{\varepsilon}}^{pl}$ is the plastic strain rate; $\dot{\bar{\varepsilon}}^{\text{pl}}$ is the equivalent plastic strain rate and $\dot{\bar{\varepsilon}}^{pl} = \sqrt{\frac{2}{3}\dot{\boldsymbol{\varepsilon}}^{pl}\dot{\boldsymbol{\varepsilon}}^{pl}}$.

The model combines isotropic hardening and kinematic hardening to describe the cyclic response of soil. The isotropic hardening behavior is controlled by the yield surface size σ^0, the evolution of σ^0 is related to the equivalent plastic strain $\bar{\varepsilon}^{\text{pl}}$, which can be expressed as

$$\sigma^0 = \sigma|_0 + Q_\infty\left(1 - e^{-b\bar{\varepsilon}^{pl}}\right) \tag{7.4}$$

where $\sigma|_0$ is the initial yield stress, that is, the yield stress (or yield surface size) at zero plastic strain; Q_∞ and b are material parameters; (Q_∞) is the maximum change in the size of the yield surface, and b defines the changing rate of the yield surface size as plastic straining increases. When $Q_\infty = 0$ or $b = 0$, $\sigma^0 = \sigma|_0$. At this case, the yield surface size remains unchanged, and the model is simplified to pure kinematic hardening model.

The kinematic hardening behavior is realized by the evolution of backstress $\boldsymbol{\alpha}$, which can be expressed as follows

$$\dot{\boldsymbol{\alpha}} = C\frac{1}{\boldsymbol{\sigma}^0}(\boldsymbol{\sigma} - \boldsymbol{\alpha})\dot{\bar{\varepsilon}}^{pl} - \gamma_r\boldsymbol{\alpha}\dot{\bar{\varepsilon}}^{pl} \tag{7.5}$$

where C is the initial kinematic hardening modulus; γ_r determine the rate at which the kinematic hardening moduli decreases with increasing plastic deformation, and $\gamma_r\boldsymbol{\alpha}\dot{\bar{\varepsilon}}^{pl}$ introduces the nonlinear characteristics into the kinematic hardening law.

Figure 7.1a and b illustrate the evolution of the isotropic and the kinematic hardening components under uniaxial and multiaxial condition, respectively. As shown in Fig. 7.1a, when the plastic strain is very large, σ tends to σ_{max} (regardless of softening of soil strength), and α tends to the maximum value α_s and $\alpha_s = C/\gamma_r$. On the π plane shown in Fig. 7.1b, $\boldsymbol{\alpha}$ represents the center of the yield surface, which is always within the circle of radius $\sqrt{2/3}\,C/\gamma_r$. The yield surface is a circle with a radius of $\sqrt{2/3}\,\sigma^0$. In addition, any stress point on the yield surface must be within the circle with a radius of $\sqrt{2/3}\,\sigma_{max}$, that is, this circle is actually the strength limit surface of the material.

For undrained clay, the maximum yield stress σ_{max} is as follows

$$\sigma_{max} = \sqrt{3}S_u \tag{7.6}$$

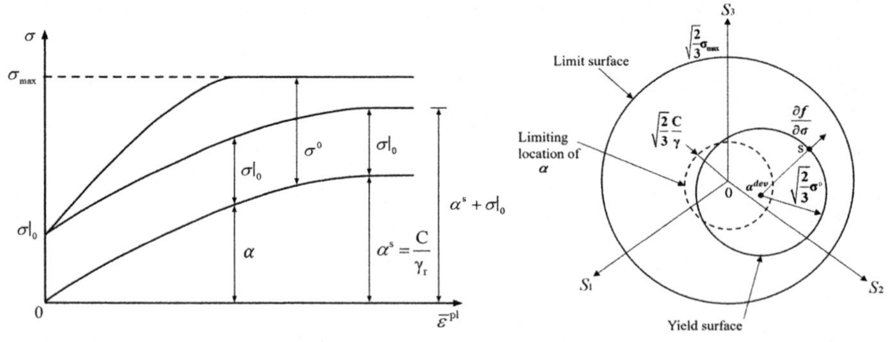

(a) uniaxial condition (b) multiaxial condition

Fig. 7.1 One and three-dimensional representation of isotropic and kinematic hardening where S_u is the undrained shear strength of clays.

7.2.2 Parameters Determination and Model Predictions

In order to simplify the calculation as much as possible, in this research, the constitutive model only considers the kinematic hardening component refer to the existing studies (Anastasopoulos et al. 2011; Kourkoulis et al. 2014). Hence, in this case, the size of the yield surface represented by Eq. (7.4) remains unchanged and satisfies $\sigma^0 = \sigma|_0$. The constitutive model contains only three model parameters: $\sigma|_0$, C and γ_r. They can be determined as follows:

① The initial yield stress $\sigma|_0$ actually represents the range of elastic domain of soil, namely the size of the initial yield surface, which can usually be represented by the following formula

$$\sigma|_0 = \lambda\sigma_{max} = \lambda\sqrt{3}S_u \tag{7.7}$$

where the empirical value of λ is 0.1–0.3.

② The initial kinematic hardening modulus C represents the initial stiffness of the soil, which is usually considered to be equal to the elastic moduli E. Empirically, E can be taken as the multiple of undrained strength S_u, that is, $C = E = \kappa S_u$, and κ is empirically taken to be 300–1800 (Anastasopoulos et al. 2011). In addition, when the shear wave velocity v_s is known, the initial maximum shear moduli G_0 can be obtained from the formula $G_0 = \rho v_s^2$ (ρ is the density of soil), and then the hardening modulus C can be determined by the formula of $C = E = 2(1 + v)G_0$ (v is Poisson's ratio).

③ γ_r determines the decreasing rate of the kinematic hardening modulus with increasing the plastic deformation, which can be determined by Eq. (7.8).

$$\gamma_r = \frac{C}{\alpha_s} = \frac{C}{\sqrt{3}S_u - \sigma|_0} \tag{7.8}$$

Cyclic simple shear tests are simulated using above constitutive model based on Abaqus software, and the simulation results are compared with test results to verify the prediction ability of the constitutive model. The finite element model for cyclic simple shear tests is shown in Fig. 7.2. The model is a cube with a side length of 10 cm. It is discretized by 8-node linear brick elements with reduced integration (C3D8R). Fixed constraints are applied to the bottom boundary. The displacement in Z and Y directions is 0 and only the movement along the shear direction (X direction) is allowed on lateral boundaries. The cyclic displacement along the X direction is applied on the top surface of the model to simulate the cyclic shearing process on the displacement control mode.

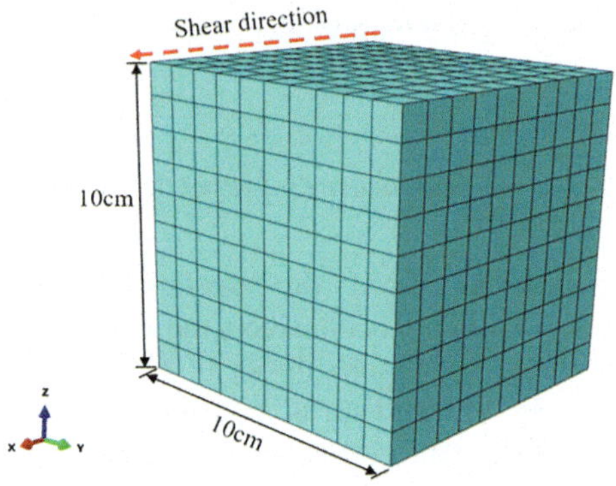

Fig. 7.2 Finite element model for cyclic simple shear tests

The undrained shear strength of clay S_u = 20 kPa, and the constitutive model parameter $\sigma|_0 = 0.1\sigma_{max} = 3.46$ kPa, $C = 1440\,S_u = 28{,}800$ kPa, $\gamma_r = 924$. Figure 7.3 shows the predicted stress–strain hysteresis curves under different amplitudes of shear strain γ in the range of 0.0001–0.01. The predictions indicate that the model can simulate the nonlinear and cyclic hysteresis characteristics of clays subjected to cyclic shearing. The curves of G/G_0–γ and ξ–γ can be obtained based on the stress–strain hysteresis curves under different strain amplitudes, as shown in Fig. 7.4. Here, G/G_0 is the shear modulus ratio, and ξ is the damping ratio. The maximum shear modulus G_0 is assumed to the value when the strain amplitude is below 10^{-6}. The corresponding test results under different strain amplitudes (Li et al. 2019) are also shown in Fig. 7.4. It can be seen that the simulation results agree with the test results in the overall. However, the predicted damping ratio is significantly larger than the test results in the case of large shear strain amplitude (such as $\gamma = 10^{-2}$), which is mainly because the predicted stress–strain hysteresis loop is relatively "fatter" and the corresponding hysteretic loop area is relatively larger. As a result, the damping ratio is overestimated under the large strain amplitude when the above constitutive model is adopted during numerical simulation.

If the undrained strength S_u is known, the parameter γ_r can be determined according to Eq. (7.8) after obtaining parameters $\sigma|_0$ and C. The value of $\sigma|_0$ and C actually depends on the empirical coefficients λ and κ, respectively. In this research, the effects of parameters $\sigma|_0$ and C on numerical simulation results are discussed by setting different λ and κ values. Figure 7.5 shows the comparison between test results and predictions of G/G_0–γ and ξ–γ curves for the same C ($C = 1440\,S_u$) and different $\sigma|_0$ ($\lambda = 0.1, 0.2, 0.3$). It can be seen that greater G/G_0 and smaller ξ can be obtained using greater $\sigma|_0$ for the same γ, which causes the predicted G/G_0–γ curves to move upward and ξ–γ curves to move down. The comparison for the same $\sigma|_0(\sigma|_0 = 0.1\sigma_{max})$ and different C ($\kappa = 500, 1000, 1800$) is shown in Fig. 7.6. Contrary to the former, smaller G/G_0 and greater ξ can be obtained using greater C for the same γ, which causes the predicted G/G_0–γ curves to move down and ξ–γ curves to move upward. Especially, the effects of parameters $\sigma|_0$ and C is most significant when the shear strain amplitude γ is within the range of 10^{-4}–10^{-3}.

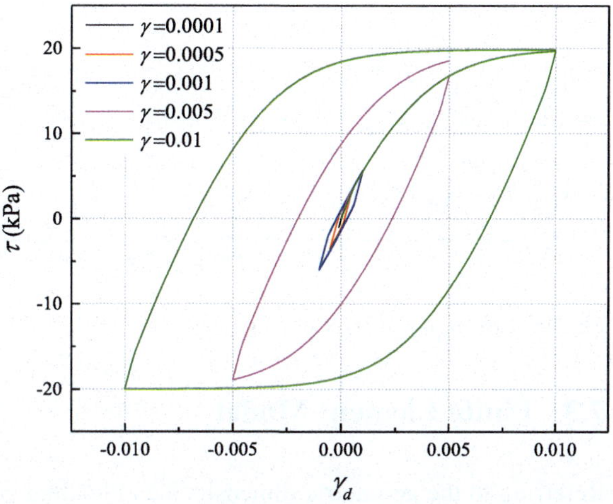

Fig. 7.3 Predicted shear stress–strain hysteresis curves under different shear strain amplitudes $\gamma =$ 0.0001–0.01

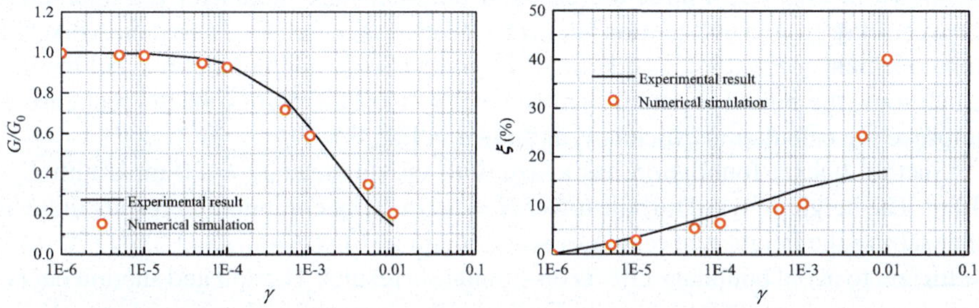

Fig. 7.4 Predicted and measured G/G_0–γ and ξ–γ curves

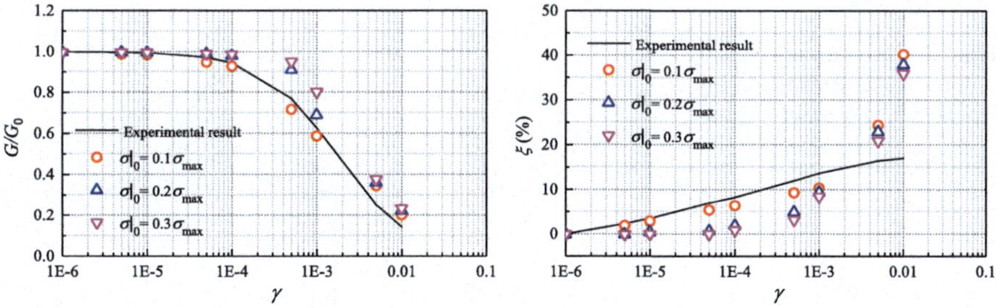

Fig. 7.5 Effects of $\sigma|_0$ on G/G_0–γ and ξ–γ curves

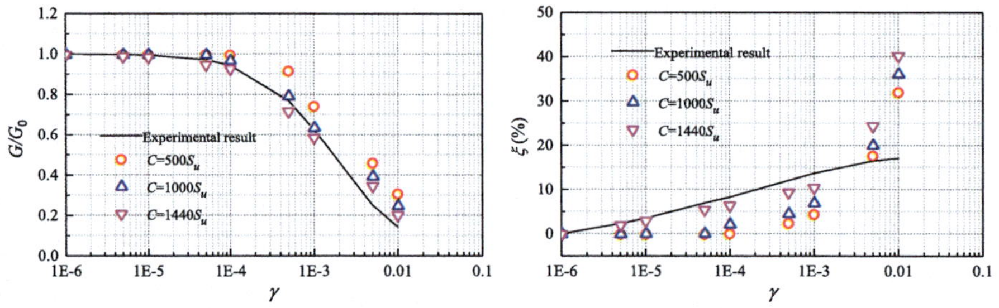

Fig. 7.6 Effects of C on G/G_0–γ and ξ–γ curves

7.3 Finite Element Model

Referring to the geometric dimensions and loading conditions of the single suction bucket foundation of a typical 3.5 MW offshore wind turbine in the North Sea as shown in Fig. 7.7, a three-dimensional finite element model for the interaction between clay ground and suction bucket foundation is established based on ABAQUS finite element software. Considering the symmetries of the geometries and loading conditions, the model is established according to a half portion of the real case to improve the calculation efficiency, as shown in Fig. 7.8.

Suction bucket foundation has a size with the diameter $D = 20$ m, length $L = 10$ m. The height of wind turbine tower $H = 30$ m. Trial computation results indicate that setting the height and length of soil domain as $6D$ (120 m) and $3.5D$ (70 m) is sufficient to avoid boundary effects on simulation results. The soil and suction bucket are discretized by 26,214 8-node linear brick elements with reduced integration (C3D8R). In order to improve calculation efficiency and ensure calculation accuracy, a finer mesh is applied within the depth of the embedment layer (where nonlinearities

Fig. 7.7 Diagram of single suction bucket foundation for typical 3.5 MW wind turbine

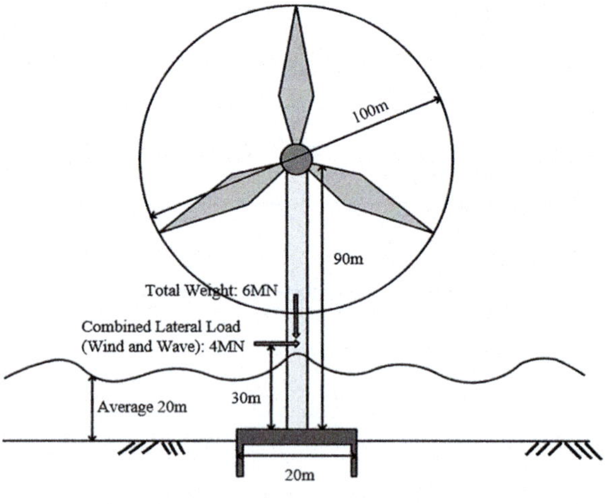

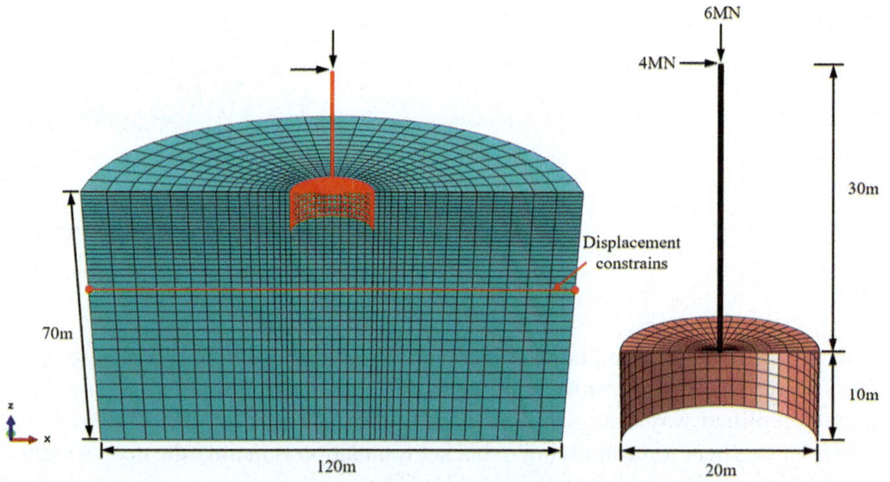

Fig. 7.8 3D finite element model for interaction between clays and single suction bucket

are more prominent), while mesh coarseness increases away from the foundation. The computer used for the simulation has Intel(R) Core(TM) i7-10,700 CPU @ 2.90 GHz, and NVIDIA GeForce GT 730 and Intel(R) UHD Graphics 630, and 16g RAM. The CPU time for each simulation case is approximately 20 h.

The stiffness of suction bucket and wind turbine tower is much greater than that of clay, especially the overlying soft clay, hence the former is set as a rigid body. The interaction between the bucket and the surrounding soil is modeled by the surface-to-surface contact in ABAQUS. In this approach, the master surface is defined as a surface that belongs to the material that either is relatively stiff or has a finer mesh geometry, and the slave surface corresponds to the softer material or material with a coarser mesh. In this study, the bucket surface was defined as the master surface, and the soil surface in contact with the bucket was defined as the slave surface. Hard contact was set in the normal direction of contact surface, which allowed separation between the interface elements when they subjected to tension. The penalty contact method was used in the tangential direction. When the two surfaces were in contact, the interface behavior was governed by Coulomb's friction theory. The critical friction shear stress τ_{crit} at the contact surface can be expressed by $\tau_{crit} = \mu \cdot p_c$ in terms of frictional coefficient μ and contact pressure p_c. $\mu = 0.3$ in this study. Tangential slip occurred when the shear stress at the contact surface exceeded τ_{crit}. In addition, the friction shear stress limit at the contact surface was set as $\tau_{max} = \alpha s_u$ to avoid unrealistic τ_{crit} due to excessive p_c. The coefficient α mainly takes into account the disturbance of suction bucket installation to the soil at the contact surface, and α is empirically taken as 0.85 (Det Norske Veritas (DNV) 2005).

Displacement boundary conditions prevent the out-of-plane movement of the vertical face of symmetry as well as the displacement of the circumferential nodes in the y direction. The base boundary of the model is simulated as rigid bedrock,

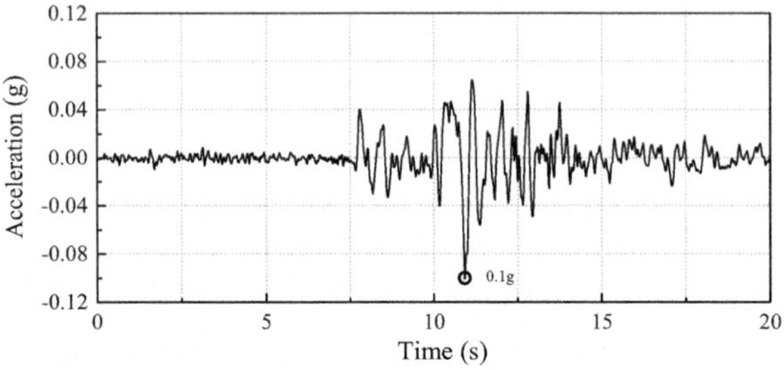

Fig. 7.9 Acceleration recorded in Loma Prieta earthquake

while for the vertical boundaries, kinematic constrains are introduced, forcing the opposite vertical sides to move simultaneously preventing any rotation, simulating in that simplified way the seismic response of infinite soil domain (Tsinidis et al. 2014). The vertical force applying to bucket foundation is mainly the dead weight of generator and tower, which is about 6 MN. The horizontal forces are mainly wave and wind loads, which are assumed to be constant as a simplification treatment. The wind load is 1 MN at 90 m above the top of suction bucket, and the wave load is 3 MN at 10 m above it. The resultant force of the two forces is 4 MN, and the action point is located at 30 m above the top of the bucket foundation. The clay site is subjected to a seismic accelerogram from the Loma Prieta earthquake record in the United States in 1989, which has a scaled peak ground acceleration (PGA) of 0.1 g, as shown in Fig. 7.9.

The dynamic explicit computation is used during the numerical simulation. The fundamental frequency of suction bucket-soil interaction system should firstly be calculated before analyzing seismic responses, and then the Rayleigh damping of soil can be determined according to the first order frequency. Subsequently, the seismic dynamic analysis is carried out according to the following three steps: ① Step 1: Geostatic stress balance analysis, that is, gravity is applied to the soil, and then the initial stress field is obtained while the initial displacement field is basically 0. ② Step 2: The horizontal and vertical concentrated loads are applied to the wind turbine tower, the responses of foundation system are analyzed using the quasi-static method. ③ Step 3: the seismic load is applied to the clay site according to the selected seismic acceleration time history, and the seismic dynamic response of foundation system is analyzed using dynamic explicit method. It should be noted that viscous damping effects result from dynamic interaction between the turbine's submerged part and the water are not considered in this study, which will lead to conservative results.

7.4 Seismic Response of Suction Bucket Foundation

The rotation angle of bucket foundation at the mudline should not exceed 0.5° (or other similar value as dictated by the turbine manufacturer) to satisfy the requirements of serviceability limit state (SLS) of offshore wind turbine (OWT) during the design life (Arany et al. 2015). Hence, it is necessary to analyze seismic dynamic responses of foundation system. In the following section, the failure mechanism of bucket foundation is firstly analyzed, and then various factors affecting seismic responses of suction bucket are analyzed systematically.

7.4.1 Rotational Bearing Mechanism

Figure 7.10 shows the bearing mechanism under seismic loads (peak acceleration scaling to 0.3 g). Obvious rotation occurs for the bucket foundation during earthquake. However, the rotation center of bucket-soil system is not fixed. At the beginning of earthquake, the rotation center is located near point A, as shown in Fig. 7.10a. And then it moves slowly to the right and lower as the earthquake continues. The rotation center vibrates at the front and back, up and down when approaching the seismic peak acceleration, and moved down obviously after the seismic peak acceleration, as shown in Fig. 7.10b and c. Subsequently, the rotation center continues to move slowly to the right and lower, and finally tends to be stable, as shown in Fig. 7.10d.

In the process of earthquake, the bucket wall near point A is subjected to active soil pressure, and gradually separates from the soil behind, forming a certain depth of gap, while the bucket wall near point C is subjected to passive soil pressure, gradually compresses the front soils. The uneven vertical displacement occurs during rotation of the bucket foundation. The point A mainly produces the upward displacement, while the points B and C mainly produce the subsidence displacement, and the settlement of point C is greater than that of point B. In the overall, a certain seismic subsidence occurs for the suction bucket foundation while rotating.

The shear stress–strain curves of three typical nodes around the suction bucket are shown in Fig. 7.11. The shear stress–strain curves show obvious nonlinear hysteresis and strain accumulation. The cumulative shear strain of Node 5294 in the passive region with a value of 5.3E−2 is more significant than that of Node 4561 in the active region with a value of 7.5E−3. For Node 710 in the bottom region, the direction of strain accumulation is opposite to the former two due to the rotational bearing mechanism. The accumulation of plastic shear strain for these nodes can well explain the accumulated rotation of the suction bucket during earthquake, and further verify the prediction ability of the soil constitutive model.

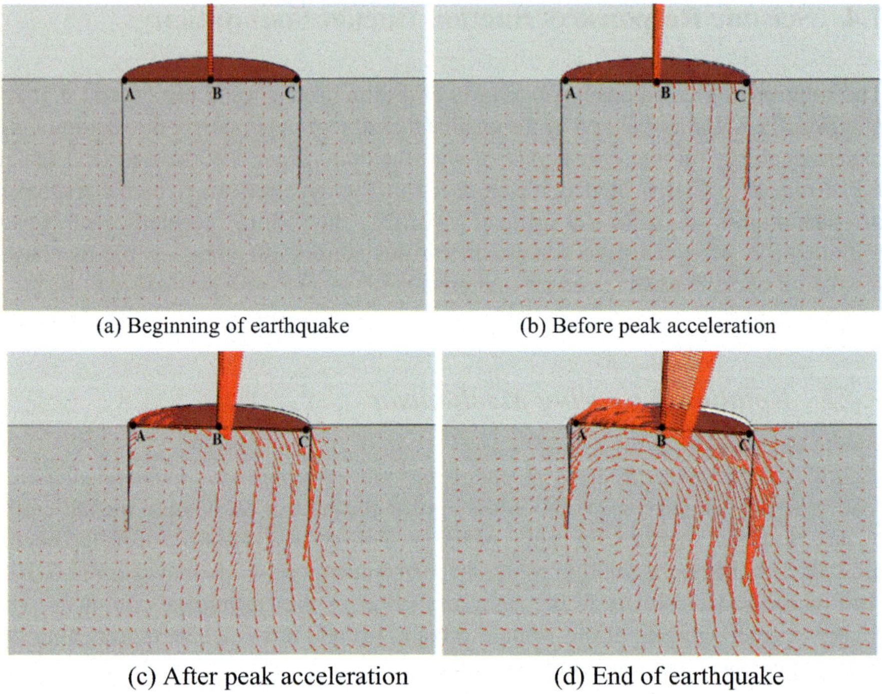

Fig. 7.10 Rotational bearing mechanism of suction bucket foundation (amplification factor of 3)

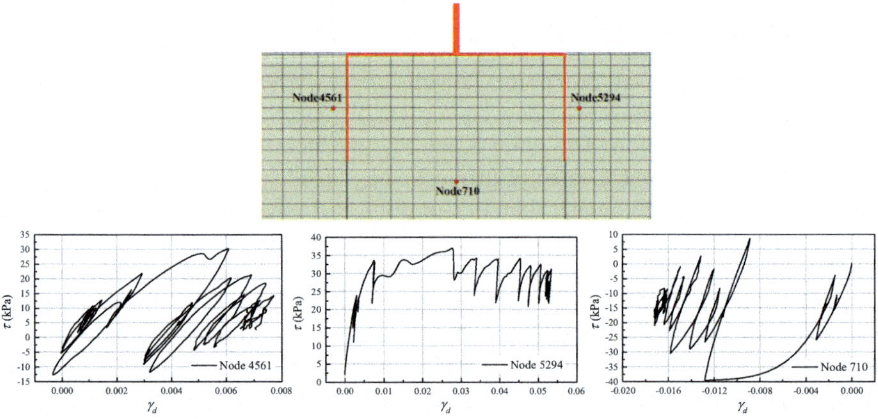

Fig. 7.11 Shear stress–strain curves of three typical nodes around the suction bucket

7.4.2 *Influence of Soil Strength*

The undrained shear strength S_u is set as five cases of 20, 30, 40, 60 and 80 kPa, and the seismic peak acceleration for all cases is 0.1 g. Figure 7.12a and b show the rotation angle-time curves and settlement-time curves of bucket, respectively, for different cases. The rotation angle and settlement gradually increase over time, and the higher the soil strength, the smaller the rotation angle and settlement. However, when S_u exceeds 60 kPa, the change of the rotation angle and settlement tends to be stable, which means that the influence of soil strength is no longer significant. The peak rotation angle and settlement at different cases are shown in Table 7.1. When the soil strength is 20 and 30 kPa, the peak rotation angle exceeds 0.5° (the requirements of serviceability limit state of OWT). In addition, in the early stage of earthquake action (about the first 5 s), the rotation angle and settlement increase rapidly, and then gradually slow down. However, abrupt increasing occurs when approaching the peak acceleration, and then slow down again. This change trend is more significant when the soil strength is lower.

The predicted peak rotation angle at different soil strength are plotted in Fig. 7.13. It can be clearly seen that the peak rotation angle monotonically decreases with increasing soil strength, and the decreasing rate is very large when the soil strength is relative low and gradually tends to be zero. The simulation results can be fitted well by the following formula

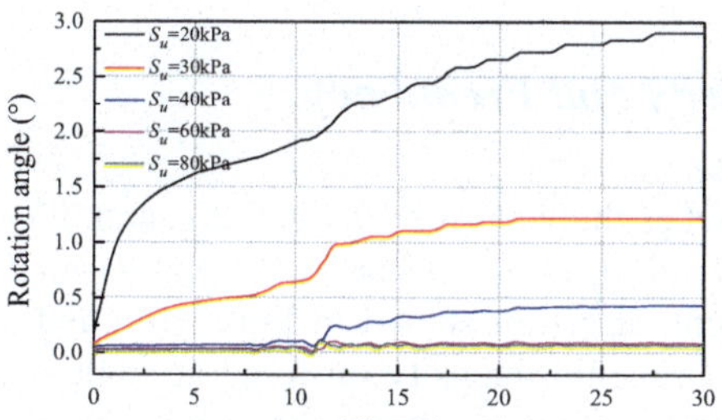

(a) Rotation angle-time curves

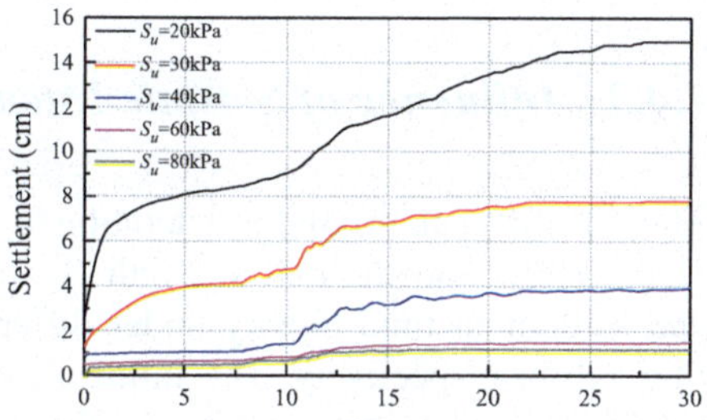

(b) Settlement-time curves

Fig. 7.12 Influence of soil strength on rotation angle and settlement of suction bucket

Table 7.1 Peak rotation angle and settlement of suction bucket under different soil strength

Soil strength (kPa)	Seismic peak acceleration (g)	Peak rotation angle (°)	Peak settlement (cm)
20	0.1	2.90	14.96
30	0.1	1.23	7.86
40	0.1	0.44	3.96
60	0.1	0.10	1.55
80	0.1	0.07	1.17

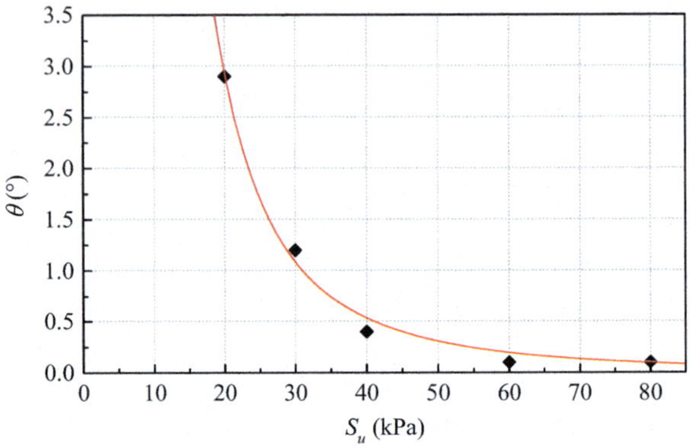

Fig. 7.13 Peak rotation angle versus soil strength

$$\theta = 4400 S_{\mathrm{u}}^{-2.44} \tag{7.9}$$

where θ is the peak rotation angle. The fitting curve represented by Eq. (7.9) is also shown in Fig. 7.13. The formula can be used to approximately estimate the peak rotation angle of suction bucket foundation of OWT in clays with different soil strength. However, the proposed formula is only applicable to those specific dimensions and loading conditions of the bucket for this study, the applicability to other cases remains to be verified.

7.4.3 Influence of Seismic Intensity and Frequency

The seismic peak ground acceleration is set as three cases of 0.1 g, 0.2 g, and 0.3 g, respectively, and the soil strength S_{u} for all cases is 40 kpa. The rotation angle and settlement time history under different cases are shown in Fig. 7.14a and in Fig. 7.14b, respectively. The influence of seismic intensity on rotation angle and settlement is not significant before the moment corresponding to the peak acceleration. After the moment, the rotation angle and settlement increase more significantly with larger seismic intensity. The values for different earthquake intensities are shown in Table 7.2. Obviously, the peak rotation angle exceeds the limit value of 0.5° when the peak acceleration is 0.2 and 0.3 g.

In order to investigate the effect of seismic frequency on seismic responses of the bucket, in addition to the Loma Prieta earthquake record above, two other seismic records are selected for comparison. The time history as input for the dynamic analyses along with their acceleration response spectra are shown in Fig. 7.15. The peak ground acceleration (PGA) of three seismic waves are scaled to 0.3 g. The predominant frequencies of seismic waves for Loma Prieta, Northern California and Kobe are 2.6 Hz, 1.6 Hz, and 0.8 Hz, respectively. The natural frequency ranges of 1p and 3p for the 3.5 MW wind turbine structure is usually less than 0.7 Hz. The rotation angle of suction bucket for three different seismic waves is shown in Fig. 7.16.

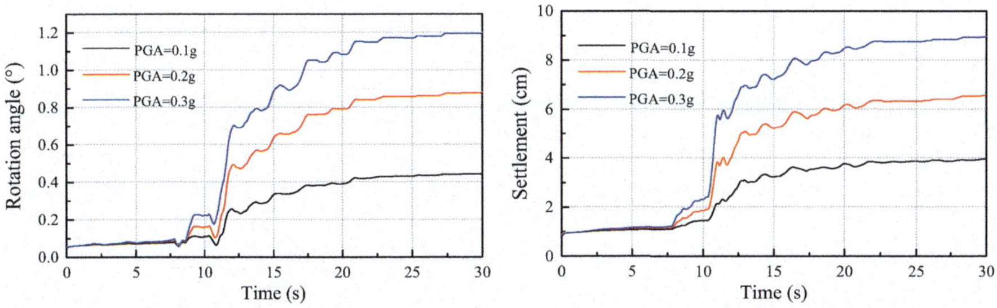

(a) Rotation angle-time curves (b) Settlement-time curves

Fig. 7.14 Influence of seismic intensity on rotation angle and settlement of suction bucket

Table 7.2 Peak rotation angle and settlement of suction bucket under different seismic intensity

Soil strength (kPa)	Seismic peak acceleration (g)	Peak rotation angle (°)	Peak settlement (cm)
40	0.1	0.44	3.96
40	0.2	0.88	6.59
40	0.3	1.20	8.98

The comparison shows the seismic frequency has a significant effect on dynamic responses of bucket. The larger cumulative rotation can occur when the seismic predominant frequency is closer to the frequency ranges of 1p and 3p for the wind turbine structure.

7.4.4 Influence of Bucket Geometry

Four suction buckets with different geometric dimensions are set to study the influence of geometry characteristics on the seismic responses, as shown in Table 7.3. The soil strength is 40 kPa and the seismic peak acceleration is 0.3 g for all cases. Comparing rotation angle-time curves of the suction bucket T1, T2, and T3 in Fig. 7.17a, it can be concluded that increasing the length of the suction bucket (or the embedment ratio) can effectively reduce the final rotation angle when the diameter of suction bucket keeps constant. Comparing with the rotation angle-time curves of the suction bucket T1 and T4, we find that final rotation angle can also be effectively reduced by increasing the diameter and length of suction bucket when the embedment ratio keeps constant. It should be noted that the geometric size of suction bucket has no significant effect on the rotation angle until reaching the peak acceleration. In addition, obvious negative rotation occurs when approaching the peak acceleration due to the shaking of the bucket in positive and negative directions, and the sloshing phenomenon is more obvious with larger embedment ratio. Comparing the rotation angle-time curves of the suction bucket T1, T2, and T4, it can be concluded that

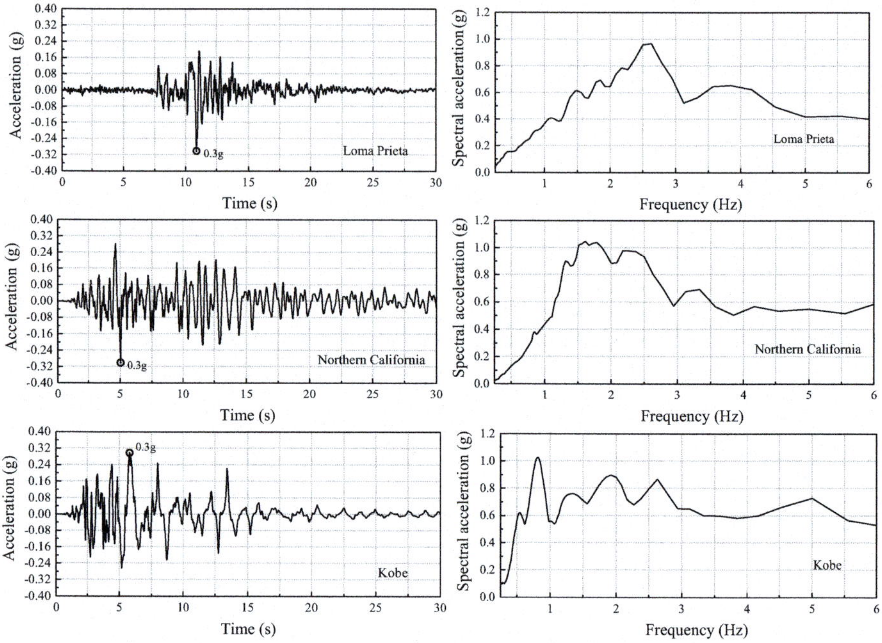

Fig. 7.15 The time history as input for the dynamic analyses along with their acceleration response spectra for three earthquake records

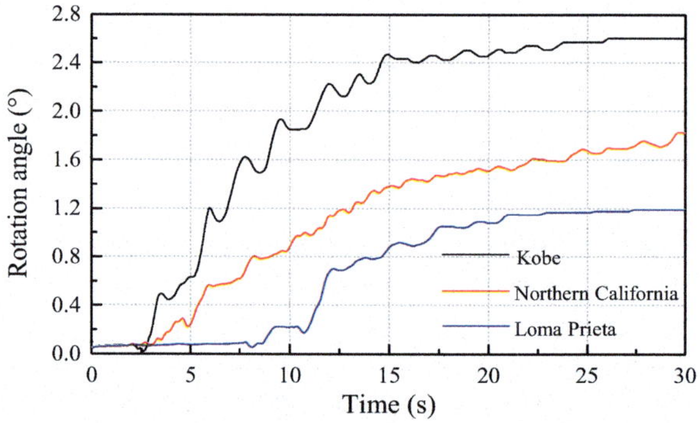

Fig. 7.16 Influence of seismic frequency on rotation angle of suction bucket

increasing the diameter appropriately ($L/D = 0.5$, $D = 20$ to $D = 25$) can reduce the final rotation angle more effectively than increasing the skirt length ($D = 20$, $L/D = 0.5$ to $L/D = 0.7$) when maintaining relative smaller embedment ratio.

It can be seen from Fig. 7.17b that increasing the length or diameter of suction bucket will not reduce the settlement, but will slightly increase it. This may be because the increase of length or diameter will enhance the stability of bucket foundation and reduce the rotation angle, however it will also increase the self-weight of bucket

Table 7.3 Peak rotation angle and settlement of suction bucket with different size

Suction bucket number	Diameter/m (*D*/m)	Length/m (*L*/m)	Embedment ratio (*L*/*D*)	Peak rotation angle (°)	Peak settlement (cm)
T1	20	10	0.5	1.20	8.98
T2	20	14	0.7	0.68	10.30
T3	20	20	1.0	0.27	10.71
T4	25	12.5	0.5	0.50	10.60

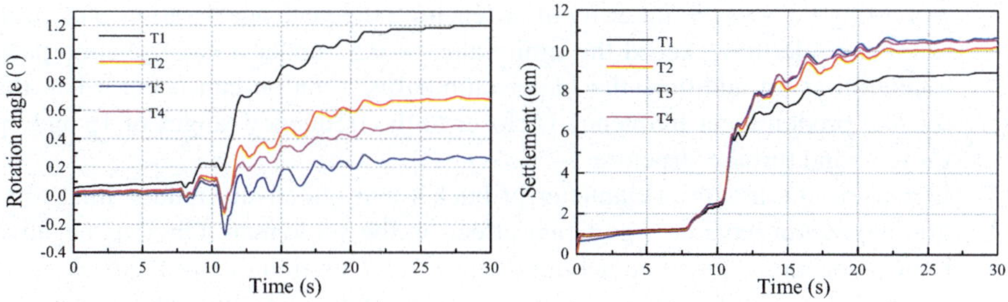

(a) Rotation angle-time curves (b) Settlement-time curves

Fig. 7.17 Influence of bucket geometry on rotation angle and settlement

foundation, resulting in slightly increased settlement. In overall, the effect of the bucket geometry on settlement is not significant.

7.5 Summary

This chapter investigated the seismic responses of suction bucket foundation for offshore wind turbine in clays by performing a series of numerical simulation. The key conclusions can be summarized as follows:

(1) A simplified kinematic hardening model available in the ABAQUS library was used to capture the undrained dynamic nonlinear behavior of clays. The method of determining model parameters was introduced, and the prediction ability of the model was validated by simulating cyclic simple shear tests. Then, a three-dimensional finite element model is developed by setting appropriate boundary and contact condition to analyze seismic responses of suction bucket in clays based on the above soil model.

(2) The rotational bearing mechanism of suction bucket foundation is revealed by the three-dimensional finite element analysis. Significant accumulated rotation occurs for the suction bucket during the earthquake. The rotation center moves continuously, and vibrates when approaching the seismic peak acceleration, and

finally tends to be stable. In the overall, a certain seismic subsidence occurs for the suction bucket when rotating. The shear stress–strain curves of soil around the bucket show obvious nonlinear hysteresis and strain accumulation which can well explain the accumulated rotation of the bucket during earthquake.

(3) The smaller rotation angle and settlement can be obtained with higher soil strength. However, when soil strength exceeds a certain value, its influence is no longer significant. It should be noted that the lower soil strength can cause the peak rotation angle to exceed the requirements of serviceability limit state of OWT.

(4) The rotation angle and settlement of suction bucket increases significantly with increasing the seismic intensity after reaching the peak acceleration. The peak rotation angle may exceed the limit value of 0.5° under a large seismic peak acceleration. In addition, the larger cumulative rotation can occur when the seismic predominant frequency is closer to the frequency ranges of 1p and 3p for the wind turbine structure.

(5) Increasing the length or diameter of bucket can effectively reduce final rotation angle, but have no significant effect on the settlement. Obvious negative rotation occurs when approaching the peak acceleration due to the shaking of the bucket in positive and negative directions, and the sloshing phenomenon is more obvious with larger embedment ratio. In addition, increasing the diameter appropriately can reduce the final rotation angle more effectively than increasing the skirt length when maintaining relative smaller embedment ratio.

At present, there are few relevant shaking table model test results of suction bucket in clays in the existing literature, so the numerical simulation results of the interaction between the bucket and soil have not been verified yet. In the follow-up work, we will try to perform relevant shaking table tests to verify the numerical method proposed by this study.

References

Ahn, J., H. Lee, and Y.T. Kim. 2014. Holding capacity of suction caisson anchors embedded in cohesive soils based on finite element analysis. *International Journal for Numerical and Analytical Methods in Geomechanics* 38 (15): 1541–1555.

Anastasopoulos, I., F. Gelagoti, R. Kourkoulis, et al. 2011. Simplified constitutive model for simulation of cyclic response of shallow foundations: Validation against laboratory tests. *Journal of Geotechnical and Geoenvironmental Engineering* 137 (12): 1154–1168.

Arany, L., S. Bhattacharya, J.H.G. Macdonald, et al. 2015. A critical review of serviceability limit state requirements for monopile foundations of offshore wind turbines. In *Offshore Technology Conference Houston.*

Byrne, B.W., and G.T. Houlsby. 2003. Foundations for offshore wind turbines. *Philosophical Transaction of the Royal Society A-Mathematical Physical and Engineering Sciences* 361: 2909–2930.

Cao, J., R. Phillips, and R. Popescu. 2003. Numerical analysis of the behavior of suction anchors in clay. *International Journal of Offshore and Polar Engineering* 13 (2): 154–159.

Chen, W., and M.F. Randolph. 2007. Uplift capacity of suction anchors under sustained and cyclic loading in soft clay. *Journal of Geotechnical and Geoenvironmental Engineering* 133 (11): 1352–1363.

Chen, X.G., T. Liu, Y.K. Jiang, et al. 2018. Stability analysis of suction bucket foundations under wave cyclic loading and scouring. *Marine Georesources and Geotechnology* 36 (7): 749–758.

Cheng, X.L., A.W. Yang, and G.N. Li. 2018. Model tests and finite element analysis for the cyclic deformation process of suction anchors in soft clays. *Ocean Engineering* 151: 329–341.

Det Norske Veritas (DNV). 2005. *Geotechnical Design and Installation of Suction Anchors in Clay*. DNV Recommended Practice RP-E303.

Faizi, K., A. Faramarzi, S. Dirar, et al. 2019. Investigating the monotonic behaviour of hybrid tripod suction bucket foundations for offshore wind towers in sand. *Applied Ocean Research* 89: 176–187.

Gelagoti, F., I. Georgiou, R. Kourkoulis, et al. 2018. Nonlinear lateral stiffness and bearing capacity of suction caissons for offshore wind-turbines. *Ocean Engineering* 170: 445–465.

Giannakos, S., N. Gerolymos, and G. Gazeta. 2012. Cyclic lateral response of piles in dry sand: Finite element modeling and validation. *Computers and Geotechnics* 44: 116–131.

Guo, Z., D.S. Jeng, W. Guo, et al. 2018. Failure mode and capacity of suction caisson under inclined short-term static and one-way cyclic loadings. *Marine Georesources and Geotechnology* 36 (1): 52–63.

Houlsby, G.T., and B.W. Byrne. 2000. Suction Caisson foundations for offshore wind turbines and anemometer masts. *Wind Engineering* 24: 249–255.

Huang, M.S., and Y. Liu. 2015. Axial capacity degradation of single piles in soft clay under cyclic loading. *Soils and Foundations* 55 (2): 315–328.

Jin, Z., Z.Y. Yin, P. Kotronis, et al. 2019. Advanced numerical modelling of caisson foundations in sand to investigate the failure envelope in the H-M-V space. *Ocean Engineering* 190: 106394.

Kim, D.J., Y.W. Choo, J.H. Kim, et al. 2014a. Investigation of monotonic and cyclic behavior of tripod suction bucket foundations for offshore wind towers using centrifuge modeling. *Journal of Geotechnical and Geoenvironmental Engineering* 140 (5): 04014008.

Kim, D.H., S.G. Lee, and I.K. Lee. 2014b. Seismic fragility analysis of 5Mw offshore wind turbine. *Renewable Energy* 65: 250–256.

Kourkoulis, R.S., P.C. Lekkakis, et al. 2014. Suction caisson foundations for offshore wind turbines subjected to wave and earthquake loading: Effect of soil–foundation interface. *Géotechnique* 64 (3): 171–185.

Li, D.Y., L.Y. Feng, and Y.K. Zhang. 2014. Model tests of modified suction caissons in marine sand under monotonic lateral combined loading. *Applied Ocean Research* 48: 137–147.

Li, H.J., H.J. Liu, and S.Y. Liu. 2017. Dynamic analysis of umbrella suction foundation embedded in seabed for offshore wind turbines. *Geomechanics for Energy and the Environment* 10: 12–20.

Li, M., J.W. Liang, Z.N. Ba, et al. 2019. Finite element analysis of seismic settlement of shield tunnel in soft soil. *China Civil Engineering Journal* 52 (S1): 240–247 (in Chinese).

Maniar, D.R. 2004. *A Computational Procedure for Simulation of Suetion Anchor Behavior Under Axial and Inclined Loads*. Austin: University of Texas at Austin.

Monajemi, H., and H.A. Razak. 2009. Finite element modeling of suction anchors under combined loading. *Marine Structures* 22 (4): 660–669.

Perveen, R., N. Kishor, and S.R. Mohanty. 2014. Off-shore wind farm development: Present status and challenges. *Renewable and Sustainable Energy Reviews* 29: 780–792.

Prevost, J.H. 1977. Mathematical modelling of monotonic and cyclic undrained clay behaviour. *International Journal for Numerical and Analytical Methods in Geomechanics* 1 (2): 195–216.

Sukumaran, B., W.O. McCarron, P. Jeanjean, et al. 1999. Efficient finite element techniques for limit analysis of suction anchors under lateral loads. *Computers and Geotechnics* 24: 89–107.

Tsinidis, G., K. Pitilakis, and A.D. Trikalioti. 2014. Numerical simulation of round robin numerical test on tunnels using a simplified kinematic hardening model. *Acta Geotechnica* 9: 641–659.

Wang, X., X. Yang, and X.W. Zeng. 2017. Seismic centrifuge modelling of suction bucket foundation for offshore wind turbine. *Renewable Energy* 114: 1013–1022.

Yang, C.B., J.M. Zhang, and R. Wang. 2017. Seismic analysis of a suction caisson foundation for offshore wind turbines. *Journal of Tsinghua University (Science and Technology)* 57 (11): 1207–1211 (in Chinese).
Yin, Z.Y., J.C. Teng, Z. Li, et al. 2020. Modelling of suction bucket foundation in clay: From finite element analyses to macro-elements. *Ocean Engineering* 210: 107577.
Zhang, B., H.L. Kang, X.H. Zhang, et al. 2019. On the responses of soil layers surrounding a suction bucket under cyclic loads. *Marine Georesources and Geotechnology* 38: 1–10.
Zhu, B., Z. Zhu, T. Li, et al. 2017. Field tests of offshore driven piles subjected to lateral monotonic and cyclic loads in soft clay. *Journal of Waterway, Port, Coastal, and Ocean Engineering* 143 (5): 05017003.

Chapter 8
Lateral Cyclic Responses of OWT Tripod Suction Bucket in Clays

Tripod suction bucket foundations are increasingly used to support offshore wind turbines (OWTs) due to their economic advantages and high overturning resistance. In the harsh marine environment, OWT tripod bucket foundations are subjected to long-term lateral cyclic loads from wind, waves and current. However, the response of tripod bucket foundations in clay to cyclic lateral loads is not well covered in the literature. Therefore, a three-dimensional (3D) numerical method is developed for analyzing their lateral cyclic response based on a simplified bounding surface model of clay. The numerical model is validated by comparing its results with scaled model test results available in the literature. The validated model is then used to investigate the rotation mechanism of tripod suction bucket under lateral monotonic and cyclic loading. The 'bonded contact' and 'separable contact' were set between the soil plug and the inner wall to simulate the upper limit and the lower limit cases of the tensile capacity of the suction bucket. The influences of the bucket-soil contact condition, loading direction and cyclic loading mode (one-way and two-way) on the cyclic behavior were analyzed. It was found that these factors have a significant impact on the cyclic responses, permanent rotation angle and failure mechanism of the tripod bucket foundation. The findings from this study can provide guidance for the design of OWT tripod bucket foundations.

8.1 Introduction

Offshore wind energy has developed rapidly in recent years due to its abundant, clean and renewable advantages. A growing number of offshore wind turbines (OWTs) are being built to produce more wind power (Perveen et al. 2014). An important consideration for OWT is its foundation as its construction and maintenance costs account for 20–30% capital cost and 12–25% life cycle cost of the project (Musial and Ram 2010). Therefore, economic and reliable foundation designs are crucial to reduce

X. Cheng et al., *Dynamic Analysis of Offshore Wind Turbine Foundations in Soft Clays*,
https://doi.org/10.1007/978-981-97-9454-6_8

construction costs and to ensure the safety of turbines. The suction bucket foundation, which can be installed by using a suction pump, is regarded as a competitive foundation type for offshore wind turbines due to its fast construction, low cost, and reusability (Byrne et al. 2002).

In the harsh marine environment, suction bucket foundations supporting OWTs are subject to the long-term lateral dynamic loads due to wind, waves, and current (Byrne and Houlsby 2003). These environmental loads induce large overturning moments on the suction bucket foundations (Houlsby and Byrne 2000), which may cause excessive rotation angle of OWTs. In current guidelines for OWTs, requirements of serviceability limit state (SLS) typically dominate the whole design (Arany et al. 2015; Liao et al. 2018). If the rotation angle exceeds the allowable value, the OWTs will not operate normally. Therefore, it is of great significance to investigate the bearing behavior of suction bucket foundations under lateral dynamic loads for the engineering design. A suction bucket foundation can be constructed in multiple configurations including the monopod and multiple-suction bucket foundations (e.g., tripod foundation), as shown in Fig. 8.1. The former has been widely used in practical engineering. Compared with the monopod bucket foundation, the multiple-suction bucket foundation has greater overturning resistance, and hence it is gradually favored by industry (Jeong et al. 2021; Houlsby et al. 2005a).

The behavior of monopod suction buckets under lateral cyclic loading was investigated extensively through: (a) prototype and reduced-scale filed tests (Houlsby et al.

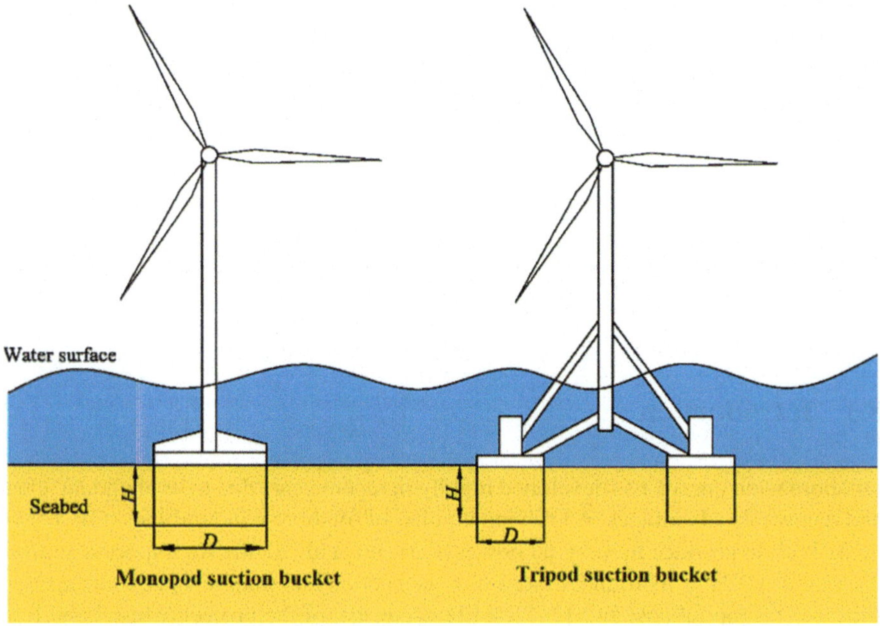

Fig. 8.1 Monopod and tripod suction bucket foundations

2005b, 2006; Barari and Ibsen 2012; Zhang et al. 2015); (b) scaled 1-g laboratory tests (Wang et al. 2006; Zhu et al. 2013, 2018; Foglia and Ibsen 2016; Hung et al. 2018); (c) centrifugal model tests (Zhang et al. 2007; Cox et al. 2014; Wang et al. 2017a, b); and (d) numerical simulations (Kourkoulis et al. 2014; Gelagoti et al. 2018; Cheng and Wang 2016; Cheng et al. 2018, 2020a; Yi et al. 2021). These studies provided good understanding of lateral cyclic behavior of the monopod suction bucket, such as the evolution of its cumulative rotation and unloading stiffness with the cyclic load amplitude, frequency and direction and number of cycles. In recent years, new types of bucket foundations have also been investigated, such as the hybrid foundation consisting of a traditional monopile and a wide-shallow bucket, and their superiority over the traditional bucket foundation has been proven (Chen et al. 2020; Li et al. 2020, 2021a, b; Ma and Yang 2020; Wang et al. 2020b; Trojnar et al. 2021).

The multiple-suction bucket foundation improves the overturning resistance of OWTs by transferring the overturning moment of the whole system into axial push–pull forces of individual buckets (Kim et al. 2014a; Zhang et al. 2016). The vertical pullout and compression behaviors of each single bucket elements were investigated to infer the cyclic behaviors of multiple-suction buckets under lateral loading (Kelly et al. 2006; Lu et al. 2007; Jeong et al. 2019). Centrifugal model studies were conducted to compare the horizontal cyclic behaviors of monopod suction bucket and tripod suction bucket foundations installed in sands (Kim et al. 2014b; Wang et al. 2018). The results revealed that the moment-rotation relationship of the tripod suction bucket exhibited an almost bilinear response, while the monopod foundation displayed a continuous hardening response. The cumulative rotation of monopod suction bucket in medium dense sand increased monotonically as the number of load cycles increased, while the rotation of tripod suction bucket foundation increased at the initial stage of cyclic loading but then gradually decreased, showing a favorable "self-healing" behavior. Jeong et al. (2021) investigated the moment-rotation responses, the cyclic stiffness, and permanent displacements of the tripod foundation in sands by performing centrifuge model tests. These tests demonstrated that the cyclic behavior of the tripod foundation was significantly affected by the loading amplitude and direction.

The literature review indicated that most of the existing studies on tripod suction bucket focus on its behavior in sands, while investigations of its response in clay are scarce. Only a few researchers studied the monotonic bearing behavior of tripod suction bucket in clays (Hung and Kim 2012; Kim et al. 2014a; Petas et al. 2016; He et al. 2021). To the best of the author's knowledge, the cyclic behavior of tripod suction bucket in clays has not been reported. This chapter develops a numerical method for analyzing the lateral cyclic behavior of tripod suction bucket in clays based on a simplified bounding surface model of clay. Subsequently, the rotation mechanism of tripod suction bucket under lateral monotonic and cyclic loading is investigated. The influences of some factors such as the bucket-soil interface contact, the loading direction, and the cyclic loading mode (one-way and two-way) on the cyclic behavior are analyzed.

8.2 Numerical Simulation Method and Verification

It is undoubtedly ideal to validate the feasibility of the numerical method through comparison with model test results of tripod suction bucket installed in clay. Regrettably, there are no model tests reported in the literature for tripod suction bucket in clay. However, since a tripod bucket foundation resists the overturning moment by vertical uplift and compression of individual buckets, it is reasonable to consider accurate prediction of the monopod bucket behavior under vertical loads is a good indicator for proper simulation of the tripod bucket behavior under lateral loads. Therefore, model tests of monopod foundations under vertical monotonic and cyclic loading reported in the existing literature are selected for numerical simulation, and the feasibility of the numerical method is validated by comparing the simulation results with the test results.

8.2.1 *Simulation of Monotonic Loading Test*

Cao et al. (2002) performed a series of centrifuge tests on a 1/100 model monopod suction bucket in clay subjected to vertical monotonic pullout load. The prototype bucket diameter $D = 5.17$ m, the skirt length $L = 24.5$ m and the bucket wall thickness $t = 0.065$ m. The undrained shear strength of soft clay varied linearly with soil depth, which was represented by $S_u = 1.14\,z$ (z is soil depth). The centrifuge test number SAT06-3 was chosen for numerical simulation. The test was carried out in two stages, installation and monotonic vertical loading. During the installation stage, 50 mm (prototype 5 m) soil plug was generated. Based on the geometric dimensions of the monopod suction bucket and soil strength profile in the model test, a three-dimensional finite element model was established employing the finite element software ABAQUS (Dassault Systemes Simulia Corp 2014). Exploiting the symmetries of the geometry and loading conditions, the model simulated only one half of the system, as shown in Fig. 8.2. The bucket and the soil were simulated using 11,160 8-node linear brick elements (C3D8). Normal horizontal constraints were applied to the vertical boundaries, and fixed constraints were applied to the bottom boundary while the top boundary was free. Vertical loads were applied to the center of the bucket top.

The interaction between the suction buckets and surrounding soil was simulated utilizing the surface-to-surface contact method. The surface of suction bucket was defined as the master surface (relatively stiff), and the soil surface in contact with suction bucket was defined as the slave surface (relatively soft). Hard contact was set in the normal direction of contact surface for the bucket outer wall, which allowed separation between the interface elements when they are subjected to tension. The penalty contact method was used in the tangential direction. When the two surfaces are in contact, the interface behavior is governed by Coulomb's friction theory. The critical friction shear stress τ_{crit} at the contact surface can be expressed by $\tau_{crit} = \mu \cdot p_c$

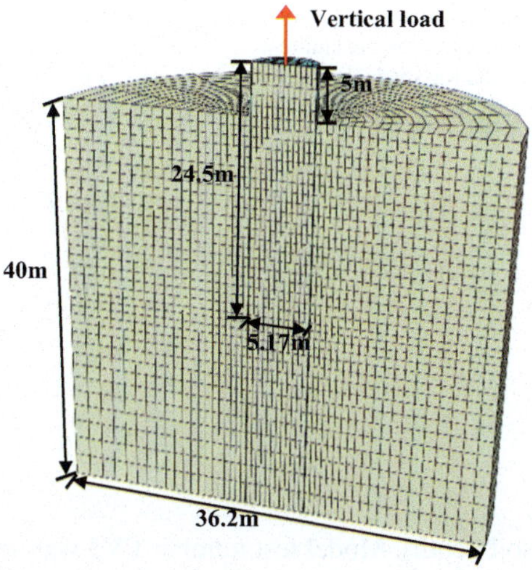

Fig. 8.2 3D finite element model of monopod bucket foundation under vertical monotonic pullout loading

in terms of frictional coefficient μ and contact pressure p_c, with $\mu = 0.25$ in this study. When the shear stress at the contact surface exceeded τ_{crit}, tangential slip occurred. The tie-contact condition was set between the soil plug and the inner wall considering that the significant passive suction makes them remain in close contact. The soil behavior was simulated by the simplified single bounding surface model described in Chap. 2 with parameters $A_0 = 2\ S_u$; $G = 500\ S_u$; $\mu = 2$; $\xi = 0$. It should be noted that parameter $\xi = 0$ is used for the numerical simulation because monotonic loading does not involve cyclic degradation. The bucket was treated as a rigid body during the calculations.

Figure 8.3 compares the calculated and measured vertical load–displacement curves of the suction bucket. The calculated results agree with the experimental results in the overall trend, which validates the rationality of the numerical method to predict the monotonic behavior of bucket. It should be noted that, as the upward displacement increases the bearing capacity of the suction bucket reaches the peak pullout force and then gradually decreases, i.e., it shows softening behavior. However, this numerical method is still unable to simulate the softening behavior of suction bucket during uplift, although it can relative accurately predict the ultimate bearing capacity.

8.2.2 Simulation of Cyclic Loading Test

Villalobos et al. (2010) performed a series of model testing for suction bucket foundation in clay subjected to vertical cyclic loading. The suction bucket diameter $D = 150$ mm, the skirt length $L = 150$ mm and the wall thickness $t = 1$ mm. The soil layer had a linear shear strength profile described by $S_u = 4z + 5.6$ kPa (z is

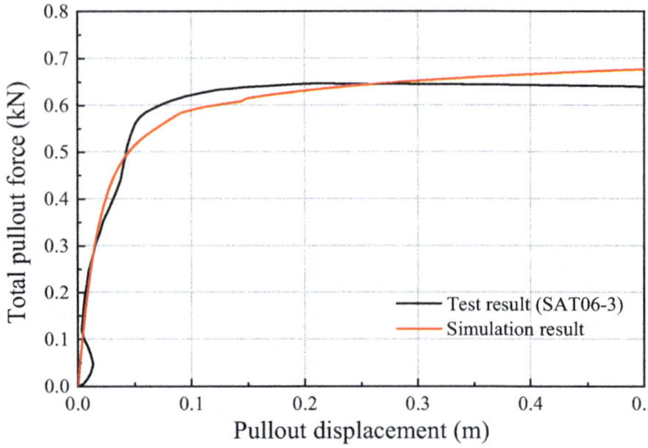

Fig. 8.3 Comparison of simulation results and test results for vertical monotonic load–displacement curve

soil depth). Model test number FV5 was selected for the numerical simulation. The suction bucket was subjected to 8 cyclic loading epochs of 10 cycles each in sequence after a pushed installation. The cyclic loading amplitudes increased from ± 47 to ± 560 N. In addition, the suction bucket was subjected to a mean vertical load of 250 N. The mean load was the maximum load to install the bucket, which is denoted as the maximum preload (V_o) experienced by the bucket prior to the cycling.

A three-dimensional finite element model was established to simulate the model test above, as shown in Fig. 8.4. The element type, boundary conditions, pile-soil contact, etc., were same as the finite element model used for the monotonic loading case. It should be noted that the pushing installation method was used in this test, i.e., the suction bucket was installed without suction assistance; hence the passive suction is insignificant during the vertical cyclic loading process. Therefore, frictional contact conditions of the tangential slip and normal separation were set between the soil plug and the bucket inner wall. The clay cyclic behavior was simulated by the simplified single bounding surface model above with parameters $\mu = 2$ and $\xi = 2$, $A_0 = 2\ S_u$ and $G = 500\ S_u$, i.e., A_0 and G increased linearly along the soil depth z.

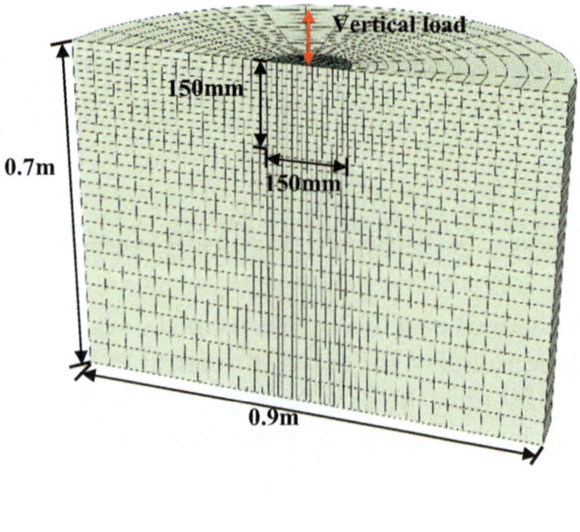

Fig. 8.4 3D finite element model of monopod bucket foundation under vertical cyclic loading

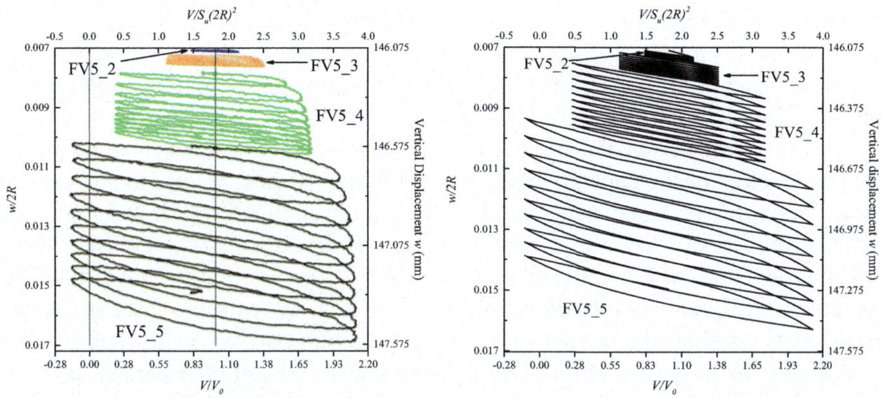

(a) Model test results (b) Numerical simulation results

Fig. 8.5 Comparison of simulation results and test results for vertical cyclic load–displacement curve

The load–displacement curves of the four loading epochs from FV5_2 to FV5_5 are presented in Fig. 8.5. The vertical load V is normalized by soil strength (S_u) at the reference point (depth of 125 mm) or by the maximum preload V_0. The displacements are normalized by the bucket diameter ($D = 2R$). It is observed that the calculated response can capture the nonlinear hysteretic characteristics of the load–displacement response, and the accumulation of foundation settlement as the number of load cycles and load amplitude increase. The calculated responses generally agree with the measured results, which validates the ability of the numerical method to predict the vertical cyclic behavior of suction buckets.

8.3 Numerical Model of Tripod Suction Bucket

A three-dimensional finite element model for a tripod bucket foundation in clays was established as shown in Fig. 8.6. The dimensions of the foundation in this study were determined by referring to the first successfully installed tripod bucket foundation at the wind farm Borkum Riffgrund 1 (Ehrmann et al. 2016). The diameter and height of each pod were both 8 m with a total weight of 850 t. The wall thickness of each pod was 0.2 m. The center-to-center distance between each pod was 30 m. The bucket was made of steel with an elastic modulus of 210 GPa and a Poisson's ratio of 0.3. Soil domain with a size of 120 m × 116 m × 37 m (length × width × height) was meshed, which was sufficient to avoid boundary effects on the obtained results. The soil and OWT structure were discretized by 34,933 8-node linear brick elements. A linear increasing soil shear strength profile described by $S_u = 2z + 2$ kPa was used in the numerical simulation. The geostatic stress balance analysis was first carried out, i.e., the gravity was applied to the soil domain to establish the initial stress field.

A vertical load of 10 MN was then applied at the tower section 33 m away from the mud surface (Point C in Fig. 8.6) to simulate the dead weight of the whole OWT structure (Tran et al. 2017). Finally, the horizontal load was applied to Point C as shown in Fig. 8.6.

As a consequence of the installation method, the suction bucket foundation is able to develop tensile capacity when it is subjected to uplift, owing to the negative excess pore pressures between the bucket lid and the confined soil. It was found that the contribution of the passive suction, generated during caisson's pull out, to the total pullout capacity is significant (Cao et al. 2002). In view of the great difficulty to accurately and quantitatively simulate passive suction, two extreme cases are usually assumed to consider the impact of the passive suction on the tensile capacity. One case is that the passive suction is zero, so the bucket lid can be separated from the soil plug inside bucket, which corresponds to the lower limit of the tensile capacity as shown in Fig. 8.7a; another case is that the passive suction is large enough, so the bucket lid remains in contact with the soil plug inside bucket, which corresponds to the upper limit of the tensile capacity as shown in Fig. 8.7c. However, the actual case should be between the two extreme cases, as shown in Fig. 8.7b. In this study, the frictional contact condition of the tangential slip and normal separation ('separable contact' for short name) was set between the soil plug and the inner wall to simulate the lower limit case of the tensile capacity; the tie-contact condition ('bonded contact' for short name) was set between the soil plug and the inner wall to simulate the upper limit case of the tensile capacity.

Offshore wind turbines may be subjected to wind and wave loads in any horizontal direction in the marine environment. Due to the asymmetric layout of the tripod bucket foundation, the loading direction will affect its bearing characteristics. The angle between the horizontal load direction and the x-axis positive direction is defined as the loading angle α, as shown in Fig. 8.8. The influence of the bucket-soil contact

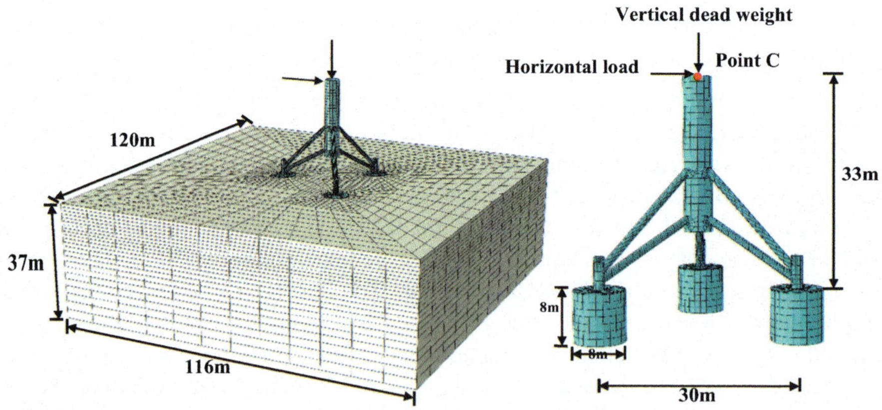

Fig. 8.6 3D finite element model for tripod suction bucket foundation in clays

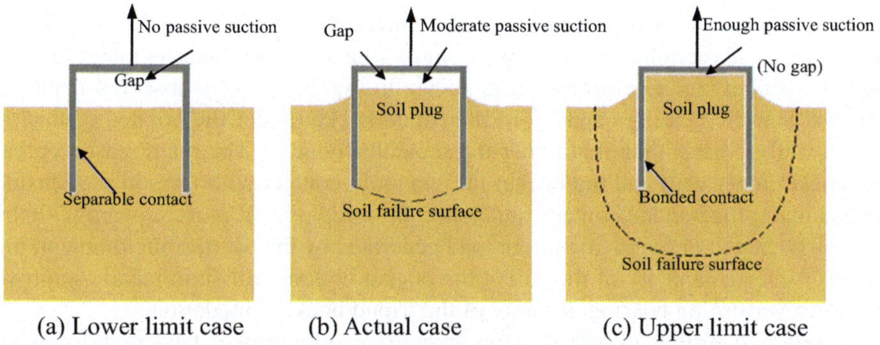

Fig. 8.7 Diagrammatic sketch of suction bucket foundation under tensile loads

Fig. 8.8 Plan view of tripod suction bucket foundation

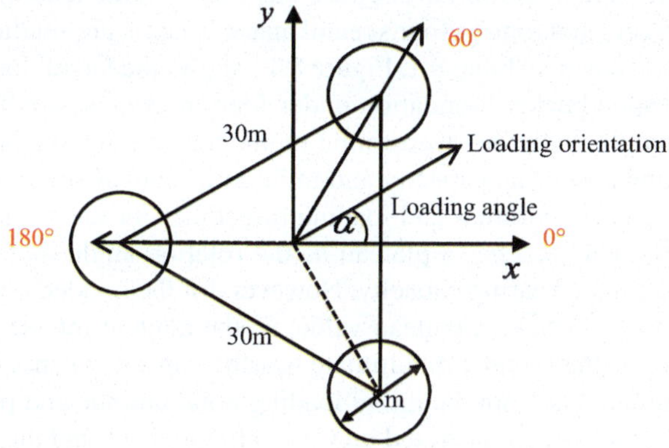

condition and the loading direction on monotonic and cyclic behavior of the tripod suction bucket foundation is analyzed in detail below.

8.4 Monotonic Loading Response of Tripod Suction Bucket

Figure 8.9 shows the horizontal ultimate bearing capacity of tripod bucket for 13 loading angles ranging from 0° to 180° (15° interval) under two different bucket-soil contact conditions, i.e., 'bonded contact' and 'separable contact'. It is observed that the variation trend of bearing capacity with the loading angle under the two contact conditions is just opposite. For the bonded contact, the bearing capacity corresponding to 0° and 120° loading directions is the highest, while that corresponding to 60° and 180° loading directions is the lowest. For the separable contact, the opposite is true. Since the plan layout of the tripod bucket foundation is an equilateral triangle as shown in Fig. 8.8, the loading conditions in the 0° and 120°directions are the same, both of which are one bucket under tension and two buckets are under compression as shown in Fig. 8.10a. Similarly, the loading conditions in the 60° and 180° directions are the same, both of which are two buckets under tension and one bucket under compression as shown in Fig. 8.10b. It is observed from the Fig. 8.8 that the bearing capacity of the tripod bucket under the bonded contact is significantly greater than that under the separable contact. The main reason is that the bucket inner wall and the bucket lid remain in contact with the soil plug inside bucket under the bonded contact, and hence the soil plug will move upward with the suction bucket under vertical pullout load generated by the overturning moment; the reverse bearing capacity of the soil at the bucket bottom will significantly improve the anti-overturning bearing capacity of the tripod bucket foundation.

In order to further clarify the comprehensive influence of bucket-soil contact conditions and loading direction on the horizontal bearing behaviors of tripod bucket foundation, the direction 0° (the most favorable direction) and 180° (the most unfavorable direction) are selected as two extreme loading cases as shown in Fig. 8.10. Correspondingly, four special analysis cases are set during finite element simulation as shown in Table 8.1. Figure 8.11 shows the lateral load-rotation angle curves of the tripod bucket foundation under four analysis cases. The foundation rotates significantly under the overturning moment caused by the horizontal load, and its rotation angle, bearing capacity and failure mechanism are closely related to the bucket-soil contact condition and loading direction. For the separable contact (Cases 1 and 2), the load reaches a plateau as the rotation angle increases, which is defined as the ultimate bearing capacity. However, for the bonded contact (Cases 3 and 4), the load does not have a plateau value, so the tangent intersection method is used to determine the foundation ultimate bearing capacity (Kim et al. 2014a). Two tangent lines are drawn from the initial loading point and the end point of the load–displacement curve, as demonstrated on Case 3 in Fig. 8.11 and the ordinate of the intersection of the two tangents define the foundation ultimate bearing capacity.

Figure 8.12 shows the foundation displacement field at failure under the four analysis cases. Comparing Cases 1 and 2 in Fig. 8.11, the ultimate bearing capacity for 0° loading is smaller than that for 180° loading under the separable contact. This

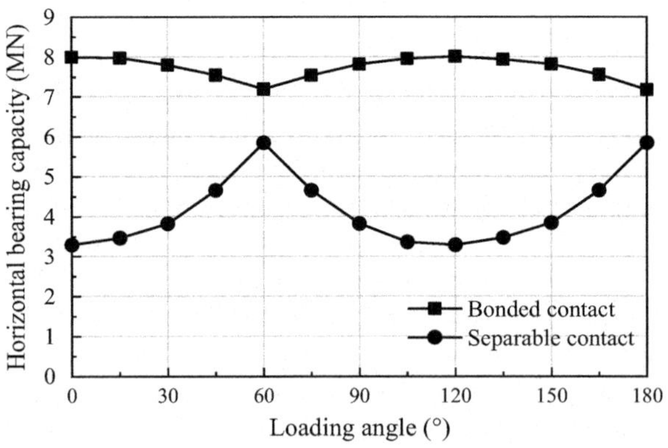

Fig. 8.9 Lateral bearing capacity of tripod bucket with various loading angle under different contact conditions

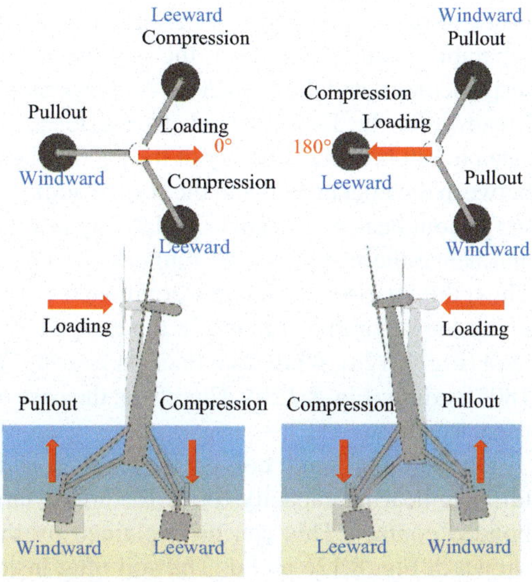

(a) 0° loading direction (b) 180° loading direction

Fig. 8.10 Diagrammatic sketch of tripod bucket under two extreme loading directions

Table 8.1 Ultimate bearing capacity for four analysis cases

Analysis cases	Loading angle α	Bucket-soil contact condition	Ultimate bearing capacity (MN)
Case 1	0°	Separable contact	3.3
Case 2	180°	Separable contact	5.8
Case 3	0°	Bonded contact	8
Case 4	180°	Bonded contact	7.2

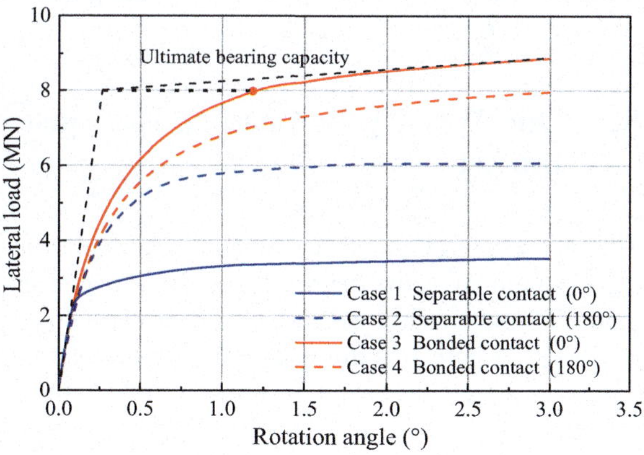

Fig. 8.11 Lateral load-rotation angle curves of the tripod bucket foundation under four analysis cases

is due to the corresponding bearing mechanism as shown in Fig. 8.12a and b. In the separable contact condition, the windward bucket is pulled with a certain upward displacement, and the bucket lid is completely separated from the internal soil plug. Meanwhile, the leeward bucket experiences compressive (downward) settlement. The pullout bearing capacity of the suction bucket is mainly provided by the friction between the bucket wall and the soil, in addition to the bucket self-weight. Therefore, the pullout bearing capacity is relatively low and upward displacement is significant, and hence the tripod bucket foundation experiences relatively large rotation angle. Thus, the bearing capacity is dominated by the windward bucket (upward bucket) under the separable contact. In the case of 0° loading, only one bucket bears uplift load (Fig. 8.12a), while two buckets bear uplift load at the same time in the case of 180° loading (Fig. 8.12b). Therefore, the case of 0° loading leads to a smaller bearing capacity.

On the other hand, comparing Cases 3 and 4 in Fig. 8.12 demonstrates that the ultimate bearing capacity for 0° loading is larger than that for 180° loading for the bonded contact. This can be explained by the corresponding bearing mechanism shown in Fig. 8.12c and d. The soil plug inside the bucket moves upward with the suction bucket under the bonded contact condition. Consequently, the total uplift bearing capacity will comprise the friction between the bucket outer wall and the surrounding soil, the self-weight of the bucket and the reverse bearing capacity of the soil at the bucket bottom. Correspondingly, the vertical uplift displacement of the windward bucket is relatively smaller under the same overturning moment and the rotation of the tripod bucket foundation is mainly caused by the downward displacement of the leeward bucket. Hence, the bearing capacity of the foundation is dominated by the compressive (downward) settlement of the leeward bucket. Two bucket bears compression load in the case of 0° loading (Fig. 8.12c), while only one bucket bears compression load at the same time in the case of 180° loading (Fig. 8.12d). Therefore, the case of 0° loading leads to larger bearing capacity.

8.5 One-Way Cyclic Loading Response of Tripod Suction Bucket

The one-way cyclic load is applied at the two critical loading angles, 0° and 180°. F_{ULS} is defined as the bearing capacity of tripod bucket in the ultimate limit state, then different cyclic load levels F_{cyc} are set with reference to F_{ULS}. That is $F_{\mathrm{cyc}} = \eta\, F_{\mathrm{ULS}}$, where η is defined as the cyclic load ratio, and set as 0.2, 0.4, 0.5 and 0.6 to investigate the effect of the amplitude of the cyclic loading on the foundation capacity. Figure 8.13 displays the scheme of half sine one-way cyclic load time histories employed in the analysis. It should be noted that F_{ULS} is different for different bucket-soil contact conditions and loading angles as shown in Table 8.1. Therefore, F_{ULS} is determined according to the specific analysis cases in Table 8.1.

Fig. 8.12 Displacement filed of tripod foundation at failure under four analysis cases (amplification factor of 1)

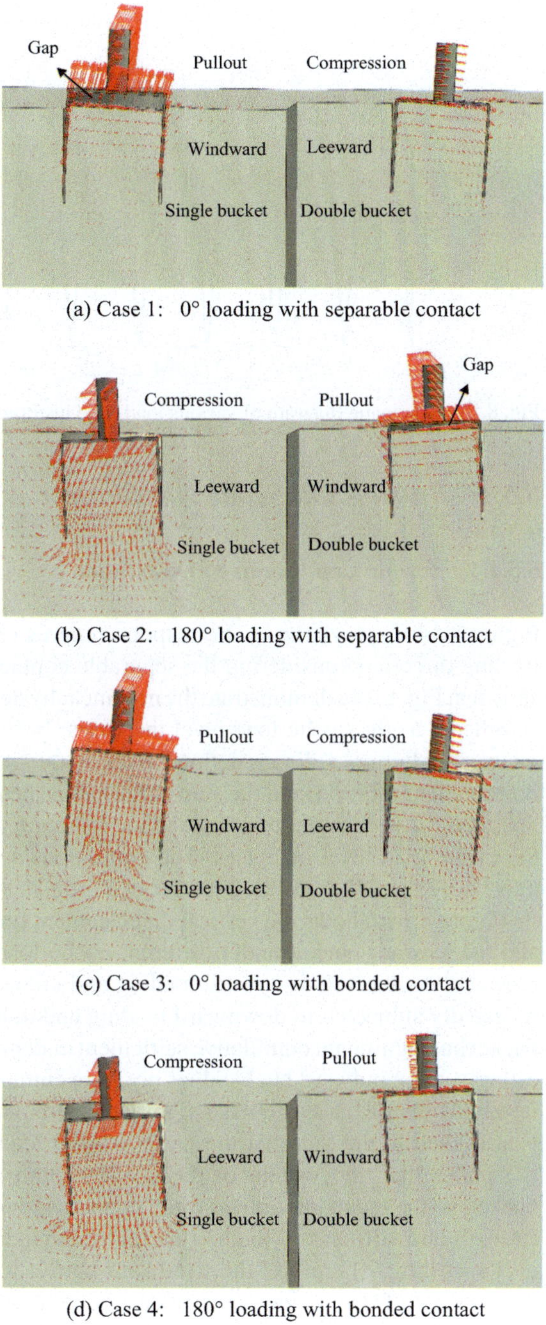

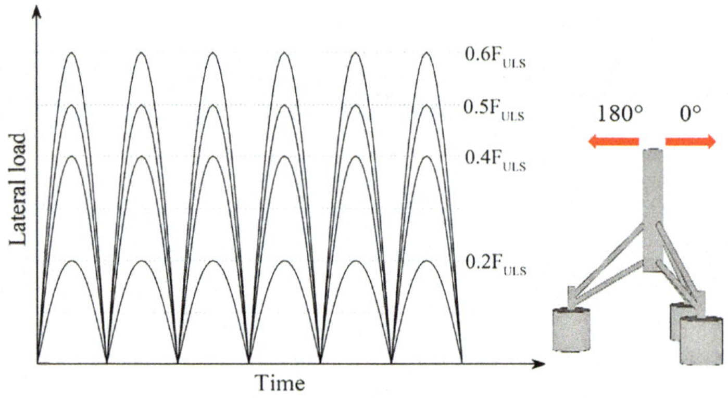

Fig. 8.13 Schematic diagram of various load-time histories for one-way cyclic loading

8.5.1 Bucket-Soil Separable Contact

8.5.1.1 Cyclic Loading in 0° Direction

Figure 8.14 displays the cyclic response curves of tripod bucket foundation for 0° loading direction considering the separable contact. The load-rotation angle curves shown in Fig. 8.14a demonstrate the nonlinear hysteretic characteristics become more significant as the cyclic load level increases. In Fig. 8.14b, the rotation that can be recovered after unloading is defined as temporary rotation, the unrecoverable rotation is permanent rotation, and the sum of the two is the peak rotation. Figure 8.14b shows that both the permanent rotation and the temporary rotation increase significantly as the cyclic load level increases. The continuous accumulation of the rotation angle may cause the OWT to reach the service limit state (SLS).

Figure 8.14c shows the vertical displacement time history of the windward bucket and the leeward bucket under different cyclic load levels. The windward bucket is repeatedly subjected to upward loading and unloading, while the leeward bucket is repeatedly subjected to downward loading and unloading. As expected, the leeward bucket has significant cumulative settlement under repeated downward loading. Interestingly, the windward bucket does not have cumulative upward displacement under repeated upward loading but has significant cumulative settlement. This phenomenon is attributed to the fact that the vertical uplift load applied to the windward bucket is smaller than the weight of the OWT system including the superstructure and bucket, so the resultant vertical force on the windward bucket is still downward. The strength and stiffness of the soil around the windward bucket are degraded during cyclic loading. Therefore, the windward bucket will have cumulative settlement as the number of cycles increases under the resultant vertical force. In addition, the temporary displacement of the windward bucket is larger than that of the leeward bucket for all cyclic load levels because the tensile bearing capacity of the windward

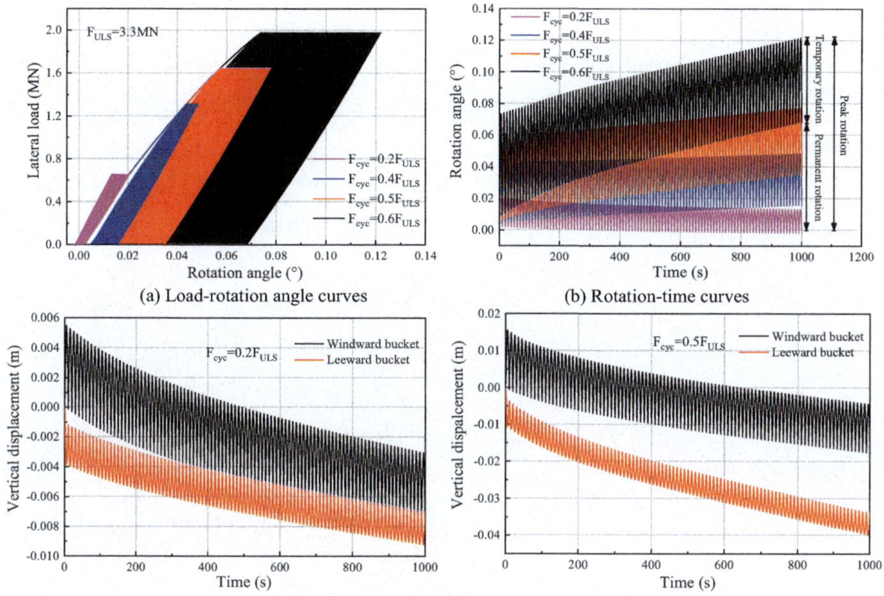

(a) Load-rotation angle curves

(b) Rotation-time curves

(c) Vertical displacement-time curves

Fig. 8.14 Cyclic response curves of tripod bucket foundation in 0° loading direction under the separable contact

bucket is usually significantly smaller than the compressive bearing capacity of the leeward bucket.

It should be noted that for relatively high cyclic load level such as 0.5 F_{ULS}, the settlement of the leeward bucket is always greater than that of the windward bucket, and the settlement difference between the two increases gradually as the number of cycles increases as shown in Fig. 8.14c. Therefore, the tripod bucket foundation gradually rotates clockwise (positive rotation angle), and the cumulative rotation angle increases gradually with the number of cycles as shown in Fig. 8.14b. However, for the relatively low cyclic load level such as 0.2 F_{ULS}, the settlement of windward bucket and leeward bucket are both small; the increasing rate of settlement of leeward bucket is smaller than the windward bucket although the total settlement of leeward bucket is greater, and hence the settlement difference decreases gradually. Therefore, the cumulative rotation angle decreases gradually or even becomes slightly negative (reverse rotation) as the number of cycles increases, as shown in Fig. 8.14b. It can be inferred that the rotation behavior of the tripod bucket foundation is closely related to the level of horizontal cyclic load and the vertical dead weight load.

Figure 8.15 presents the displacement field at the instant of peak load for different cycle number ($N = 1, 25$, and 100), which can explain the increase in foundation rotation. Figure 8.15a shows that the windward bucket at the initial loading has a certain upward displacement, the bucket lid is separated from the internal soil plug with a certain gap, while the leeward bucket has a certain downward displacement. As the

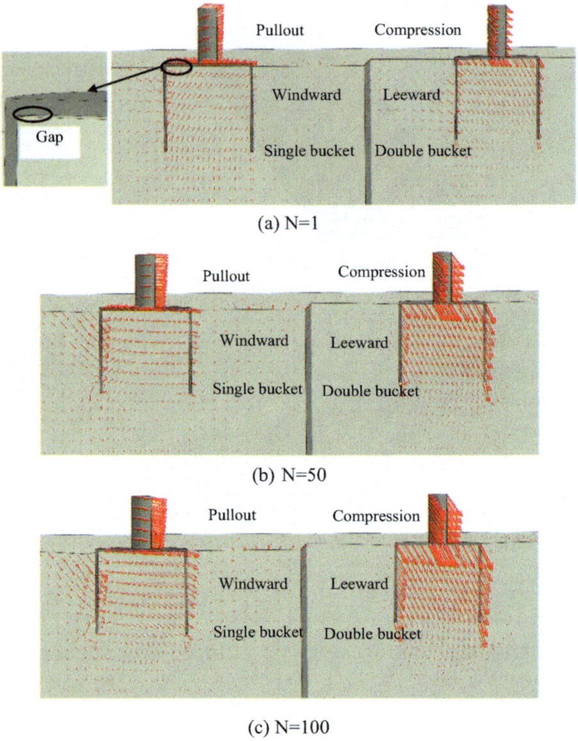

Fig. 8.15 Displacement field of tripod bucket foundation at instant of peak load for different cycle number (0.5 F_{ULS}, amplification factor of 5)

number of cycles increases, the displacement of windward bucket gradually changes from upward to downward due to cumulative settlement as shown in Fig. 8.15b and c.

8.5.1.2 Cyclic Loading in 180° Direction

Figure 8.16 shows the cyclic response curves of tripod bucket foundation in 180° loading direction considering the separable contact. Figure 8.16b shows that for the same cyclic load ratio, the cumulative rotation angle of foundation caused by cyclic loading in 180° direction is significantly greater than that in 0° direction shown in Fig. 8.14b. The permanent rotation increases as the number of load cycles increases for all cyclic load levels and it is much higher than the temporary rotation. Therefore, the gradual increase in permanent rotation is the main reason for reaching the SLS during the design life of the OWT. The foundation shows 'cyclic shakedown' behavior when the cyclic load level is relatively small such as 0.2 F_{ULS}, and the 'cyclic degradation' behavior when the cyclic load level is relatively large such as 0.6 F_{ULS}. The permanent rotation exceeds the limit of 0.25° in the SLS of OWT when the cyclic load level exceeds 0.2 F_{ULS}.

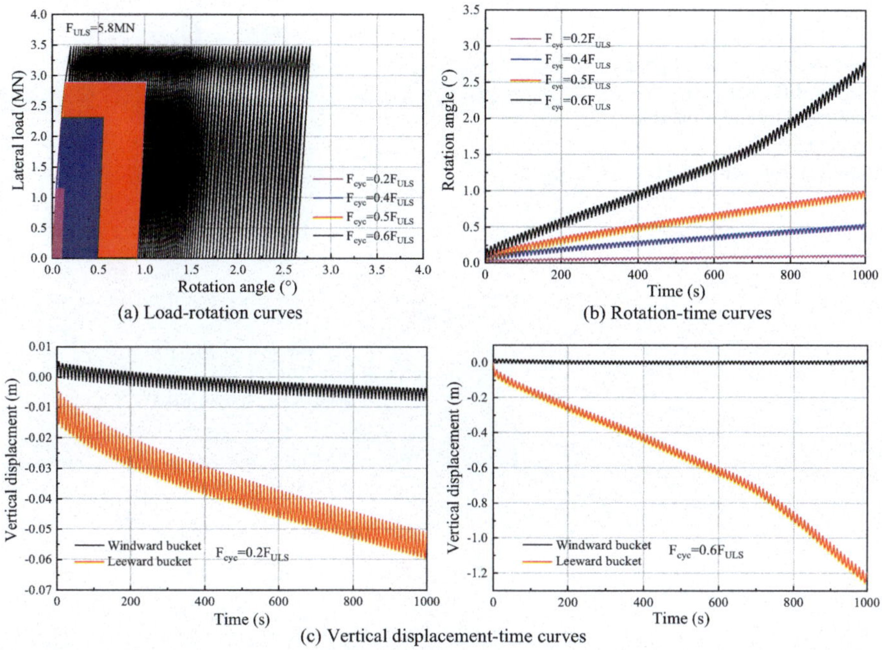

Fig. 8.16 Cyclic response curves of tripod bucket foundation in 180° loading direction under separable contact

Figure 8.16c shows that the cumulative settlement of the windward bucket is very small compared with the leeward bucket. In addition, contrary to the case in 0° direction, the temporary displacement of the windward bucket is less than that of the leeward bucket for all cyclic load levels. This is attributed to the much larger tensile bearing capacity of two windward bucket (180° direction) than that of one windward bucket in the case of 0° direction. On the other hand, the compressive bearing capacity of one leeward bucket for loading in the 180° direction is significantly smaller than that of two leeward buckets in the 0° direction. The tripod bucket foundation gradually rotates counterclockwise as the number of cycles increases due to the increasing difference of settlement as gleaned from the displacement field presented in Fig. 8.17. It can be concluded that the significant cumulative settlement of leeward bucket caused by relatively low compressive bearing capacity is the main reason for the foundation rotation.

Fig. 8.17 Displacement field of tripod bucket foundation at instant of peak load for different cycle number under the separable contact (0.5 F_{ULS}, amplification factor of 1)

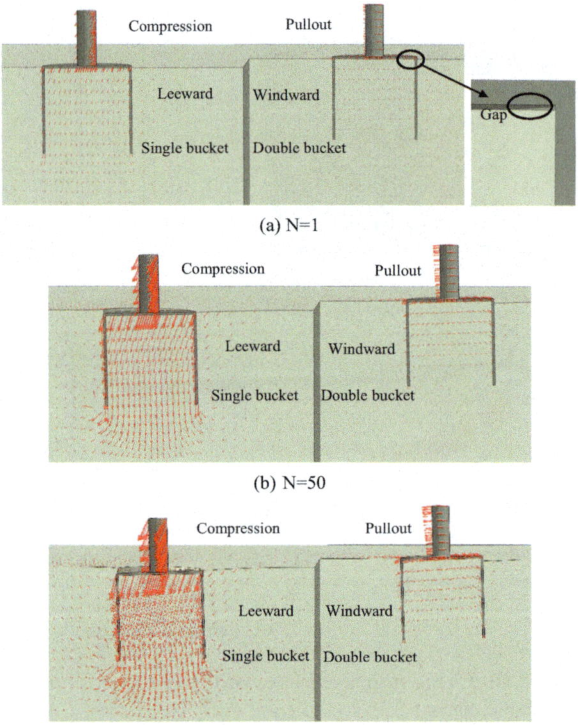

8.5.2 *Bucket-Soil Bonded Contact*

8.5.2.1 Cyclic Loading in 0° Direction

Figure 8.18 shows that the cyclic response curves of tripod bucket foundation in 0° loading direction under the bonded contact is significantly different from that for the separable contact. For the same cyclic load ratio, the cumulative rotation angle of foundation in the bonded contact condition is greater than that in the separable contact condition. The difference can be well explained by the vertical displacement time history of the windward bucket and the leeward bucket in Fig. 8.18c. For relatively high cyclic load levels such as 0.5 F_{ULS}, the windward bucket experiences cumulative upward displacement under repeated upward loading, which is different from the corresponding displacement time history in the separable contact condition. Because the ultimate bearing capacity of the foundation under the bonded contact is higher than that under the separable contact, the cyclic load level applied to the foundation is larger for the same cyclic load ratio. When the vertical uplift load applied to the windward bucket is larger than the dead weight of the OWT system, the resultant vertical force acting on the windward bucket is upward and causes cumulative upward

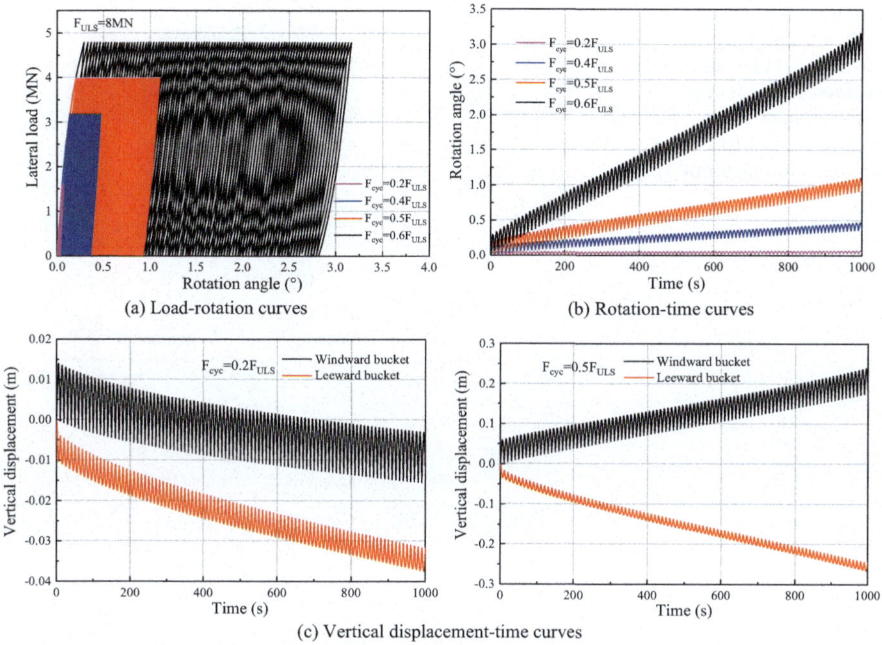

Fig. 8.18 Cyclic response curves of tripod bucket foundation in 0° loading direction under bonded contact

displacement. However, for the relatively low cyclic load level such as 0.2 F_{ULS}, the settlement occurs for both the windward bucket and the leeward bucket because the vertical uplift load applied to the windward bucket is smaller than the dead weight of the OWT system. Figure 8.19 displays the foundation displacement field, which reveals that windward bucket experiences significant upward displacement and the leeward bucket experiences significant downward displacement. The windward bucket lid remains in contact with the internal soil plug due to the bonded contact. The increasing difference of vertical displacement between windward and leeward buckets accelerates the rotation of the tripod bucket foundation.

8.5.2.2 Cyclic Loading in 180° Direction

Figure 8.20 shows the cyclic response curves of tripod bucket foundation in 180° loading direction under the bonded contact, which are similar to that under the separable contact condition. However, the foundation permanent rotation for the bonded contact is relatively larger than that for the separable contact for the same cyclic load ratio. Figure 8.21 shows the displacement field of tripod bucket foundation at the instant of peak load for different cycle numbers when the cyclic load level is 0.5 F_{ULS}. The cumulative settlement of leeward bucket is larger than the upward

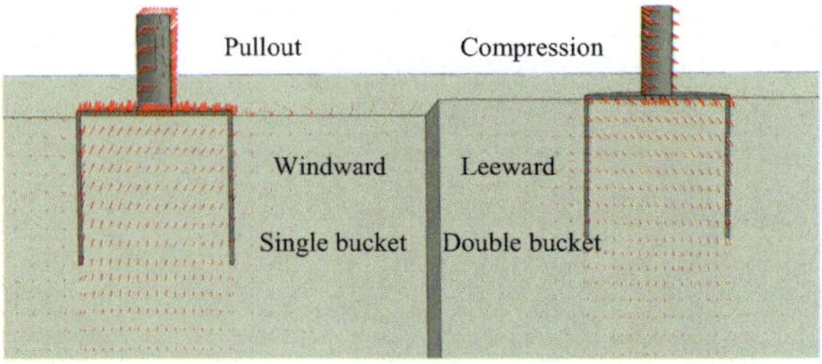

(a) N=1

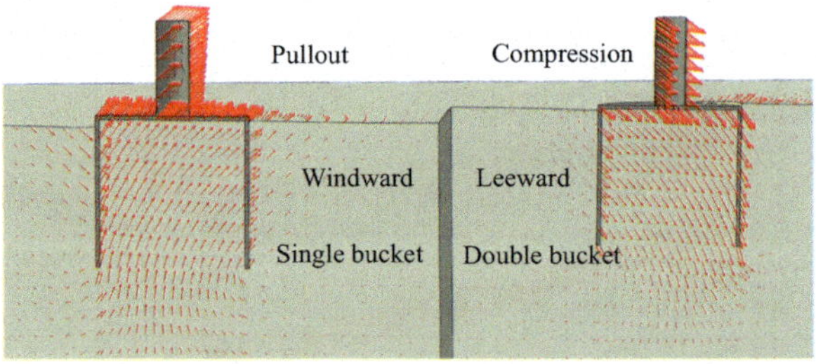

(b) N=50

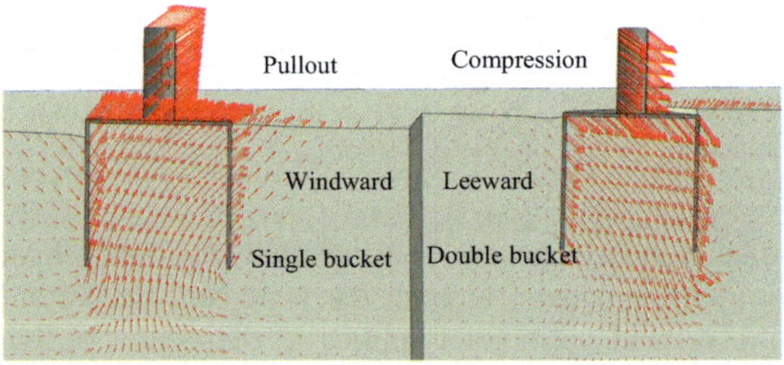

(c) N=100

Fig. 8.19 Displacement field of tripod bucket foundation at instant of peak load for different cycle number under the bonded contact (0.5 F_{ULS}, amplification factor of 1)

displacement (almost 0) of windward bucket, and hence it causes the rotation of the tripod bucket foundation. The bearing behavior of foundation is similar to that under separation contact but the windward bucket lid is not separated from the internal soil plug due to the bonded contact condition.

8.6 Two-Way Cyclic Loading Response of Tripod Suction Bucket

Two-way cyclic loading is applied alternately at two loading angles of 0° (positive load value) and 180° (negative load value) as per the schematic diagram shown in Fig. 8.22. Because the ultimate bearing capacity is different for the two loading angles in both bucket-soil contact types (separable or bonded), F_{ULS} is defined as the

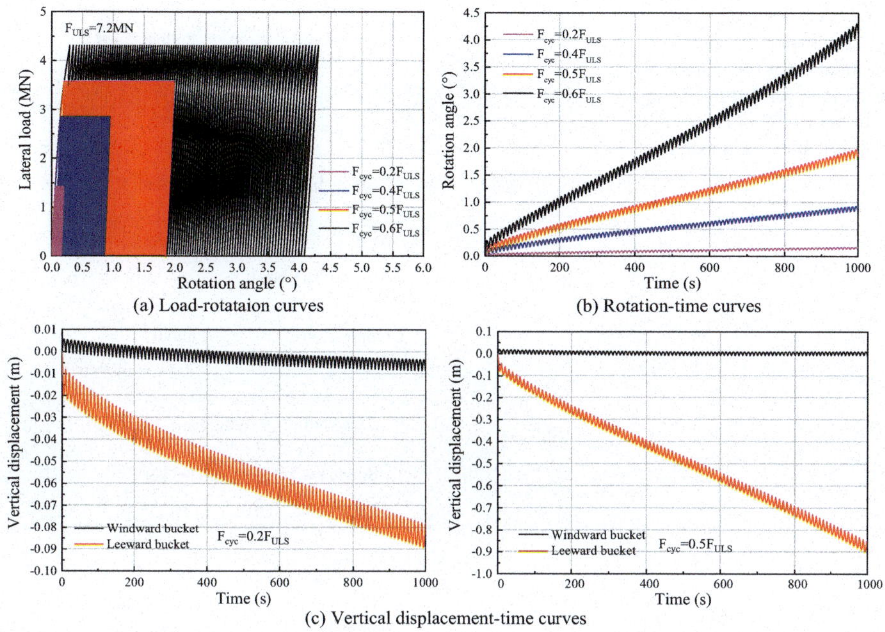

Fig. 8.20 Cyclic response curves of tripod bucket foundation in 180° loading direction under bonded contact

smaller value of the two ultimate capacity values. Different cyclic load levels F_{cyc} are then set with reference to F_{ULS}, i.e., $F_{\text{cyc}} = 0.2$, 0.4, and 0.6 F_{ULS}. Additionally, displacement-controlled two-way cyclic loadings are performed to further clarify the bearing mechanism of tripod bucket foundation.

8.6.1 Bucket-Soil Separable Contact

Figure 8.23 shows the cyclic response curves of tripod bucket foundation subjected to two-way cyclic loading considering separable contact. Interestingly, the foundation rotation angle gradually accumulates counterclockwise, even though it is subjected to symmetrical two-way cyclic load. The accumulated settlement occurs for both the 'single bucket' and the 'double bucket'; however, the single bucket settlement is much larger than the double buckets which results in gradual increase in foundation counterclockwise rotation. This phenomenon is mainly attributed to the fact that the compressive bearing capacity of 'single bucket' is significantly smaller than that of 'double buckets', and hence the former is more prone to settle than the latter under the same vertical cyclic loads. Figure 8.24 shows the displacement filed of foundation at the instants of positive and negative peaks of load for $N = 1$ and $N = 100$ when the cyclic load level is 0.6 F_{ULS}. Figure 8.24a shows that the 'single bucket' at the

Fig. 8.21 Displacement field of tripod bucket foundation at the instant of peak load for different cycle number (0.5 F_{ULS}, amplification factor of 1)

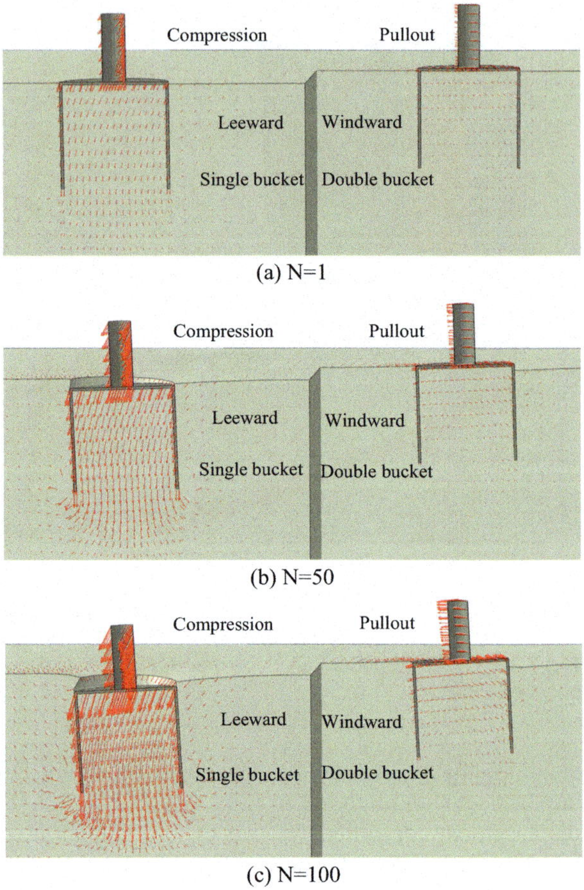

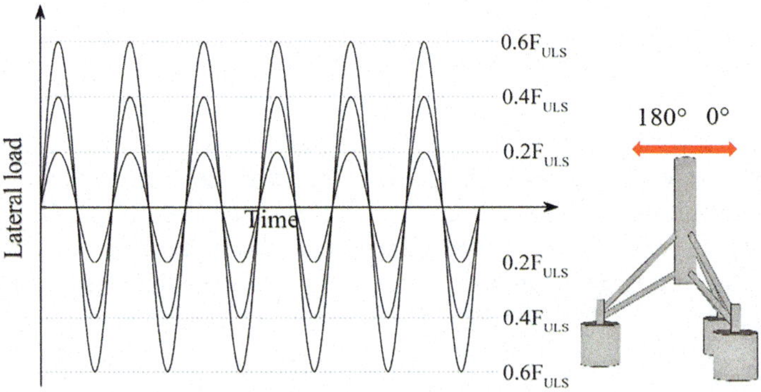

Fig. 8.22 Schematic diagram of load-time histories used for two-way cyclic loading

positive peak load has a slightly upward displacement, while the 'double bucket' has a slightly downward displacement; hence, the foundation rotates slightly clockwise (positive rotation angle). By contrast, the foundation rotates slightly counterclockwise (negative rotation angle) at the first negative peak load as shown in Fig. 8.24b. The 'single bucket' and the 'double buckets' experience large permanent settlements as the number of cycles increases, and large counterclockwise rotation occurs at the instant of both positive and negative peak load for $N = 100$ as shown in Fig. 8.24c and d.

Figure 8.25 presents the load-rotation angle curve of displacement-controlled cyclic loading of rotation angle amplitude is 0.25°, which can explain the difference of bearing capacity under different loading directions. The decrease in foundation stiffness (tangent slope of load-rotation angle curve) is larger under clockwise rotation (positive rotation angle) compared to the counterclockwise rotation (negative rotation angle). The foundation reaches its ultimate bearing capacity (the value for Case 1 in Table 8.1) when rotating clockwise, but it is far from reaching its ultimate bearing capacity (the value for Case 2 in Table 8.1) when rotating counterclockwise. Interestingly, cyclic hardening occurs for clockwise rotation, i.e., the bearing capacity gradually increases as the cycle number increases, while cyclic degradation occurs for counterclockwise rotation, i.e., the bearing capacity gradually decreases as the cycle number increases. This is attributed to the accumulation of permanent rotation in the counterclockwise direction (negative rotation angle) during cyclic loading,

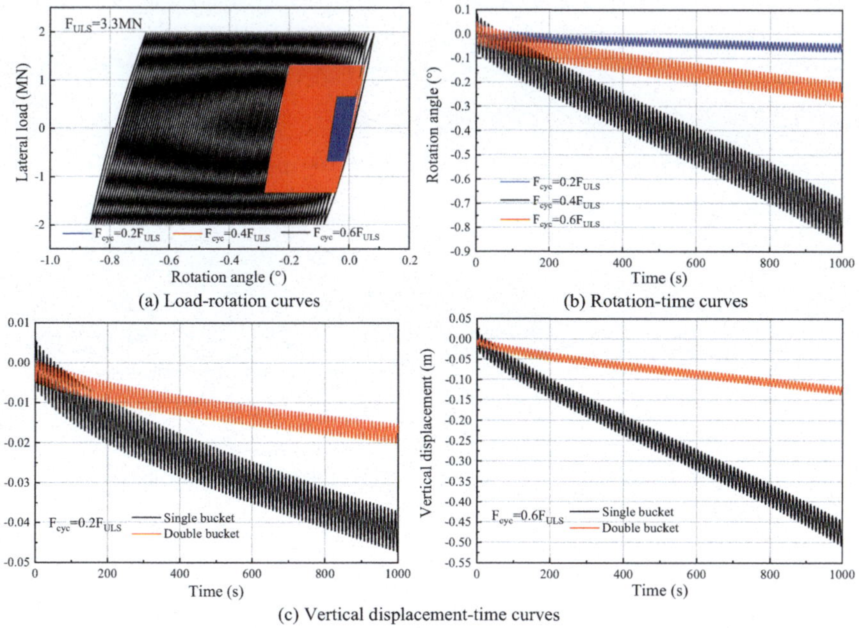

Fig. 8.23 Cyclic response curves of tripod bucket foundation subjected to two-way cyclic loading under separable contact

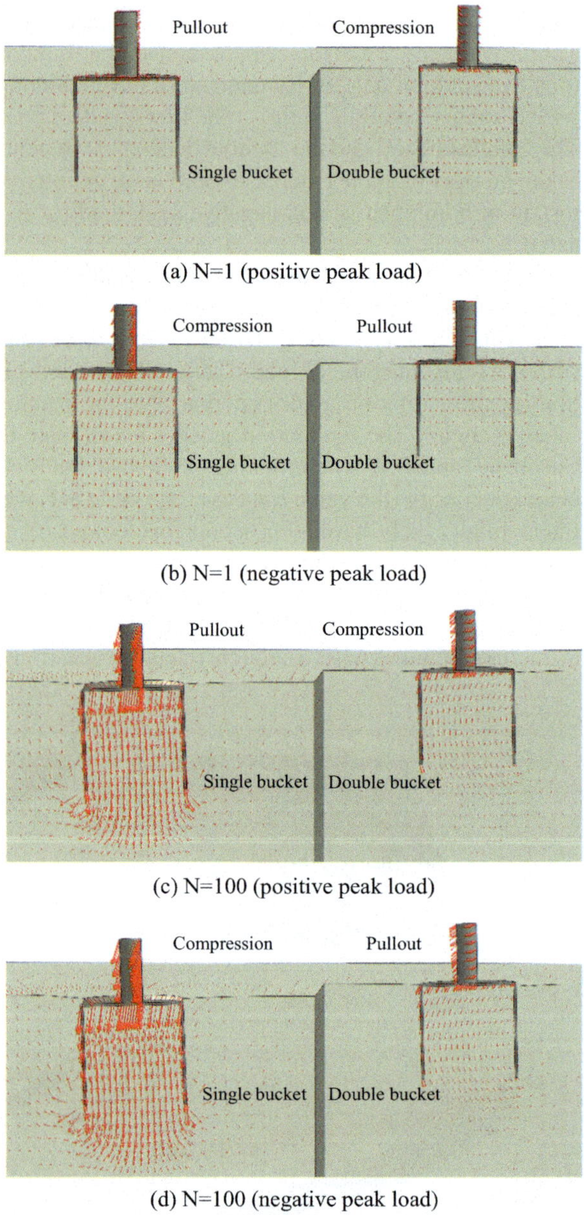

Fig. 8.24 Displacement field of tripod bucket foundation at the instant of positive and negative peak load for different cycle number (0.6 F_{ULS}, amplification factor of 1)

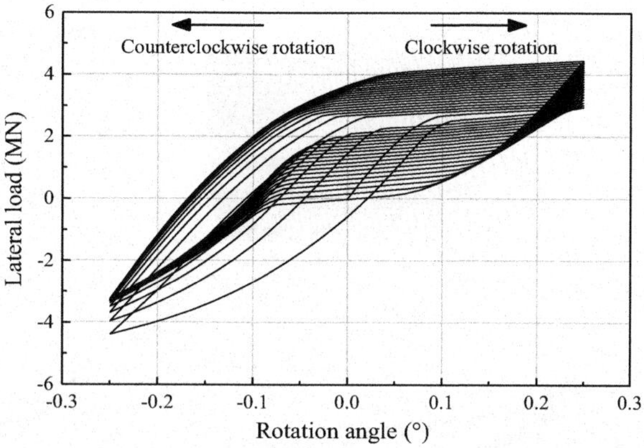

Fig. 8.25 Lateral load-rotation angle curves of tripod bucket foundation for displacement-controlled cyclic loading under separable contact condition

which increases the bearing capacity in the 0° loading direction and decreases the bearing capacity in the 180° loading direction.

8.6.2 Bucket-Soil Bonded Contact

Figure 8.26 shows the cyclic response curves of tripod bucket foundation subjected to two-way cyclic loading considering bonded contact. Same as the separable contact, the tripod bucket foundation gradually rotates counterclockwise. However, the permanent rotation angle is larger for the same cyclic load ratio, and it exceeds the SLS value of the OWT at relatively high cyclic load level. On the other hand, for small cyclic load (i.e., 0.2 F_{ULS}), the foundation exhibits 'cyclic shakedown' behavior. Figure 8.27 presents the foundation displacement field at the instant of positive and negative peaks of load for $N = 1$ and $N = 100$ at 0.4 F_{ULS}, which explains the large permanent counterclockwise rotation. The soil flow in mud surface is very significant due to large settlement of the 'single bucket'.

Figure 8.28 shows the load-rotation angle curve of displacement-controlled cyclic loading with a rotation angle amplitude of 0.25°. Same as the separable contact case, cyclic hardening occurs for clockwise rotation, while cyclic degradation occurs for counterclockwise rotation. However, the peak load for clockwise rotation is larger than that for counterclockwise rotation because the ultimate bearing capacity in the 0° loading direction is greater than that in the 180° loading direction for the bonded contact. The foundation does not reach its ultimate bearing capacity in both the 0° and 180° loading directions; therefore, the nonlinearity of load-rotation angle curve is not as obvious as that in Fig. 8.25.

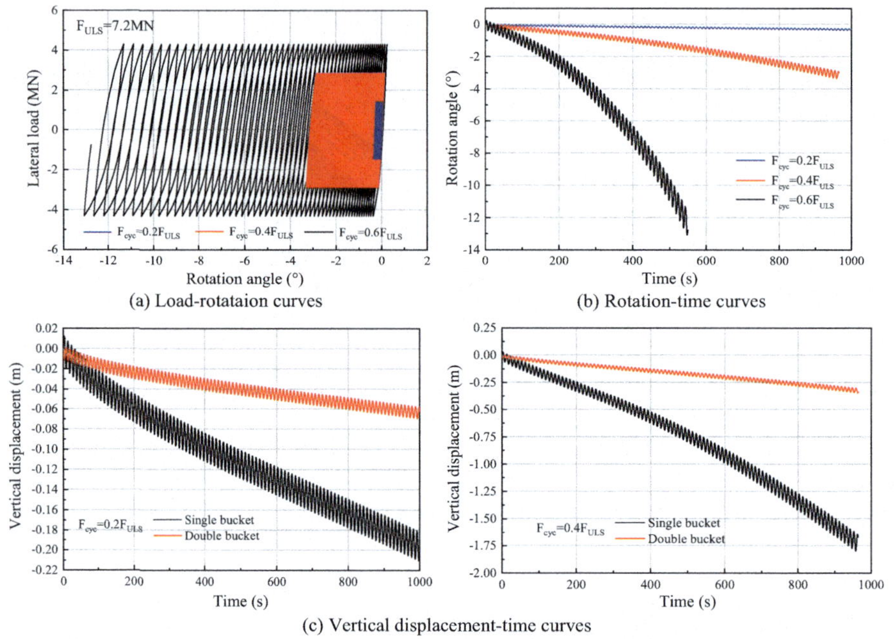

Fig. 8.26 Cyclic response curves of tripod bucket foundation subjected to two-way cyclic loading under bonded contact

8.7 Discussion

The obtained results and discussion correspond to the rotation behavior of tripod suction bucket foundation under the two extreme conditions of 'bonded contact' and 'separable contact', which represent the upper and lower limit cases of the tensile capacity of the suction bucket. However, for both the monotonic and cyclic loading, the actual case should be between the two extreme cases. The tensile capacity of suction bucket is closely related to the negative excess pore pressure inside the bucket, and the magnitude of the negative pressure depends on the loading rate. For the short-term loads, such as earthquake load or strong storm load, the negative excess pore pressure will develop rapidly, and the bucket-soil contact is closer to the bonded contact condition. For the long-term loads, such as wind, waves and current loads during the whole service life of offshore wind turbine, the negative excess pore pressure will gradually dissipate with time, and the bucket-soil contact is closer to the separable contact condition. Therefore, the present numerical simulation results under the bonded and separable contact conditions can provide some guidance for the evaluation of the short-term and long-term bearing behavior of the tripod suction bucket foundation, respectively. Given that it is difficult to quantitatively simulate the development of negative excess pore pressure and accurately evaluate the bearing

Fig. 8.27 Displacement field of tripod bucket foundation at the instant of positive and negative peak load for different cycle number (0.4 F_{ULS}, amplification factor of 1)

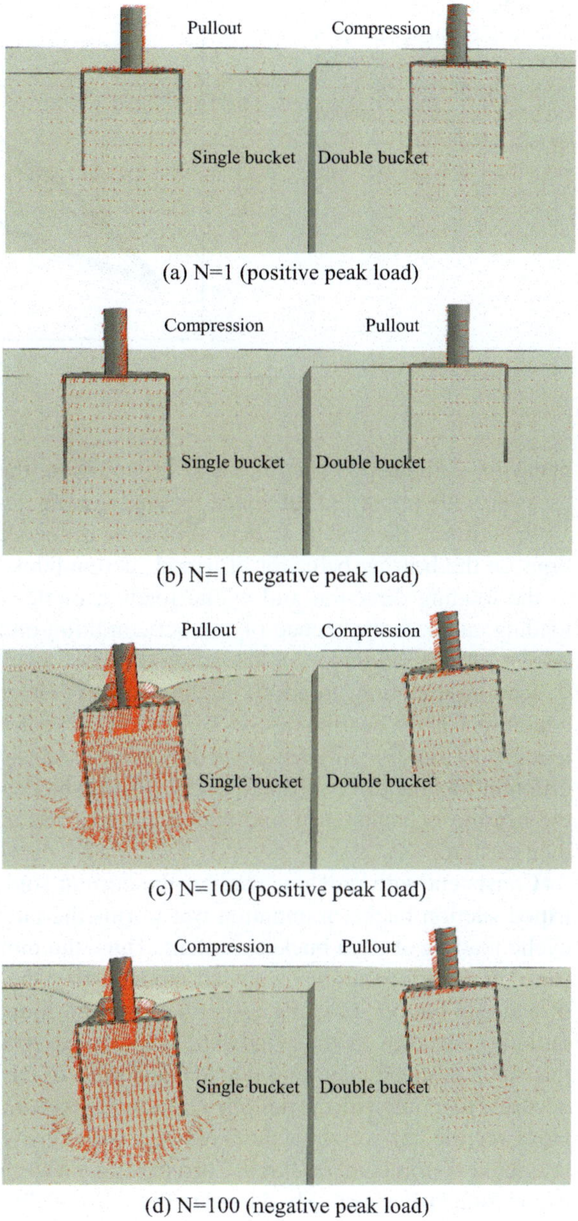

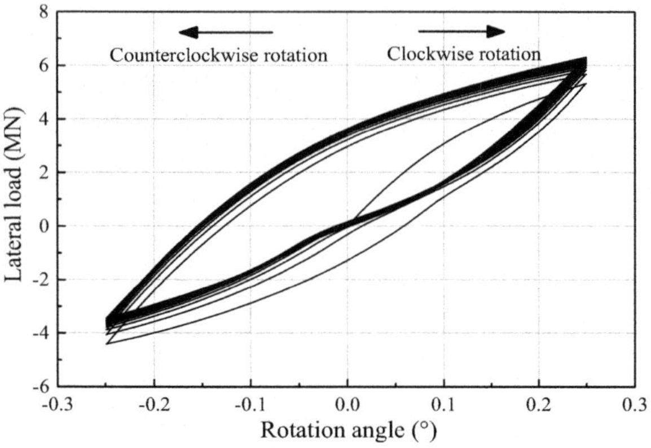

Fig. 8.28 Lateral load-rotation angle curves of tripod bucket foundation for displacement-controlled cyclic loading under bonded contact condition

behavior of the tripod suction bucket foundation, the findings for the two extreme cases provide important reference for engineering design.

In addition, the results indicated that the influence of bucket-soil contact conditions on the bearing behavior of tripod suction bucket foundation is closely related to the loading direction and cyclic loading mode. For the monotonic and cyclic loading cases, the influence of contact condition on the bearing behavior of tripod bucket foundation under 180° loading direction is more significant than that under 0° loading direction. In addition, the results revealed that the influence of contact conditions on the bearing behavior of tripod bucket foundation under two-way cyclic loading is more significant than that under one-way loading. These findings for different loading directions and cyclic loading modes enhance our understanding of the bearing behavior of tripod bucket foundations and provide some guidance for their design.

Consistent with previous studies, the current study once again confirms that the tripod suction bucket foundation transforms the lateral cyclic loads into a vertical cyclic loading at each bucket element. Thus, the moment-rotation responses of the tripod foundation can be evaluated by analyzing the compression-pullout behavior of a single bucket. Existing tests on the tripod foundation in sands found that the pullout resistance of the windward bucket was critical to the overall capacity of the tripod system under lateral loads, and most of the large rotational displacements occurred together with the pullout displacements (Kim et al. 2014b; Jeong et al. 2021). However, the self-weight of the OWT superstructure was usually not considered in the previous experimental research. Current research has found that the cyclic response of tripod foundation is closely related to the self-weight of the superstructure and cyclic load level. The vertical uplift load applied to the windward bucket is smaller than the weight of the OWT system when the lateral cyclic load level is relatively small, so the resultant vertical force on the windward bucket is still downward. Therefore, the windward bucket experiences cumulative settlement under the resultant vertical force during cyclic loading due to the degradation of the clay strength and stiffness around the bucket. In many cases, it is the differential settlement between the windward bucket and the leeward bucket that causes the rotation of the tripod foundation in clay. The influence of the vertical load level on the lateral cyclic response of the tripod foundation should be studied in more detail in the follow-up work.

8.8 Summary

This chapter investigated the rotation behaviors of tripod suction bucket under lateral monotonic and cyclic loading by performing three-dimensional finite element analysis. The key conclusions can be summarized as follows:

(1) The numerical method developed based on a simplified bounding surface model of clay was validated by comparing its predictions with experimental results available in the literature.
(2) The ultimate bearing capacity for 0° loading (single bucket under uplift) is smaller than that for 180° loading (double buckets under uplift) under separable contact condition because the bearing capacity is dominated by the windward bucket (upward bucket). The ultimate bearing capacity for 0° loading (double buckets under compression) is larger than that for 180° loading (single bucket under compression) under bonded contact condition because the bearing capacity is dominated by the leeward bucket (compression bucket).
(3) For the one-way cyclic loading in 0° direction for the separable contact condition, both the leeward and windward buckets experience different cumulative settlements, which causes the foundation rotation. The rotation behavior is found to be closely related to the level of horizontal cyclic load and the vertical dead weight load of the OWT and buckets.
(4) For the one-way cyclic loading in 180° direction under the separable contact, the foundation shows 'cyclic shakedown' behavior for small cyclic load levels, and 'cyclic degradation' behavior for large cyclic load level. The cumulative settlement of leeward bucket is large due to its low compressive bearing capacity, which causes the foundation rotation.
(5) For the one-way cyclic loading in 0° direction under bonded contact, the foundation cumulative rotation is greater than that under the separable contact for the same cyclic load ratio. For high cyclic load levels, the windward bucket experiences upward displacement and the leeward bucket experiences downward displacement, which accelerates the foundation rotation.
(6) For the one-way cyclic loading in 180° direction under the bonded contact, the bearing behavior of foundation is same as that under the separation contact although the permanent rotation angle is larger than the separation contact condition. This is because the windward bucket lid is not separated from the internal soil plug due to the bonded contact condition.

(7) For the two-way cyclic loading under separable contact, both the 'single bucket' and 'double buckets' experience cumulative settlement but the settlement of the single bucket is larger, which causes gradual increase of foundation counterclockwise rotation. For the displacement-controlled cyclic loading, cyclic hardening occurs for clockwise rotation, while cyclic degradation occurs for counterclockwise rotation.
(8) For the two-way cyclic loading under bonded contact, the foundation exhibits 'cyclic shakedown' behavior for small cyclic load levels, and 'cyclic degradation' behavior for large cyclic load levels. For displacement-controlled cyclic loading, the load-rotation angle curve is similar to that under separable contact; however, the peak load for clockwise rotation is larger than that for counterclockwise rotation.

The present research has confirmed that the cyclic behavior of the tripod foundation in clays is affected significantly by the bucket-soil contact conditions, the loading direction and cyclic loading modes. These findings in this study can enhance our understanding of the bearing behavior of tripod foundations and provide some guidance for their design under complex load environment.

References

Arany, L., S. Bhattacharya, J.H.G. Macdonald et al. 2015. A critical review of serviceability limit state requirements for monopile foundations of offshore wind turbines. In *Offshore Technology Conference Houston*.

Barari, A., and L.B. Ibsen. 2012. Undrained response of bucket foundations to moment loading. *Applied Ocean Research* 36: 12–21.

Byrne, B.W., and G.T. Houlsby. 2003. Foundations for offshore wind turbines. *Philosophical Transaction of the Royal Society A-Mathematical Physical and Engineering Sciences* 361: 2909–2930.

Byrne, B., G. Houlsby, C. Martin, and P. Fish. 2002. Suction caisson foundations for offshore wind turbines. *Wind Engineering* 26 (3): 145–155.

Cao, J., J.M.E. Audibert, Z. Al-Khafaji, et al. 2002. Numerical analysis of the behavior of suction caissons in clay. In *The Twelfth International Offshore and Polar Engineering Conference (ISOPE)*. International Society of Offshore and Polar Engineers.

Chen, D., P. Gao, S. Huang, et al. 2020. Static and dynamic loading behavior of a hybrid foundation for offshore wind turbines. *Marine Structures* 71: 102727.

Cheng, X.L., and J.H. Wang. 2016. An elastoplastic bounding surface model for the cyclic undrained behaviour of saturated soft clays. *Geomechanics and Engineering* 11 (3): 325–343.

Cheng, X.L., A.W. Yang, and G.N. Li. 2018. Model tests and finite element analysis for the cyclic deformation process of suction anchors in soft clays. *Ocean Engineering* 151: 329–341.

Cheng, X.L., X.L. Du, D.C. Lu, et al. 2020. A simple single bounding surface model for undrained cyclic behaviours of saturated clays and its numerical implementation. *Soil Dynamics and Earthquake Engineering* 139: 106389.

Cox, J.A., C.D. O'Loughlin, M. Cassidy, et al. 2014. Centrifuge study on the cyclic performance of caissons in sand. *International Journal Physical Modelling Geotechnics* 14: 99–115.

Dassault Systemes Simulia Corp. 2014. Abaqus Analysis User's Manual Version 6.14. Providence.

Ehrmann, A., N. Penner, C.G. Gebhardt, et al. 2016. Offshore support structures with suction buckets: parameter fitting of a simplified foundation model. In *The 26th International Ocean and Polar Engineering Conference.*

Foglia, A., and L.B. Ibsen. 2016. Monopod bucket foundations under cyclic lateral loading. *International Journal Offshore and Polar Engineering* 26 (2): 109–115.

Gelagoti, F., I. Georgiou, R. Kourkoulis, et al. 2018. Nonlinear lateral stiffness and bearing capacity of suction caissons for offshore wind-turbines. *Ocean Engineering* 170: 445–465.

He, B., J. Jiang, J. Cheng, et al. 2021. The capacities of tripod bucket foundation under uniaxial and combined loading. *Ocean Engineering* 220: 108400.

Houlsby, G.T., and B.W. Byrne. 2000. Suction Caisson foundations for offshore wind turbines and anemometer masts. *Wind Engineering* 24: 249–255.

Houlsby, G.T., Ibsen, L.B., Byrne, B.W. 2005a. Suction caissons for wind turbines. In *Frontiers in Offshore Geotechnics*. Perth, WA: ISFOG, 75–93.

Houlsby, G.T., R.B. Kelly, J. Huxtable, et al. 2005b. Field trials of suction caissons in clay for offshore wind turbine foundations. *Géotechnique* 55 (4): 287–296.

Houlsby, G.T., R.B. Kelly, J. Huxtable, et al. 2006. Field trials of suction caissons in sand for offshore wind turbine foundations. *Géotechnique* 56 (1): 3–10.

Hung, L.C., and S.R. Kim. 2012. Evaluation of vertical and horizontal bearing capacities of bucket foundations in clay. *Ocean Engineering* 52: 75–82.

Hung, L.C., S.H. Lee, S. Vicent, et al. 2018. An experimental investigation of the cyclic response of bucket foundations in soft clay under one-way cyclic horizontal loads. *Applied Ocean Research* 71: 59–68.

Jeong, Y.H., J.H. Kim, H.J. Park, et al. 2019. Cyclic behavior of unit bucket for tripod foundation system supporting offshore wind turbine via model tests. *Wind Energy* 22 (2): 257–268.

Jeong, Y.H., K.W. Ko, D.S. Kim, et al. 2021. Studies on cyclic behavior of tripod suction bucket foundation system supporting offshore wind turbine using centrifuge model test. *Wind Energy* 24: 515–529.

Kelly, R.B., G.T. Houlsby, and B.W. Byrne. 2006. Transient vertical loading of model suction caissons in a pressure chamber. *Geotechnique* 56 (10): 665–675.

Kim, S.R., L.C. Hung, and M. Oh. 2014a. Group effect on bearing capacities of tripod bucket foundations in undrained clay. *Ocean Engineering* 79: 1–9.

Kim, D.J., Y.W. Choo, J.H. Kim, et al. 2014b. Investigation of monotonic and cyclic behavior of tripod suction bucket foundations for offshore wind towers using centrifuge modeling. *Journal of Geotechnical and Geoenvironmental Engineering* 140 (5): 04014008.

Kourkoulis, R.S., P.C. Lekkakis, et al. 2014. Suction caisson foundations for offshore wind turbines subjected to wave and earthquake loading: Effect of soil–foundation interface. *Géotechnique* 64 (3): 171–185.

Li, X., X. Zeng, and X. Wang. 2020. Feasibility study of monopile-friction wheel-bucket hybrid foundation for offshore wind turbine. *Ocean Engineering* 204: 107276.

Li, L., H. Liu, and W. Wu. 2021a. Investigation on the behavior of hybrid pile foundation and its surrounding soil during cyclic lateral loading. *Ocean Engineering* 240: 110006.

Li, X., X. Zeng, X. Yu, et al. 2021b. Seismic response of a novel hybrid foundation for offshore wind turbine by geotechnical centrifuge modeling. *Renewable Energy* 172: 1404–1416.

Liao, W.M., J.J. Zhang, J.B. Wu, et al. 2018. Response of flexible monopile in marine clay under cyclic lateral load. *Ocean Engineering* 147: 89–106.

Lu, X., Y. Wu, B. Jiao, et al. 2007. Centrifugal experimental study of suction bucket foundations under dynamic loading. *Acta Mechanica Sinica* 23 (6): 689–698.

Ma, H., and J. Yang. 2020. A novel hybrid monopile foundation for offshore wind turbines. *Ocean Engineering* 198: 106963.

Musial, W., and B. Ram. 2010. *Large-Scale Offshore Wind Power in the United States: Assessment of Opportunities and Barriers*, 1–221. Golden: National Renewable Energy Laboratory.

Perveen, R., N. Kishor, and S.R. Mohanty. 2014. Off-shore wind farm development: Present status and challenges. *Renewable and Sustainable Energy Reviews* 29: 780–792.

Petas, V., G. Kolotsios, K. Georgiadis, et al. 2016. Effect of horizontal loading on the bearing capacity of tripod skirted foundations on clay. In *Proceedings of the 26th International Ocean and Polar Engineering Conference*, Rhodes, vol. I, pp. 16–230.

Tran, N.X., L.C. Huang, and S.R. Kim. 2017. Evaluation of horizontal and moment bearing capacities of tripod bucket foundations in sand. *Ocean Engineering* 140: 209–221.

Trojnar, K. 2021. Simplified design of new hybrid monopile foundations for offshore wind turbines. *Ocean Engineering* 219: 108046.

Villalobos, F.A., B.W. Byrne, and G.T. Houlsby. 2010. Model testing of suction caissons in clay subjected to vertical loading. *Applied Ocean Research* 32 (4): 414–424.

Wang, Y.H., X.B. Lu, S.Y. Wang, et al. 2006. The response of bucket foundation under horizontal dynamic loading. *Ocean Engineering* 33: 964–973.

Wang, X., X. Yang, and X.W. Zeng. 2017a. Seismic centrifuge modelling of suction bucket foundation for offshore wind turbine. *Renewable Energy* 114: 1013–1022.

Wang, X., X. Yang, and X.W. Zeng. 2017b. Lateral response of improved suction bucket foundation for offshore wind turbine in centrifuge modelling. *Ocean Engineering* 141: 295–307.

Wang, L.Z., H. Wang, B. Zhu, et al. 2018. Comparison of monotonic and cyclic lateral response between monopod and tripod bucket foundations in medium dense sand. *Ocean Engineering* 155: 88–105.

Wang, X., X. Zeng, X. Li, et al. 2020. Liquefaction characteristics of offshore wind turbine with hybrid monopile foundation via centrifuge modelling. *Renewable Energy* 145: 2358–2372.

Yi, J.T., F. Liu, and T.B. Zhang. 2021. Determination of the ultimate consolidation settlement of jack-up spudcan footings embedded in clays. *Ocean Engineering* 236: 109509.

Zhang, J.H., L.M. Zhang, and X.B. Lu. 2007. Centrifuge modeling of suction bucket foundations for platforms under ice-sheet-induced cyclic lateral loadings. *Ocean Engineering* 34 (8–9): 1069–1079.

Zhang, P., Y. Guo, Y. Liu, et al. 2015. Model tests on sinking technique of composite bucket foundations for offshore wind turbines in silty clay. *Journal of Renewable and Sustainable Energy* 7 (3): 033113.

Zhang, P.Y., J.C. Shi, H.Y. Ding, et al. 2016. Design of offshore wind power foundation with multi-bucket. *Transactions of Tianjin University* 22 (6): 502–507.

Zhu, B., B.W. Byrne, and G.T. Houlsby. 2013. Long-term lateral cyclic response of suction caisson foundations in sand. *Journal of Geotechnical and Geoenvironmental Engineering* 139 (1): 73–83.

Zhu, F.Y., C.D. O'Loughlin, B. Bienen, et al. 2018. The response of suction caissons to long-term lateral cyclic loading in single-layer and layered seabeds. *Géotechnique* 68 (8): 729–741.

Chapter 9
Seismic Responses of OWT Tripod Suction Bucket in Clays

Tripod suction buckets offer many advantages as foundations for offshore wind turbines including fast and economic installation and high overturning resistance. In this application, tripod suction buckets are subjected to dynamic loads such as wind and waves as well as earthquakes. However, their dynamic and seismic performance characteristics in cohesive soils are not well covered in the literature. This chapter investigates the dynamic behavior of tripod foundation installed in clay employing advanced three-dimensional nonlinear finite element analysis. Its rotation mechanism is analyzed considering different load cases including environmental loads, seismic load or combined action of seismic and environmental loads. The dynamic response was evaluated considering ground motions with varying intensity and frequency content, and environmental loads with varying magnitudes. It was found that even though the OWTs system may not experience instability failure due to the combined action of seismic and environmental loads, it is likely to fail to meet the requirements of SLS due to large permanent rotation associated with large settlement of the leeward bucket compared to that of the windward bucket. It was found that the frequency content of the ground motion has a significant effect on dynamic response of tripod foundation; larger permanent rotation can occur when the seismic predominant frequency is close to the frequency ranges of 1p and 3p for the wind turbine structure. The rotation angle of wind turbine structure depends on both the predominant frequency and intensity of the ground motion. These observations should be considered in the seismic design of tripod foundations of OWTs.

9.1 Introduction

Offshore wind turbine (OWT) foundations are subject to dynamic loads due to wind, waves and earthquakes (Byrne and Houlsby 2003). These environmental loads induce large overturning moments on the OWT foundations (Houlsby and Byrne 2000), and

X. Cheng et al., *Dynamic Analysis of Offshore Wind Turbine Foundations in Soft Clays*,
https://doi.org/10.1007/978-981-97-9454-6_9

consequently, the resistance of OWT foundations to lateral loads and associated moments is more critical than their resistance to vertical loads. In this regard, tripod suction buckets represent a viable foundation option for OWTs. In addition to their high overturning resistance, they have simple structure with convenient installation at low construction cost, and potential for reuse. However, their resistance to environmental and seismic loading is a challenging design consideration (Perveen et al. 2014).

A suction bucket foundation can be constructed in multiple configurations including the monopod and multiple-suction bucket foundations (e.g., tripod foundation). The multiple-suction bucket foundation has the potential to improve the overturning resistance of OWTs by transferring the overturning moment of the whole system into axial push–pull force of individual buckets. However, most available studies on the seismic response of suction buckets-supported OWTs focused on the performance of monopod suction bucket foundations installed in sand. A special emphasis was placed on the accumulation of pore pressure in sand, and the effects of skirt length, bucket diameter and dead weight on the seismic rotation and settlement of monopod suction bucket (Yamazaki et al. 2003; Yu et al. 2014; Olalo et al. 2016; Wang et al. 2017a; Zayed et al. 2019). These studies indicated that increasing the diameter, skirt length or dead weight of the monopod suction bucket can enhance its anti-seismic liquefaction capacity, and hence effectively reduce its rotation and settlement. In recent years, some new types of bucket foundation have been researched, such as the hybrid foundation consisting of a traditional monopile and a wide-shallow bucket, and their superiority over the traditional bucket foundation has been proved (Chen et al. 2020; Li et al. 2021a, 2022). In addition to researching more reasonable foundation forms to improve the bearing capacity of the OWT system subjected to marine environmental loads such as wind, wave and current loads, vibration control has also gaining focuses to avoid excessive vibration affecting the normal service of OWT as the turbine tower becomes taller and more slender (Zuo et al. 2017; Sun and Jahangiri 2018; Chen et al. 2021).

Offshore wind turbines located in seismic active regions are susceptible to earthquakes, in addition to wind and wave dynamic loads. Therefore, the combined action of environmental loads and seismic forces should be considered when evaluating the seismic response of OWT suction bucket foundation. Several researchers investigated the bearing capacity and deformation characteristics of monopod suction buckets under lateral cyclic loading caused by winds and waves (Houlsby et al. 2005b; Wang et al. 2006, 2017b; Zhu et al. 2013, 2018; Cox et al. 2014; Foglia and Ibsen 2016; Hung et al. 2018; Cheng et al. 2018, 2020). These studies provided good understanding of lateral cyclic behavior of the monopod suction bucket, such as the evolution of its cumulative rotation and unloading stiffness with the load cyclic load amplitude, frequency, direction and number of cycles. Many of the findings of these studies are relevant to the seismic response of monopod suction buckets.

The bearing mechanism of multiple-suction bucket under overturning moment is different from that of the monopod bucket. Multiple-bucket foundation resists the overturning moment by vertical uplift and compression behaviors of individual buckets, while monopod suction bucket mainly depends on rigid rotation (Kim et al. 2014c; Zhang et al. 2016). Therefore, some studies evaluated the behavior of monopod bucket under vertical loading instead of considering multiple-suction buckets under lateral loading or overturning moment (Kelly et al. 2006; Lu et al. 2007; Jeong et al. 2019). Centrifugal model studies were conducted to compare the horizontal cyclic behaviors of monopod suction bucket and tripod suction bucket foundations. Kim et al. (2014a) reported that the moment-rotation relationship of the tripod suction bucket exhibited an almost bilinear response, while that of the monopod one displayed a continuous hardening response. Wang et al. (2018a) observed that the cumulative rotation of monopod suction bucket in medium dense sand increased monotonically as the number of load cycles increased, while the rotation of tripod suction bucket increased at the initial stage of cyclic loading but then gradually decreased, showing a favorable "self-healing" behavior. Jeong et al. (2021) confirmed that the cyclic behavior of the tripod foundation was affected by the load level and direction. These studies demonstrated that the performance characteristics of monopod suction bucket foundations differ from those of the tripod suction bucket foundation. However, there are few reports on the behavior of tripod suction bucket under earthquakes.

Dynamic soil-structure interaction (SSI) of foundations has major influence on the seismic response of OWTs system and has been extensively investigated through numerical simulation. Dynamic Winkler models involving springs and dampers have been utilized (Zhao and Maisser 2006; Zhong and Huang 2014; Kim et al. 2014b; Zuo et al. 2018; Wang et al. 2018b). However, the dynamic Winkler model is overly simplified to accurately simulate the complex bucket-soil dynamic interaction as it ignores the soil continuity. In addition, it is difficult to consider the effect of soil stiffness degradation and deformation accumulation on the dynamic response of the OWTs system under earthquake, especially considering the flow mechanism of soil around foundation. On the other hand, time domain numerical analyses employing nonlinear soil constitutive models are more suitable for simulating dynamic bucket-soil interaction (Kourkoulis et al. 2014; Cerfontaine et al. 2016; Li et al. 2017; Asheghabadi et al. 2019; Cheng et al. 2024). However, the availability of proper constitutive models of complex nonlinear dynamic soil response and the computational effort and cost hinder the development of this method.

The literature review indicated scarce information on the dynamic response of tripod suction bucket foundations installed in cohesive soil and subjected to earthquake and environmental loads. Unlike sandy soil foundation, cohesive soil foundation is not prone to liquefaction, but significant total settlement and uneven settlement may occur under earthquake, which will affect the normal operation of the OWTs. Therefore, this study investigates the dynamic behavior of tripod suction bucket foundations installed in cohesive soils under seismic and environmental loading (Cheng et al. 2022). Three-dimensional finite element analyses of OWT system are performed considering the soil nonlinearity using a simplified kinematic hardening soil constitutive model. The rotation mechanism of tripod foundations in clays is analyzed, and various factors affecting their seismic responses are discussed.

9.2 Numerical Method

9.2.1 Finite Element Model

A three-dimensional finite element model is established employing the finite element software ABAQUS (Dassault Systemes Simulia Corp 2014) to simulate the interaction between cohesive soil and tripod suction buckets considering the specifications of a typical Vestas V90 3MW wind turbine (see Table 9.1). A schematic of the main environment of the OWT is shown in Fig. 9.1a. The OWT system consists of three-blade rotor, nacelle, tower, substructure, triple suction buckets, and soil as shown schematically in Fig. 9.1b. The wind turbine and suction buckets are made of steel with Young's modulus = 210 GPa, Poisson's ratio = 0.27 and density = 7850 kg/m^3. The assembly of rotor blades, nacelle, and hub weighs 10^5 kg (Nam et al. 2019). Exploiting the symmetries of the geometry and loading conditions, the model simulates only one half of the system, as shown in Fig. 9.2a, to improve the computational efficiency, while Fig. 9.2b displays the overall outline of the whole OWT system for clarity. A plan view of the system is shown in Fig. 9.2c. A sensitivity study was conducted to establish the length, width and height of soil domain in order to eliminate the boundary effects on the results. Soil domain with different sizes are meshed to calculate dynamic responses (such as rotation angle-time history) of suction bucket under earthquake. The size is considered to be appropriate when it has little effect on calculation results. It was found that mesh size of 62 m × 30 m × 20 m (length × width × height) is sufficient to avoid boundary effects on the obtained results. The soil and OWT structure were discretized by 43,092 8-node linear brick elements with reduced integration (C3D8R), while the blades, hub and nacelle were discretized using 3453 quadratic tetrahedral elements (C3D10). To improve computational accuracy and efficiency, a finer mesh was applied within the depth of embedment layer where nonlinearity is expected to be more prominent, while coarser mesh was applied away from the foundation.

The interaction between the tripod suction buckets and surrounding soil was simulated utilizing the surface-to-surface contact method. The surface of suction buckets was defined as the master surface (relatively stiff), and the soil surface in contact with suction buckets was defined as the slave surface (relatively soft). Hard contact was set in the normal direction of contact surface (both inner and outer walls of bucket), which allowed separation between the interface elements when they are subjected to tension. The penalty contact method was used in the tangential direction. When the two surfaces are in contact, the interface behavior is governed by Coulomb's friction theory. The critical friction shear stress τ_{crit} at the contact surface can be expressed by $\tau_{\text{crit}} = \mu \cdot p_{\text{c}}$ in terms of frictional coefficient μ and contact pressure p_{c}, with $\mu = 0.3$ in this study. When the shear stress at the contact surface exceeded τ_{crit}, tangential slip occurred.

Normal horizontal constraints were applied to the vertical face of symmetry to prevent the out-of-plane movement. The base boundary of the model was simulated as rigid bedrock, and the top boundary was fully free. For the vertical boundaries,

Table 9.1 Dimensions of OWT model

Property	Symbol	Dimension
Tower diameter	D_{tw}	4 m
Tower length	L_{tw}	60 m
Tower wall thickness	t_{tw}	4 cm
Main substructure diameter	D_m	4 m
Main substructure length	L_m	30 m
Main substructure wall thickness	t_m	12 cm
Suction buckets diameter	D_{sb}	6 m
Suction buckets length	L_{sb}	3 m
Suction buckets wall thickness	t_{sb}	5 cm
Supporting structure diameter	D_{ss}	1.5 m
Supporting structure wall thickness	t_{ss}	5 cm
Substructure center to suction buckets distance	D_{cs}	10 m
Center-to-center distance between the buckets	D_{cc}	17.3 m

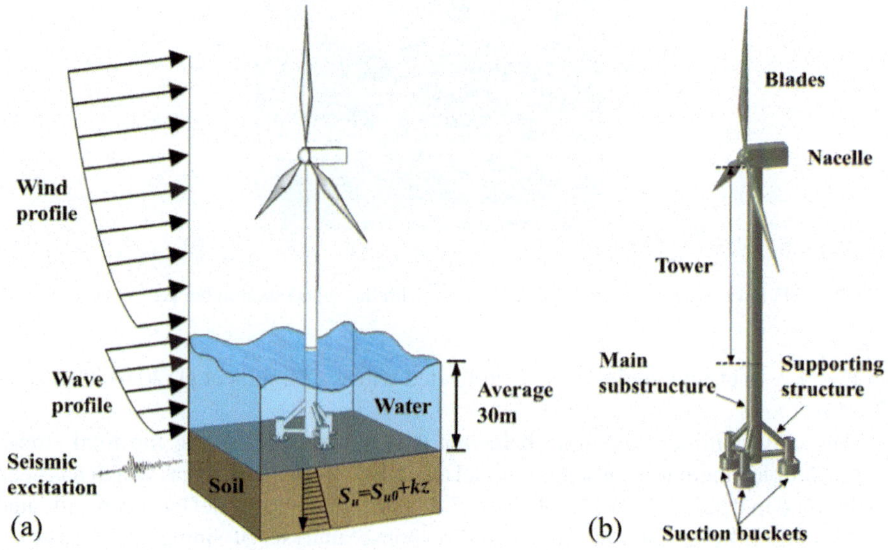

Fig. 9.1 Schematic of OWT system with the tripod suction bucket: **a** main environment of the OWT and **b** components of the OWT

equivalent displacement constrains were introduced to simulate the seismic response of infinite horizontal strata. The nodes with the same height on both sides of the analytical region are bound together (the displacement is always consistent), forcing the opposite vertical sides to move simultaneously preventing any rotation, as shown

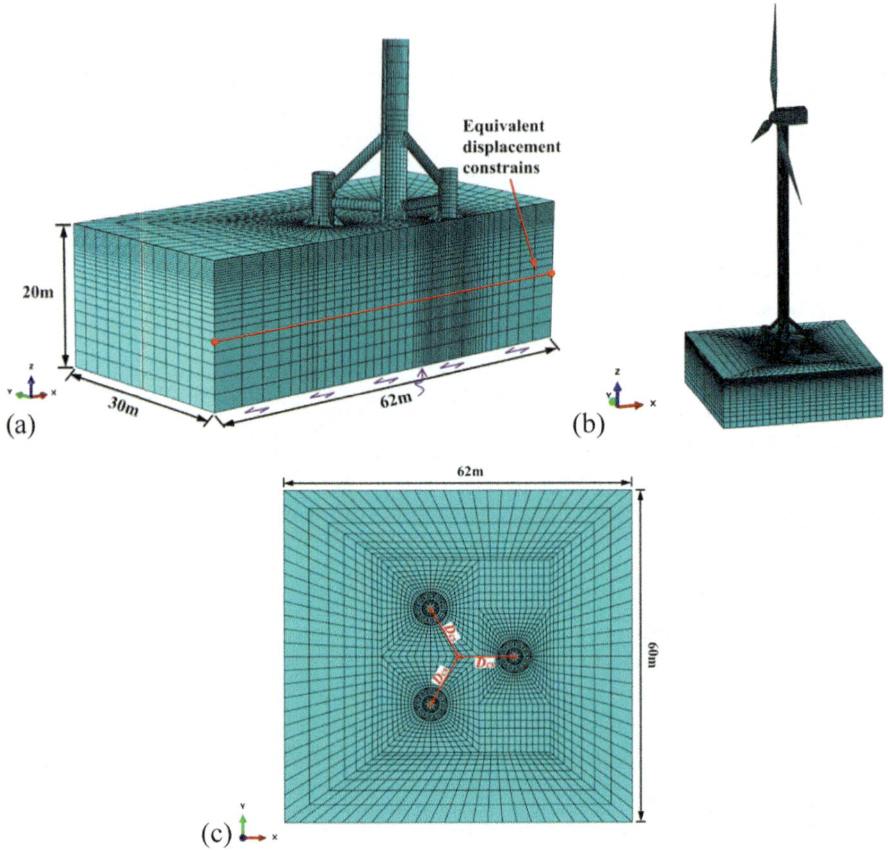

Fig. 9.2 3D finite element model of OWT system with the tripod suction bucket in clays

in Fig. 9.2, to eliminate the effect of reflected waves (Tsinidis et al. 2014; Gao et al. 2021).

The clay nonlinear dynamic behavior is simulated employing the total stress-based simple kinematic hardening model available in Abaqus library, which has been elaborated in detail in Sect. 7.2 of Chap. 7. In particular, the model has been validated and successfully applied to the seismic response analysis of single suction bucket (Zhang et al. 2021). The constitutive model contains only three model parameters, i.e., $\sigma_{|0}$, C and γ_r, which are related to soil shear strength, S_u. The initial yield stress $\sigma_{|0}$ actually represents the range of elastic domain of soil, which may be determined by $\sigma_{|0} = \lambda\sqrt{3}S_u$, and λ is 0.1 in this study. The initial kinematic hardening modulus C represents the initial stiffness of the soil, which is considered to be equal to the elastic modulus E. Empirically, E can be taken as the multiple of undrained strength S_u, that is, $C = E = \kappa S_u$, in this study, κ is taken to be 1440. Finally, γ_r determines the decreasing rate of kinematic hardening modulus as plastic deformation increases, which can be determined by $\gamma_r = C/\left(\sqrt{3}S_u - \sigma_{|0}\right)$. In this study, the undrained shear strength S_u is assumed to increase linearly with depth, i.e., $S_u = S_{u0} + kz$, z is the depth of soil layer, k is the shear strength increment per unit depth and S_{u0} is the undrained shear strength at mud surface, as shown in Fig. 9.1. $S_{u0} = 20\,\text{kPa}$, and $k = 3$. Therefore, the model parameters are actually related to the soil depth, which can be realized through the user-defined field variables in ABAQUS.

9.2.2 *Loads on OWT System*

9.2.2.1 Wind Loads

The wind load acting on the OWTs can be divided into two components: load acting on turbine tower and load acting on turbine blades. The wind load acting on the turbine tower is a distributed force that can be given by (Binh et al. 2008)

$$F_{sh}(z) = 0.5\rho_a C_D D(V + v)^2 \tag{9.1}$$

where ρ_a is the air density which can be assumed as 1.225 at 15 °C and 1 atm; D is the tower diameter; $V(z)$ and $v(z, t)$ are the mean and fluctuating wind velocity along the tower, respectively; C_D is the drag coefficient, which depends on Reynold's number and surface roughness (Det Norske Veritas (DNV) 2010), and $C_D = 1.2$ in this study. The wind loads at different locations along the tower are different. In order to simplify the analysis, the tower is divided into six segments, each with a length of 10 m, and drag force is assumed to be uniform within each segment. In the finite element model, a reference point is defined at the midpoint of each segment, coupled with the cross-section of the corresponding segment, and then wind loads are applied on these reference points.

During OWT operation, the blades rotate and a thrust force is created in the wind direction at the tower top, which can be approximated by Bernoulli's equation as (Lee et al. 2010), i.e.,

$$F_T = 0.5\rho_a \pi R_T^2 V_s^2 (1 + 2v_s/V_s) C_T \tag{9.2}$$

where V_s and v_s are the mean and fluctuating wind velocity at the hub height, respectively; R_T is rotor radius, which is 45.6 m (Nam et al. 2019); C_T is thrust coefficient and $C_T = 4\alpha(1 - \alpha)$, where α is axial induction factor, with a value range of [0–1] and taken $\alpha = 0.5$ in this study, which renders maximum C_T (Hansen 2008). F_T is applied on the hub center in the form of longitudinal point load.

The Davenport wind speed spectrum (Davenport 1961, 1967) and the autoregressive (AR) model in linear filtering method are used to simulate the fluctuating wind velocity (Iannuzzi and Spinelli 1987). Three mean wind velocities at height of 10 m are selected in this study, i.e., $V_{10} = 8$, 10 and 12 m/s. Figure 9.3a shows the wind

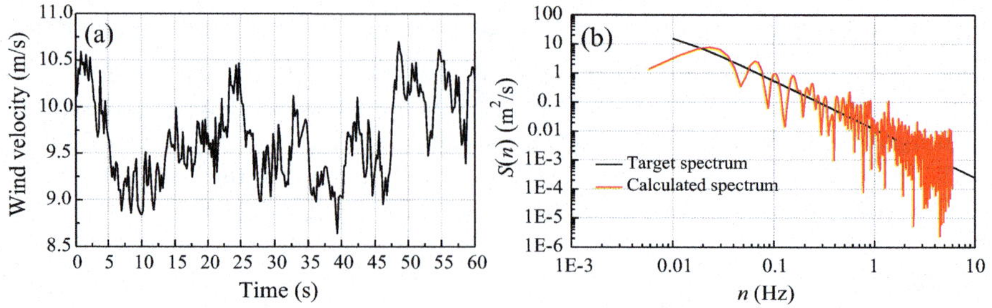

Fig. 9.3 **a** Wind velocity time history; **b** Calculated and target spectra

Fig. 9.4 Wind load time history at the hub

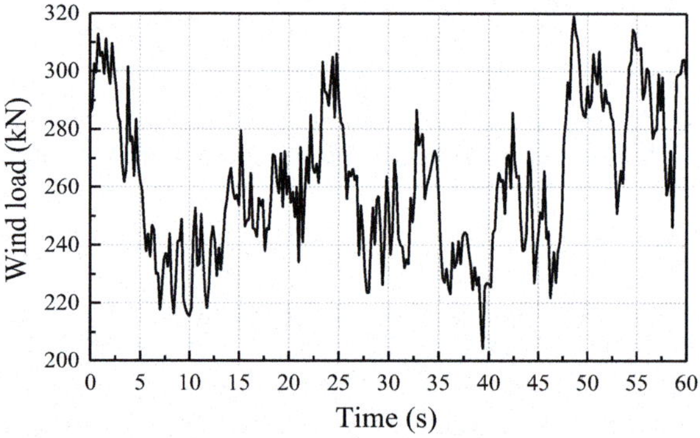

velocity time history at the hub for $V_{10} = 8\,\text{m/s}$ and Fig. 9.3b compares its power spectral density (PSD) with the target spectrum. The calculated spectrum agrees well with the target spectrum as shown in Fig. 9.3b. For $V_{10} = 8\,\text{m/s}$, the simulated wind load time history at the hub is shown in Fig. 9.4.

9.2.2.2 Wave Loads

The stochastic wave elevation can be established by using the empirical equation on the PSD of the wave elevation proposed by Pierson and Moskowitz (1964; Yu 1981)

$$S_{\eta\eta}(\omega) = 0.78\omega^{-5} \exp\left(-3.11\omega^{-4} H_{1/3}^{-2}\right) \tag{9.3}$$

where $H_{1/3}$ is the significant wave height, ω is the angular frequency.

The sea surface elevation $\eta(\mathrm{t})$ in the time domain can then be simulated as (Zhao and Hu 2012):

$$\eta(t) = \sum_{i=1}^{m} \alpha_i \cos(k_i x - \omega_i t + \varepsilon_i) \tag{9.4}$$

where $\eta(t)$ is the instantaneous height of fluctuating water surface relative to static water surface; x indicates position, namely $x = 0$; α_i, k_i and ω_i are the amplitude, wave number and angular frequency of the ith component wave, respectively; t is time and ε_i is an independent random variable representing phase of each component wave uniformly distributed in the range of $[0, 2\pi]$; α_i can be calculated using Eq. (9.5) (Ren and Wang 2004)

$$\alpha_i = \sqrt{2S_{\eta\eta}(\overline{\omega}_i)\Delta\omega_i} \tag{9.5}$$

in which the exact meaning of symbol $\overline{\omega}_i$ and $\Delta\omega_i$ can be found in (Yu 1981).

In this study, three significant wave heights $H_{1/3} = 5$ m, 10 m and 14 m are selected. Figure 9.5 compares the calculated and target power spectral density for $H_{1/3} = 5$ m. It can be seen that the calculated spectrum agrees well with the target spectrum.

The Morison formula can be used to calculate the sea wave load on the tower (Det Norske Veritas (DNV) 2010). Accordingly, the horizontal sea wave load df per length dz can be calculated as (Cao et al. 2020):

$$df = df_D + df_I = \frac{1}{2}\rho C_D D v_x |v_x| dz + \frac{\pi}{4}\rho C_M D^2 a_x dz \tag{9.6}$$

where df_D is the drag force and df_I is the inertia force; the water density, $\rho = 1030$ kg/m^3; C_D, C_M are the drag and inertia coefficients, respectively, and $C_D = 1.2$, $C_M = 2.0$ in the simulation; D is the tower diameter; v_x and a_x are the wave induced velocity and acceleration of water, respectively, in the horizontal direction (Yu and Kang 1982; Zuo et al. 2018).

The tower part in the water is divided into three segments; each is 10 m. In the finite element model, a reference point is defined in each segment and coupled with the cross-section of the corresponding segment, and then sea wave loads are applied on these reference points. Figure 9.6 shows the simulated typical sea wave load time history (significant wave height $H_{1/3} = 5$ m).

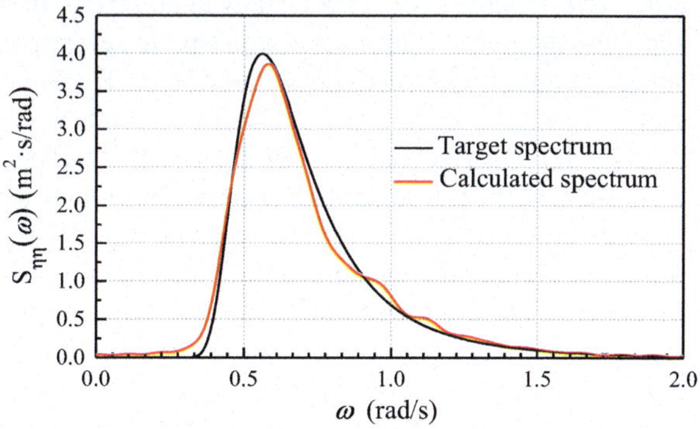

Fig. 9.5 Calculated and targeted power spectral density of wave load acting on structure

Fig. 9.6 Simulated typical sea wave load time history

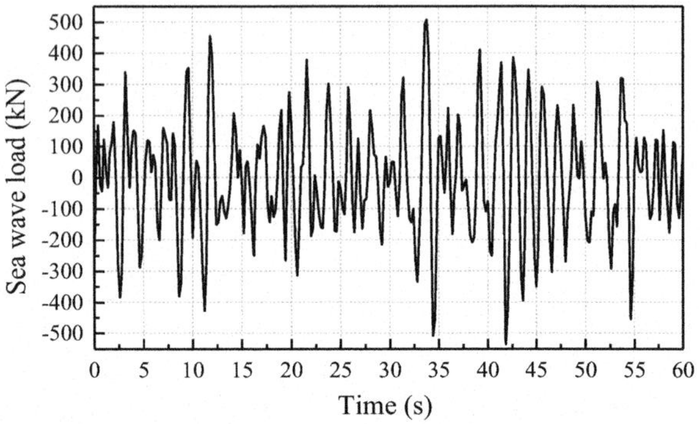

9.2.2.3 Hydrodynamic Pressure Caused by Earthquake

A vibrating tower in sea water will bear additional force due to hydrodynamic pressure associated with its motion. The interaction between water and tower can be simulated by the equivalent mass of water, that is, the additional mass method (Liaw and Chopra 1974). The added mass of tower can be expressed by (Det Norske Veritas (DNV) 2010)

$$m_{\mathrm{a}} = C_{\mathrm{a}} A_{\mathrm{p}} \rho_{\mathrm{w}} \tag{9.7}$$

where C_{a} is the added mass coefficient and $C_{\mathrm{a}} = 1.0$ in the simulation (Cao et al. 2020); ρ_{w} is the sea water density; A_{p} is the section area of the tower. In dynamic analysis, the calculated mass of the tower part in water should be the sum of the added mass and the actual mass of the tower.

9.2.2.4 Seismic Loads

The natural frequency ranges of 1p and 3p for the Vestas V90 3 MW wind turbine structure is usually < 0.93 Hz (Arany et al. 2016). In order to investigate the effect of frequency content of ground motion on the seismic response of OWT system, three seismic records with different predominant frequencies are selected. The predominant frequencies for Kobe, Loma Prieta and Northern California seismic waves are 0.8 Hz, 1.1 Hz, and 1.6 Hz, respectively. The acceleration time history and response spectra for these three ground motion records are shown in Fig. 9.7. The peak ground acceleration (PGA) of ground motion is scaled to be 0.3 g. The duration of an earthquake signal is defined as the time between the first exceedance of a certain acceleration threshold (taken as 0.05 g in the study) and the last exceedance of this threshold (Bolt 1973), as shown in Fig. 9.7. The bottom of the finite element model is considered to be rigid bedrock, and the input ground motion is applied as vertically propagating shear wave (SV) represented by the acceleration time history at the bottom of the model, as demonstrated in Fig. 9.2.

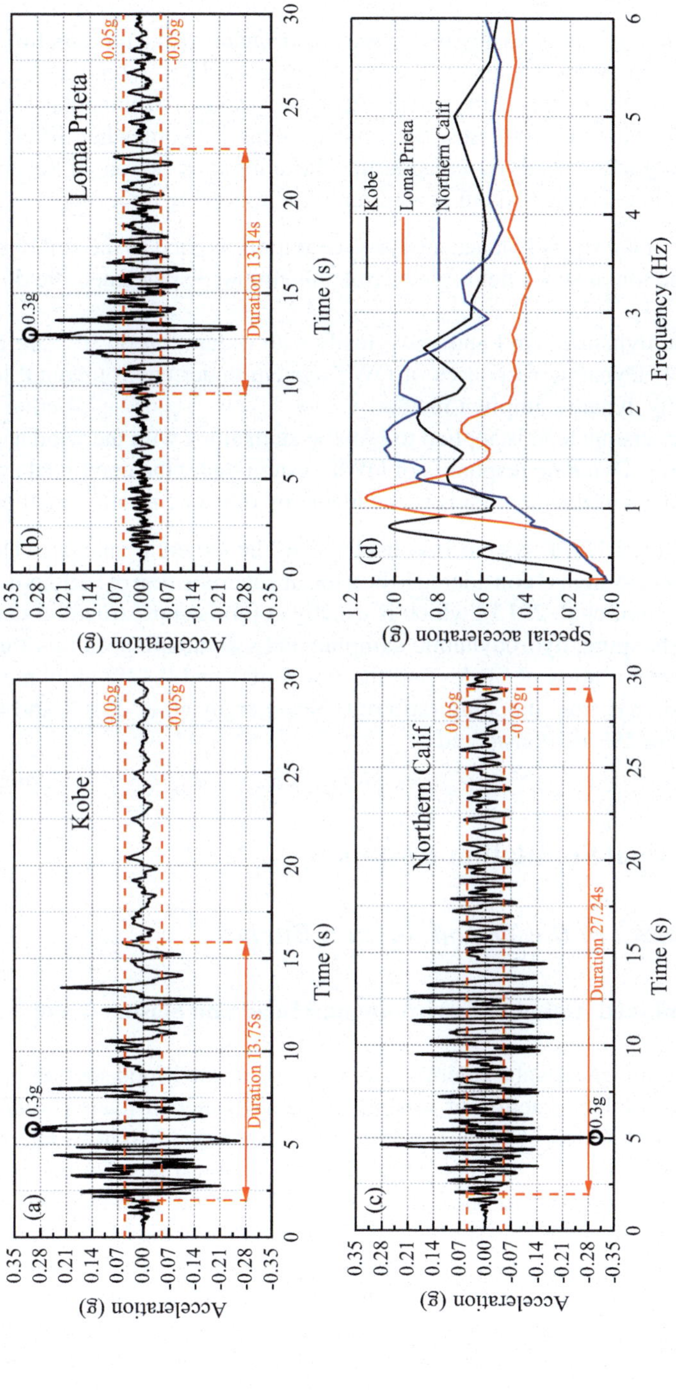

Fig. 9.7 Acceleration time history and response spectra for three earthquake records

9.2.3 Numerical Analysis Steps

In order to systematically investigate the bearing mechanism of tripod suction bucket foundation in cohesive soils, three load cases are considered: combined action of seismic and environmental loads, seismic load only and environmental load only. Depending on different load cases, the analysis steps are somewhat different. For example, the dynamic analysis considering the combined action of seismic and environmental loads is carried out according to the following three steps.

Step 1: Geostatic stress balance analysis. Gravity is applied to the soil domain to establish the initial stress field. Meanwhile, the initial displacement field is set to 0.

Step 2: The dynamic wind and wave loads with a duration of 30 s are applied to the OWT. Dynamic responses of OWT system under environmental loads is calculated by dynamic implicit method.

Step 3: The seismic load is applied to OWT system along with the wind and wave loads for 30 s. Dynamic responses of OWT system under the combined action of seismic and environmental loads is calculated by dynamic implicit method.

The damping of OWT system is considered in the dynamic analysis. The total damping of OWT system is considered to be a linear superposition of various damping sources (Damgaard et al. 2013), which is usually comprised of structural damping, aerodynamic damping, hydrodynamic damping and soil damping. In this study, the structural damping ratios of blades and the tower of 1 and 0.43% (Jonkman et al. 2009), the hydrodynamic damping ratio of 0.23% (Arany et al. 2016), and the soil damping ratio of 5% are considered.

9.3 Numerical Simulation Results

9.3.1 Rotation of the Tripod Suction Bucket

9.3.1.1 Combined Action of Environmental Load and Seismic Load

Figure 9.8 presents the displacement vector diagram of the tripod suction bucket at the end of the combined action of seismic and environmental loads (Loma Prieta earthquake record scaled to PGA $= 0.3$ g, significant wave height m $H_{1/3} = 5\,\text{m}$ and mean wind velocity $V_{10} = 8\,\text{m/s}$). Figure 9.8 demonstrates that the bucket on the windward side (windward bucket) and the bucket on the leeward side (leeward bucket) experienced significant settlement. The settlement of the leeward bucket was more significant than that of the windward bucket, which caused the tripod foundation to rotate in the downwind direction. Figure 9.9a shows the rotation time history of the windward and leeward buckets. The rotation of the two were almost identical, and either can represent the rotation of the tripod foundation as a whole. During the action of wind and wave load (first 30 s), there was no obvious accumulation of tripod foundation rotation, and only slight shaking occurred. During the combined action of seismic and environmental loads (30 s–60 s), both the vibration rotation angle and permanent rotation angle increased significantly. Here, we define the rotation that can be recovered after unloading as vibration rotation, the unrecoverable rotation as permanent rotation, and the maximum value of rotation as the peak rotation as shown in Fig. 9.9a. At the end of the earthquake record, the permanent rotation angle has exceeded the allowable rotation angle of 0.25° in the serviceability limit state (SLS) of OWTs. Therefore, during the combined action of seismic and environmental loads, even if there is no instability failure of the OWTs system, it is very likely to fail to meet the requirements of SLS due to large permanent rotation. Figure 9.9b shows the vertical displacement time history of the windward and leeward buckets, which further confirmed quantitatively that the settlement of the leeward bucket was much greater than that of the windward bucket. The uneven settlement of the buckets on the windward and leeward sides was the direct cause of the rotation of the tripod foundation.

Figure 9.10a and b compare the horizontal displacement time history of the bucket and soil for Point A on the windward side and Point B on the leeward side marked

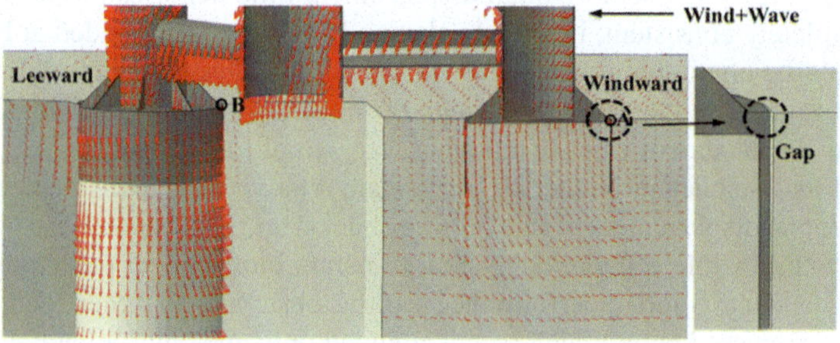

Fig. 9.8 Displacement vector diagram of tripod suction bucket at the end of combined action of seismic and environmental loads (amplification factor of 3)

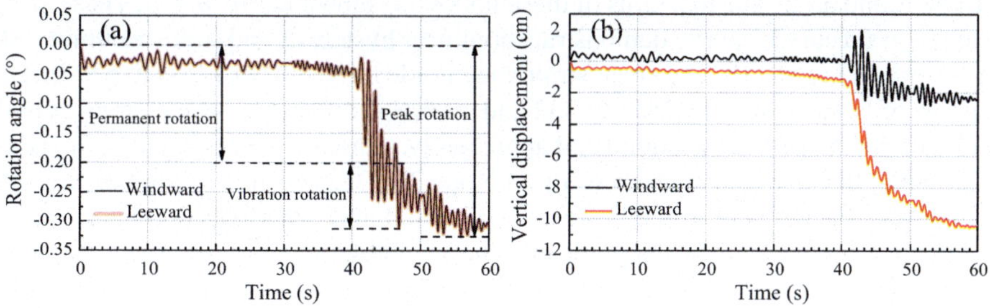

Fig. 9.9 Rotation and vertical displacement time history of the windward and leeward buckets: **a** rotation angle-time history; **b** vertical displacement–time history

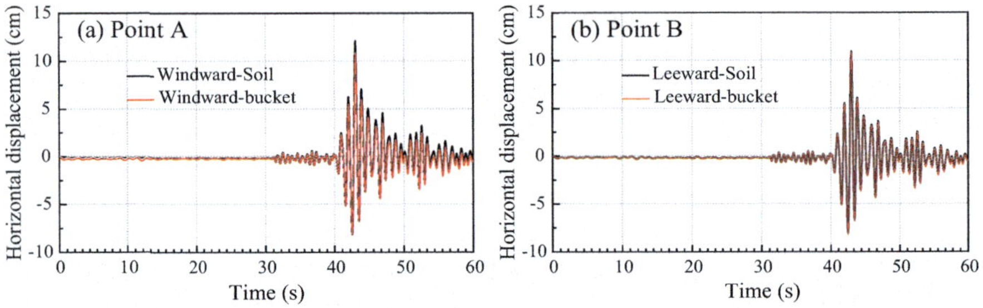

Fig. 9.10 Comparison of horizontal displacement time history of bucket and soil: **a** Point A on the windward side; **b** Point B on the leeward side

in Fig. 9.8. During the combined action of seismic and environmental loads, significant horizontal vibration displacement occurred for the bucket on the windward and leeward sides, but there was no significant horizontal permanent displacement at the end of the earthquake. It can be seen from Fig. 9.10a that the horizontal displacement of the soil at Point A was slightly larger than that of the bucket. Hence, at the end of the earthquake, a gap opened at the bucket-soil interface at Point A, which can be observed in Fig. 9.8. The horizontal displacement time histories of bucket and soil were completely consistent, i.e., the bucket and soil were still bonded at Point B.

It should be noted that the absolute horizontal displacement caused by the horizontal movement of the soil layer under earthquake is significant, but the relative horizontal displacement of bucket-soil is very small, as shown in Fig. 9.10. For the tripod suction bucket foundation, significant relative vertical displacement rather than horizontal displacement occurs between bucket and soil, because the overturning moment of the whole foundation system transfers into vertical push–pull force of individual bucket. Similar performance can be observed for other loading conditions to be discussed below, such as seismic load only and environmental load only. Limited to space, the following sections mainly discusses the vertical displacement of each bucket.

Figure 9.11 shows shear stress–strain curves of some typical nodes around the tripod foundation. The locations of these nodes are shown in Fig. 9.11a. These shear stress–strain curves clearly demonstrate nonlinear hysteresis and strain accumulation as shown in Fig. 9.11b. The shear strain magnitude ranged from 1 E–3 to 6 E–2. The curves of Nodes 9400, 13,285, 30,332 and 10,608 show that the shear strain at both sides of the bucket wall accumulated appreciably, although in opposite directions. In particular, the cumulative shear strain of Node 30,332, which is located on the side of leeward bucket wall in the downwind direction and experienced greater passive earth pressure, was the most significant. The accumulation of plastic shear strain for these nodes can well explain the accumulated rotation of the suction bucket during earthquake. It should be noted that the shear stress–strain curves of Nodes 2351 and 2422 had good symmetry, with relatively large vibration strain and small permanent strain, which is mainly due to the symmetry of the shear stress suffered by the two nodes in a special position just below the bucket.

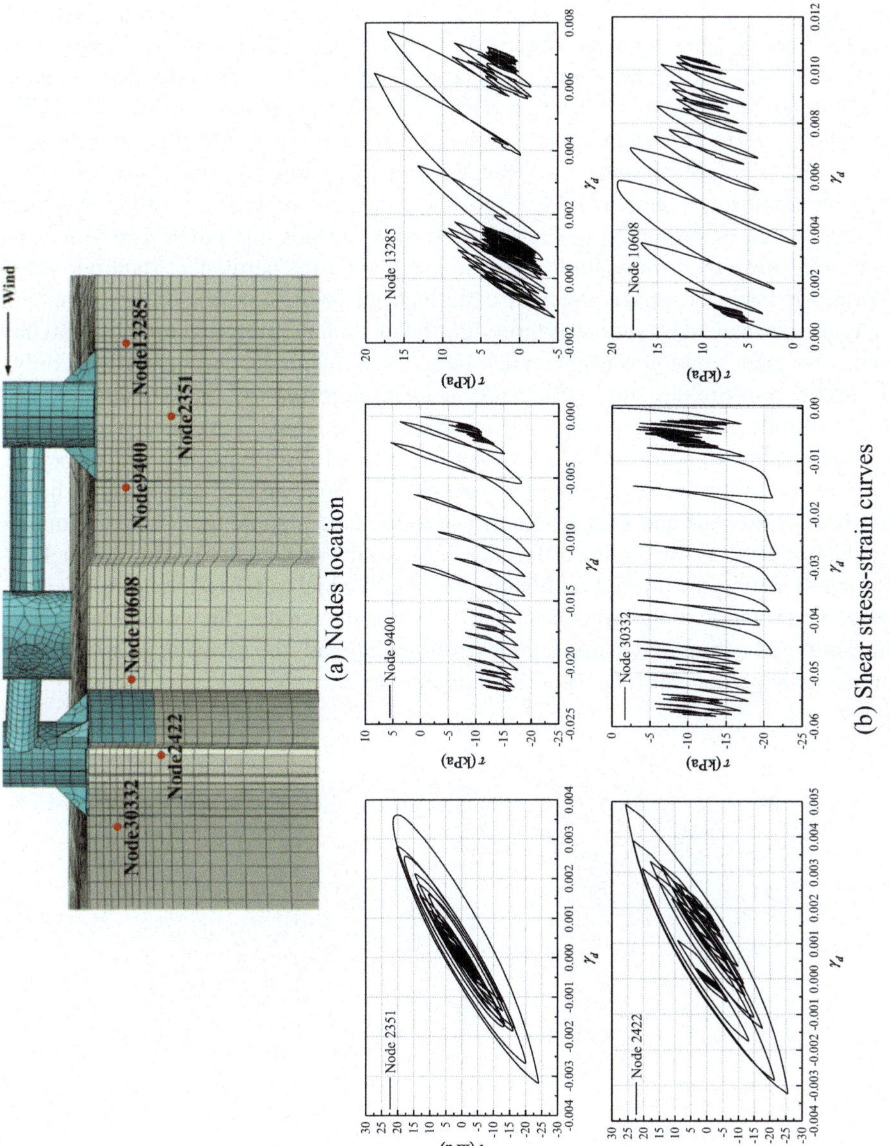

Fig. 9.11 Shear stress–strain curves of some typical nodes around the tripod foundation

9.3.1.2 Seismic Load Only

The earthquake record, wind and wave loads used in this section are the same as those in Sect. 9.3.1.1. The wind and wave loads were also applied in the first 30 s, but they were removed and only the seismic load was applied in the last 30 s. In order to highlight the dynamic response of the OWT under seismic load, Fig. 9.13 only gives the time history during the seismic load period. Figure 9.12 shows the displacement vector diagram of the tripod suction buckets under seismic load only. It should be noted that the tripod foundation rotated in the windward direction under the seismic load alone, which is just opposite to the result of the combined action of seismic and environmental loads shown in Fig. 9.8. The comparison of rotation time histories for the two cases presented in Fig. 9.13a further verifies this point. The vibration rotation of the tripod foundation was also larger than the permanent rotation under seismic load only. However, under the combined action of seismic and environmental loads, the permanent rotation was more significant than the vibration rotation, and the final permanent rotation was also much larger than that under earthquake load only, but in the opposite direction. Therefore, we can infer that the environmental loads would significantly increase the rotation angle of the tripod foundation and change its rotation direction. Figure 9.13b shows the vertical displacement time history of the buckets on the windward and leeward sides in both cases. Figure 9.13b shows that both windward and leeward buckets experienced significant settlement under the seismic load. However, the difference between the settlements of the two buckets was small, that is, the uneven settlement was small, which resulted in relatively small permanent rotation of the tripod foundation. When the wind turbine was subjected to the combined action of seismic and environmental load, the uneven settlement was large, so the permanent rotation was relatively large.

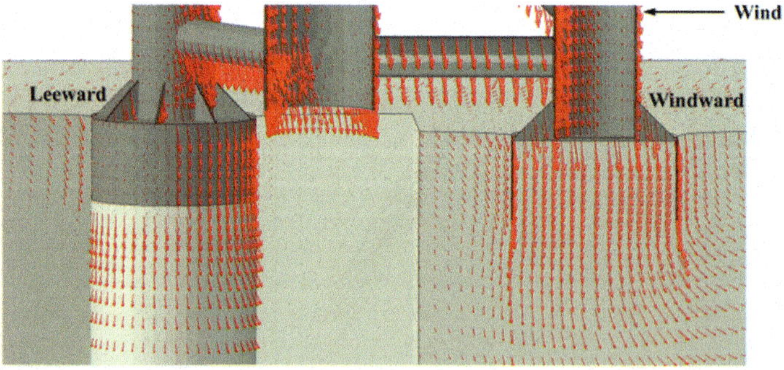

Fig. 9.12 Displacement vector diagram of tripod suction bucket at the end of seismic load only (amplification factor of 3)

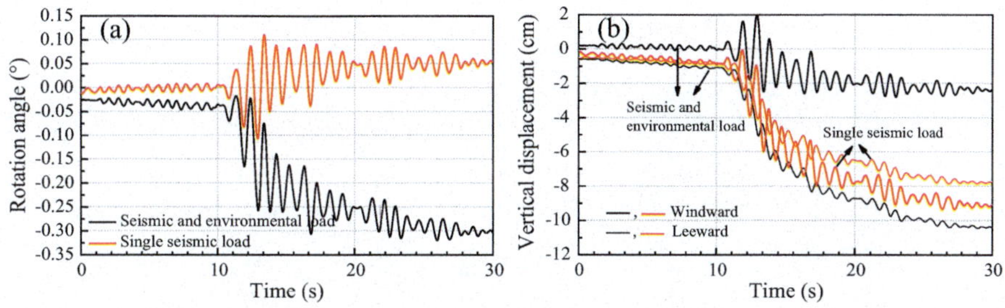

Fig. 9.13 Comparison of dynamic response under seismic load only and combined action of seismic and environmental loads: **a** rotation angle-time history; **b** vertical displacement–time history

9.3.1.3 Environmental Load Only

Figure 9.14 displays the displacement vector diagram of the tripod suction bucket under environmental loads only. The wave and wind loads were calculated for significant wave height $H_{1/3} = 5\,\text{m}$ and mean wind velocity $V_{10} = 12\,\text{m/s}$. Figure 9.14 demonstrates that the windward bucket experienced upward displacement and the leeward bucket experienced downward displacement, which caused the tripod foundation to rotate in the downwind direction. This phenomenon is also different from that observed behavior for the combined action of seismic and environmental loads (Fig. 9.8), where the both windward and leeward buckets experienced downward settlement and the tripod foundation also rotated in the downwind direction. Figure 9.15a and b, respectively, show the rotation angle and vertical displacement time histories of the tripod foundation under different wind speeds (total duration of 30 s). As can be seen from Fig. 9.15b, the upward displacement of the windward bucket was greater than the downward displacement of the leeward bucket. This finding is consistent with the results of cyclic horizontal loading tests on model tripod suction reported by Jeong et al. (2021). This is because the pullout bucket capacity was less than its compressive capacity. In addition, as the wind speed increased, the upward displacement of the windward bucket increased significantly, and the difference in vertical displacement between the windward and leeward buckets increased. As a result, the rotation of the tripod foundation became more significant as shown in Fig. 9.15a.

9.3.2 Influence of Seismic Intensity and Frequency Content

Four analyses were conducted considering different peak ground acceleration (PGA) to investigate the impact of earthquake intensity on the dynamic response of the wind turbine system subjected to the combined action of seismic and environmental loads. For these four cases, Loma Prieta wave was used as the input ground motion. The

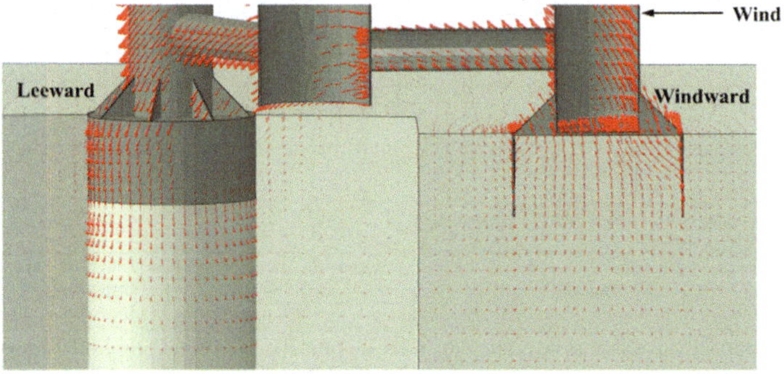

Fig. 9.14 Displacement vector diagram of tripod suction bucket under environmental loads only (amplification factor of 3)

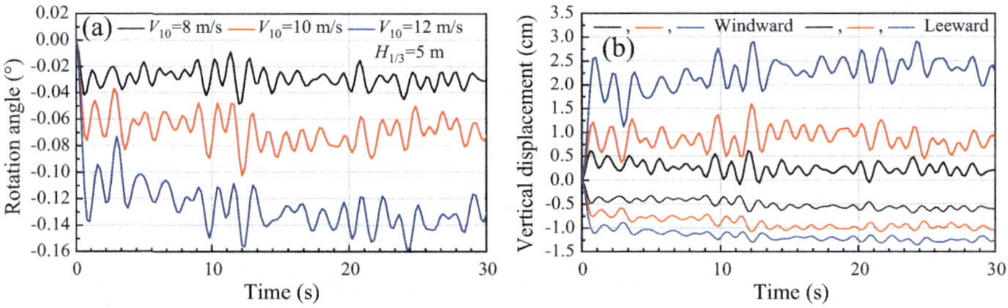

Fig. 9.15 Dynamic response of tripod suction bucket under environment loads only: **a** rotation angle-time history; **b** vertical displacement -time history

wave and wind loads were evaluated considering significant wave height $H_{1/3} = 5\,\mathrm{m}$ and mean wind speed $V_{10} = 8\,\mathrm{m/s}$.

Figure 9.16a demonstrates that, as expected, both the vibration and permanent rotations of the tripod foundation increased as PGA increased. In particular, the vibration rotation increased significantly at the instant of peak acceleration. For the soil conditions and geometric dimensions of the foundation considered in this study, when PGA is 0.3 g and 0.4 g, the permanent rotation angle exceeds the limit of 0.25° in the serviceability limit state (SLS) of OWTs. Figure 9.16b displays the vertical displacement time histories of windward and leeward buckets under different PGA. Figure 9.16b shows that the vertical displacement of both buckets increased as PGA increased. However, the increase in vertical displacement of leeward bucket was higher. This significant increase of the leeward bucket settlement was the main reason for the increase of permanent rotation of the tripod foundation as seismic intensity increased.

In order to investigate the effect of ground motion frequency content on seismic response of the tripod suction bucket, numerical simulations were performed for three different seismic records shown in Fig. 9.7, i.e., Northern California, Kobe and Loma Prieta considering PGA = 0.3 g. Wave and wind loads were evaluated

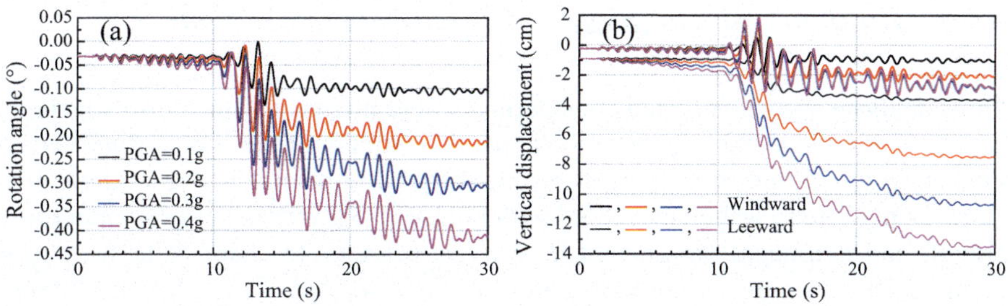

Fig. 9.16 Dynamic response of the tripod suction bucket in clay under different PGA: **a** Rotation angle-time curves; **b** Settlement-time curves

considering $H_{1/3} = 5\,\text{m}$ and $V_{10} = 8\,\text{m/s}$. The time histories of rotation angle of tripod suction bucket for the three seismic records are shown in Fig. 9.17. It can be seen that the rotation angle was much higher for Kobe earthquake compared with the other two records because the predominant frequency of the Kobe record (0.8 Hz) was closer to the 3P frequency of the 3 MW wind turbine structure, causing some resonance. However, the rotation angle of the wind turbine structure does not depend solely on the predominant frequency of the seismic wave. It is also closely related to the duration of ground motion, which can be confirmed by comparing the response of Loma Prieta earthquake record and Northern Calif earthquake record in Fig. 9.17. The duration of Loma Prieta earthquake record is significantly shorter than that of Northern Calif record as shown in Fig. 9.7; thus, the rotation angle of the former is smaller than that of the latter, although the predominant frequency of the former is closer to the 3P frequency of the wind turbine structure than that of the latter. The comparison demonstrates that the duration of the ground motion has a significant effect on the tripod foundation seismic response, and the rotation angle depends on the intensity, predominant frequency and duration of the ground motion.

Fig. 9.17 Rotation angle-time history of tripod suction bucket for three different seismic waves

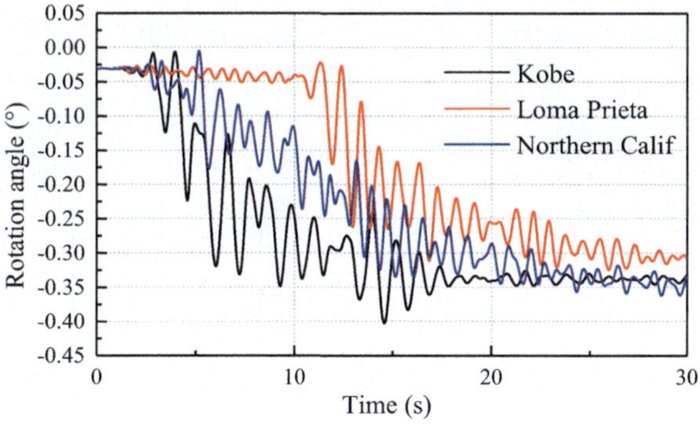

9.3.3 Influence of Magnitude of Environmental Loads

Three analyses were conducted considering different wind speeds to investigate the impact of wind load magnitude on the dynamic response of the wind turbine system subjected to the combined action of seismic and environmental loads. For these three cases, Loma Prieta earthquake record with PGA of 0.3 g was used as the input round motion, and significant wave height $H_{1/3} = 5\,\text{m}$ was used to establish the wave load. Figure 9.18 shows the dynamic response of the tripod foundation for the different loading conditions. The results clearly demonstrate that the increase in wind speed resulted in a significant increase in the foundation permanent rotation. When the mean wind speed V_{10} increased from 8 m/s to 12 m/s, the permanent rotation almost doubled as shown in Fig. 9.18a. In addition, Fig. 9.18b shows that the wind speed has a greater impact on the vertical displacement of the windward bucket than that of the leeward bucket due to relative low pullout capacity of suction bucket compared with its compressive capacity.

It should be noted that the windward bucket moved downward under earthquake loading only as shown in Fig. 9.12, but the windward bucket moves upward under environmental loads only as shown in Fig. 9.14. When earthquake and environmental loads act together, the environmental load magnitude, especially the wind load magnitude, has a significant impact on the vertical displacement of windward bucket. Figure 9.19 shows the displacement vector diagram of the tripod foundation when the mean wind speed $V_{10} = 12\,\text{m/s}$ under the combined action of seismic and environmental loads. Compared with the results for $V_{10} = 8\,\text{m/s}$ shown in Fig. 9.8, the movement of the windward bucket changed from downward movement to upward movement due to the increase of wind load. This behavior can be confirmed by the vertical displacement time history in Fig. 9.18b. Therefore, the wind load magnitude is a very important factor to be considered in the seismic design of tripod foundation.

In order to investigate the impact of wave load magnitude on the dynamic response of the wind turbine system subjected to the combined action of seismic and environmental loads, three different significant wave heights were considered to represent different wave load magnitude, meanwhile the same seismic load and wind load

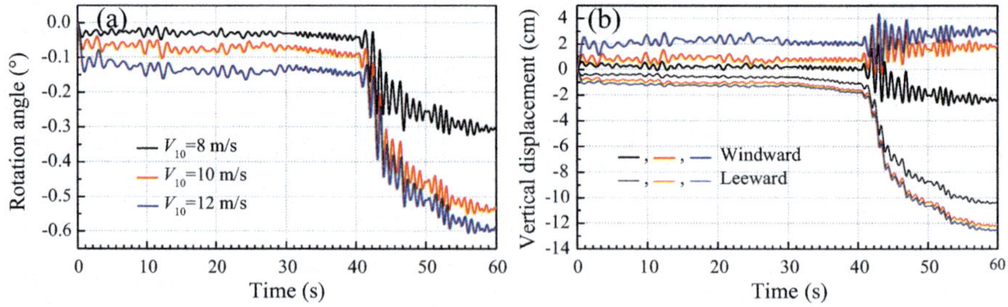

Fig. 9.18 Dynamic response of the tripod suction bucket under different wind speed: **a** rotation angle-time curves; **b** vertical displacement–time curves

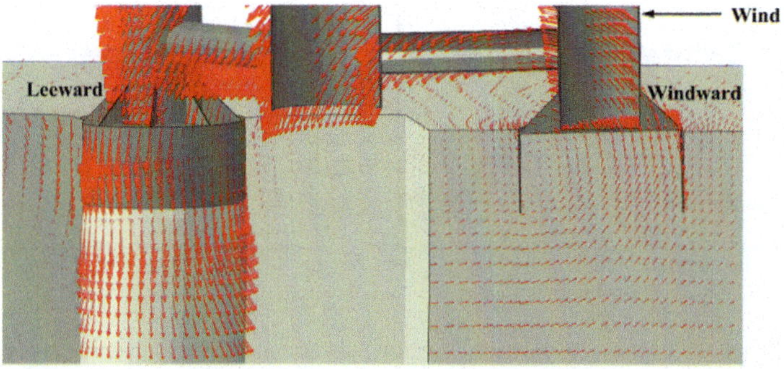

Fig. 9.19 Displacement vector diagram of the tripod suction bucket in clay under the combined action of seismic and environmental loads when the wind speed $V_{10} = 12\,\mathrm{m/s}$ (amplification factor of 3)

were applied (i.e., Loma Prieta earthquake record with PGA of 0.3 g, the mean wind speed $V_{10} = 8\,\mathrm{m/s}$). Figure 9.20 shows dynamic response of the tripod foundation under different significant wave heights. Obviously, increasing the wave load magnitude significantly increased the vibration rotation of the tripod foundation. However, interestingly, unlike wind loads, increasing the wave load magnitude does not always result in an increase of permanent rotation, as the cases of $H_{1/3} = 5$ m and 10 m shown in Fig. 9.20a. The constant mean wind load is dominant, although the wind load can be decomposed into a constant mean wind load and a fluctuating component. Therefore, the wind load can be approximately regarded as one-way dynamic load (no reverse load as shown in Fig. 9.4). The constant mean wind load mainly causes permanent rotation, while the fluctuating wind load mainly causes vibration rotation. The increase of wind speed will increase the constant mean wind magnitude more significantly, resulting in a significant increase in permanent rotation. However, the wave load is typically a two-way dynamic load, and the mean load is almost null as shown in Fig. 9.6. Therefore, increasing the wave load magnitude will significantly increase the vibration rotation. Seismic load is actually also a two-way dynamic load. When seismic and wave loads are coupled, they are not simple linear superposition, but may weaken or enhance each other, which depends on their loading time history and spectrum characteristics. Hence, compared with the wind load, the influence of wave load on permanent rotation of wind turbine during earthquake is more complex, and further research is still needed.

9.4 Summary

This chapter investigated the dynamic behavior of tripod suction bucket foundation installed in clays supporting offshore wind turbine under seismic and environmental loading by performing three-dimensional finite element analysis. The key conclusions can be summarized as follows:

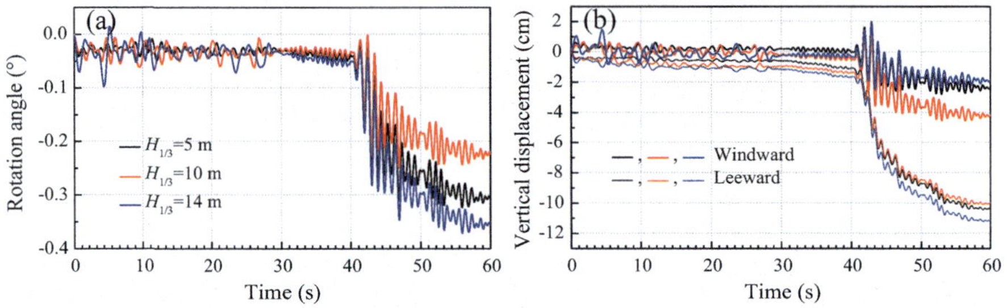

Fig. 9.20 Dynamic response of the tripod suction bucket under different significant wave heights: **a** rotation angle-time curves; **b** vertical displacement–time curves

(1) When OWTs system is subjected to the combined action of seismic and environmental loads, even if there is no instability failure of the OWTs system, it is likely to fail to meet the requirements of SLS due to large permanent rotation. The settlement of the leeward bucket is much greater than that of the windward bucket. The uneven settlement of the bucket on the windward and leeward sides is the direct cause of the rotation of the tripod foundation.
(2) The vibration rotation of the tripod foundation is larger than the permanent rotation under single seismic load. However, under the combined action of seismic and environmental loads, the permanent rotation is more significant than the vibration rotation. The environmental load will significantly increase the permanent rotation of the tripod foundation and change its rotation direction.
(3) When OWTs system is subjected to environment loads only, the windward bucket moves upward and the leeward bucket moves downward, which causes the tripod foundation to rotate in the downwind direction. The main reason is that the bucket uplift capacity is less than its compressive capacity.
(4) With the increase of PGA, the vibration and permanent rotations of the tripod foundation increase significantly. Compared with the windward bucket, the impact of earthquake intensity on leeward bucket is more significant. The significant increase of the leeward bucket settlement is the main reason for the increase of permanent rotation of the tripod foundation with seismic intensity.
(5) The seismic frequency content has a significant effect on dynamic responses of tripod foundation. Larger permanent rotation occurs when the seismic predominant frequency is closer to the frequency ranges of 1p and 3p for the wind turbine structure. The rotation angle of the wind turbine structure depends on the intensity, predominant frequency and duration of the ground motion
(6) The constant mean wind load mainly causes permanent rotation, while the fluctuating wind load mainly causes vibration rotation. The increase of wind speed will increase the constant mean wind more significantly, resulting in a significant increase in permanent rotation of tripod suction bucket foundation.
(7) Seismic load and wave loads are both two-way dynamic loads. When seismic and wave loads are coupled, they are not simple linear superposition, but may weaken or enhance each other, which depends on their loading time history

and spectrum characteristics. Hence, compared with wind load, the influence of wave load on permanent rotation of wind turbine during earthquake is more complex.

It should be noted that the efficiency of individual bucket units is affected by spacing due to the overlapping of the resisting zones. Spacing between buckets affects the bearing capacity and dynamic responses of the soil foundation system under seismic and environmental loads. Although this point is not the focus of this chapter, it should be paid attention to in engineering design, and needs to be deeply discussed in future research work.

References

Arany, L., S. Bhattacharya, J.H. Macdonald, et al. 2016. Closed form solution of Eigen frequency of monopile supported offshore wind turbines in deeper waters incorporating stiffness of substructure and SSI. *Soil Dynamics and Earthquake Engineering* 83: 18–32.

Asheghabadi, M.S., M. Sahafnia, A. Bahadori, et al. 2019. Seismic behavior of suction caisson for offshore wind turbine to generate more renewable energy. *International Journal of Environmental Science and Technology* 16: 2961–2972.

Binh, L., T. Ishihara, P. Phuc, et al. 2008. A peak factor for non-Gaussian response analysis of wind turbine tower. *Journal of Wind Engineeing and Industerial Aerodynamics* 96: 2217–2227.

Bolt, B.A. 1973. Duration of strong ground motion. In *The Proceedings of the 5th World Conference on Earthquake Engineering*, vol. 1, pp. 1304–1313.

Byrne, B.W., and G.T. Houlsby. 2003. Foundations for offshore wind turbines. *Philosophical Transaction of the Royal Society A-Mathematical Physical and Engineering Sciences* 361: 2909–2930.

Cao, G., Z. Chen, C. Wang, et al. 2020. Dynamic responses of offshore wind turbine considering soil nonlinearity and wind-wave load combinations. *Ocean Engineering* 217 (2): 108155.

Cerfontaine, B., F. Collin, and R. Charlier. 2016. Numerical modelling of transient cyclic vertical loading of suction caissons in sand. *Géotechnique* 66 (2): 121–136.

Chen, D., P. Gao, S. Huang, et al. 2020. Static and dynamic loading behavior of a hybrid foundation for offshore wind turbines. *Marine Structures* 71: 102727.

Chen, D., S.S. Huang, C.G. Huang, et al. 2021. Passive control of jacket-type offshore wind turbine vibrations by single and multiple tuned mass dampers. *Marine Structures* 77: 102938.

Cheng, X.L., A.W. Yang, and G.N. Li. 2018. Model tests and finite element analysis for the cyclic deformation process of suction anchors in soft clays. *Ocean Engineering* 151: 329–341.

Cheng, X.L., P.G. Wang, N. Li, et al. 2020. Predicting the cyclic behaviour of suction anchors based on a stiffness degradation model for soft clays. *Computers and Geotechnics* 122: 103552.

Cheng, X.L., W.L. Cheng, P.G. Wang, et al. 2022. Response of offshore wind turbine tripod suction bucket foundation to seismic and environmental loading. *Ocean Engineering* 257: 111708.

Cheng, X.L., Y. Li, K. Mu, et al. 2024. Seismic response of tripod suction bucket foundation for offshore wind turbine in sands. *Soil Dynamics and Earthquake Engineering* 177: 108353.

Dassault Systemes Simulia Corp. 2014. Abaqus Analysis User's Manual Version 6.14. Providence.

Cox, J.A., C.D. O'Loughlin, M. Cassidy, et al. 2014. Centrifuge study on the cyclic performance of caissons in sand. *International Journal Physical Modelling Geotechnics* 14: 99–115.

Damgaard, M., L.B. Ibsen, L.V. Andersen, et al. 2013. Cross-wind modal properties of offshore wind turbines identified by full scale testing. *Journal of Wind Engineeing and Industerial Aerodynamics* 116: 94–108.

Davenport, A.G. 1961. The application of statistical concepts to the wind loading of structures. *Proceedings of the Institution Civil Engineers* 19 (4): 449–472.

Davenport, A.G. 1967. The dependence of wind load upon meteorological parameters. In *Proceedings of the International Research Seminar on Wind Effects on Building and Structures*, vol. 1, pp. 19–82. University of Toronto Press, Toronto.

Det Norske Veritas (DNV). 2010. *DNV-RP-C205: Environmental Conditions and Environmental Loads*. Norway: DNV.

Foglia, A., and L.B. Ibsen. 2016. Monopod bucket foundations under cyclic lateral loading. *International Journal Offshore and Polar Engineering* 26 (2): 109–115.

Gao, B., G. Ye, Q. Zhang, et al. 2021. Numerical simulation of suction bucket foundation response located in liquefiable sand under earthquakes. *Ocean Engineering* 235: 109394.

Hansen, M.O.L. 2008. *Aerodynamics of Wind Turbines*, 2nd ed. London, UK: Earthscan.

Houlsby, G.T., and B.W. Byrne. 2000. Suction Caisson foundations for offshore wind turbines and anemometer masts. *Wind Engineering* 24: 249–255.

Houlsby, G.T., R.B. Kelly, J. Huxtable, et al. 2005. Field trials of suction caissons in clay for offshore wind turbine foundations. *Géotechnique* 55 (4): 287–296.

Hung, L.C., S.H. Lee, S. Vicent, et al. 2018. An experimental investigation of the cyclic response of bucket foundations in soft clay under one-way cyclic horizontal loads. *Applied Ocean Research* 71: 59–68.

Iannuzzi, A., and P. Spinelli. 1987. Artificial wind generation and structural response. *Journal of Structural Engineering* 113 (12): 2382–2398.

Jeong, Y.H., J.H. Kim, H.J. Park, et al. 2019. Cyclic behavior of unit bucket for tripod foundation system supporting offshore wind turbine via model tests. *Wind Energy* 22 (2): 257–268.

Jeong, Y.H., K.W. Ko, D.S. Kim, et al. 2021. Studies on cyclic behavior of tripod suction bucket foundation system supporting offshore wind turbine using centrifuge model test. *Wind Energy* 24: 515–529.

Jonkman, J.M., Butterfield, S., Musial, W., et al. 2009. Definition of a 5MW reference wind turbine for offshore system development. Office of Scientific & Technical Information Technical Reports.

Kelly, R.B., G.T. Houlsby, and B.W. Byrne. 2006. Transient vertical loading of model suction caissons in a pressure chamber. *Geotechnique* 56 (10): 665–675.

Kim, S.R., L.C. Hung, and M. Oh. 2014a. Group effect on bearing capacities of tripod bucket foundations in undrained clay. *Ocean Engineering* 79: 1–9.

Kim, D.H., S.G. Lee, and I.K. Lee. 2014b. Seismic fragility analysis of 5 Mw offshore wind turbine. *Renewable Energy* 65: 250–256.

Kourkoulis, R.S., P.C. Lekkakis, et al. 2014. Suction caisson foundations for offshore wind turbines subjected to wave and earthquake loading: Effect of soil–foundation interface. *Géotechnique* 64 (3): 171–185.

Lee, S., H. Kim, and S. Lee. 2010. Analysis of aerodynamic characteristics on a counter-rotating wind turbine. *Current Applied Physics* 10: S339–S342.

Li, H.J., H.J. Liu, and S.Y. Liu. 2017. Dynamic analysis of umbrella suction foundation embedded in seabed for offshore wind turbines. *Geomechanics for Energy and the Environment* 10: 12–20.

Li, L., H. Liu, and W. Wu. 2021. Investigation on the behavior of hybrid pile foundation and its surrounding soil during cyclic lateral loading. *Ocean Engineering* 240: 110006.

Li, L.C., M.Y. Zheng, X. Liu, et al. 2022. Numerical analysis of the cyclic loading behavior of monopile and hybrid pile foundation. *Computers and Geotechnics* 144: 104635.

Liaw, C.Y., and A.K. Chopra. 1974. Dynamics of towers surrounded by water. *Earthquake Engineering and Structural Dynamics* 3 (1): 33–49.

Lu, X., Y. Wu, B. Jiao, et al. 2007. Centrifugal experimental study of suction bucket foundations under dynamic loading. *Acta Mechanica Sinica* 23 (6): 689–698.

Nam, W., K.Y. Oh, and B.I. Epureanu. 2019. Evolution of the dynamic response and its effects on the serviceability of offshore wind turbines with stochastic loads and soil degradation. *Reliability Engineering and System Safety* 184: 151–163.

Olalo, L.T., Y.W. Choo, K.T. Bae. 2016. Influence of the skirt on the seismic response of bucket foundations for offshore wind tower using dynamic centrifuge model tests. In *Proceedings of the ASME 2016 35th International Conference on Ocean, Offshore, and Arctic Engineering (OMAE 2016)*, June 19–24. Busan, South Korea.

Perveen, R., N. Kishor, and S.R. Mohanty. 2014. Off-shore wind farm development: Present status and challenges. *Renewable and Sustainable Energy Reviews* 29: 780–792.

Pierson, W.J., and L. Moskowitz. 1964. A proposed spectral form for fully developed wind seas based on the similarity theory of SA Kitaigorodskii. *Journal of Geophysical Research Atmospheres* 69 (24): 5181–5190.

Ren, B., and Y.X. Wang. 2004. Numerical simulation of random wave slamming on structures in the splash zone. *Ocean Engineering* 31: 547–560.

Sun, C., and V. Jahangiri. 2018. Bi-directional vibration control of offshore wind turbines using a 3D pendulum tuned mass damper. *Mechanical Systems and Signal Processing* 105: 338–360.

Tsinidis, G., K. Pitilakis, and A.D. Trikalioti. 2014. Numerical simulation of round robin numerical test on tunnels using a simplified kinematic hardening model. *Acta Geotechnica* 9: 641–659.

Wang, Y.H., X.B. Lu, S.Y. Wang, et al. 2006. The response of bucket foundation under horizontal dynamic loading. *Ocean Engineering* 33: 964–973.

Wang, X., X. Yang, and X.W. Zeng. 2017a. Seismic centrifuge modelling of suction bucket foundation for offshore wind turbine. *Renewable Energy* 114: 1013–1022.

Wang, X., X. Yang, and X.W. Zeng. 2017b. Lateral response of improved suction bucket foundation for offshore wind turbine in centrifuge modelling. *Ocean Engineering* 141: 295–307.

Wang, L.Z., H. Wang, B. Zhu, et al. 2018a. Comparison of monotonic and cyclic lateral response between monopod and tripod bucket foundations in medium dense sand. *Ocean Engineering* 155: 88–105.

Wang, P., M. Zhao, and X. Du. 2018b. Wind, wave and earthquake responses of offshore wind turbine on monopile foundation in clay. *Soil Dynamics and Earthquake Engineering* 113: 47–57.

Yamazaki, H., Y. Morikawa, and F. Koike. 2003. Study on design method of suction foundation using model tests. In *Soft Ground Engineering in Coastal Areas*, pp. 419–422.

Yu, Y.X. 1981. Application of ocean wave simulation. *J Dalian U Tech* 20 (2): 84–90 (in Chinese).

Yu, Y.X., and H.G. Kang. 1982. The digital simulation of irregular wave forces on an isolated pile. *Journal of Dalian University of Technology* 21 (2): 111–118 (in Chinese).

Yu, H., X.W. Zeng, and J.J. Lian. 2014. Seismic behavior of offshore wind turbine with suction caisson foundation. In *Geo-Congress 2014: Geo-characterization and Modeling for Sustainability*, vol. 234, pp. 1206–1214.

Zayed, M., K. Kim, and A. Elgamal. 2019. Seismic response of suction caisson in large-scale shake table test. In *VII International Conference on Earthquake Geotechnical Engineering*, pp. 17–20.

Zhang, P.Y., J.C. Shi, H.Y. Ding, et al. 2016. Design of offshore wind power foundation with multi-bucket. *Transactions of Tianjin University* 22 (6): 502–507.

Zhang, J.X., W.L. Cheng, X.L. Cheng, et al. 2021. Seismic responses analysis of suction bucket foundation for offshore wind turbine in clays. *Ocean Engineering* 232: 109159.

Zhao, X.Z., and C.H. Hu. 2012. Numerical and experimental study on a 2-D floating body under extreme wave conditions. *Applied Ocean Research* 35: 1–13.

Zhao, X., and P. Maisser. 2006. Seismic response analysis of wind turbine towers including soil-structure interaction. *Proceedings of the Institution of Mechanical Engineers, Part k: Journal of Multi-Body Dynamics* 220 (1): 53–61.

Zhong, R., and M. Huang. 2014. Winkler model for dynamic response of composite caisson-piles foundations: Seismic response. *Soil Dynamics and Earthquake Engineering* 66: 241–251.

Zhu, B., B.W. Byrne, and G.T. Houlsby. 2013. Long-term lateral cyclic response of suction caisson foundations in sand. *Journal of Geotechnical and Geoenvironmental Engineering* 139 (1): 73–83.

Zhu, F.Y., C.D. O'Loughlin, B. Bienen, et al. 2018. The response of suction caissons to long-term lateral cyclic loading in single-layer and layered seabeds. *Géotechnique* 68 (8): 729–741.

Zuo, H.R., K.M. Bi, and H. Hao. 2017. Using multiple tuned mass dampers to control offshore wind turbine vibrations under multiple hazards. *Engineering Structures* 141: 303–315.
Zuo, H., K. Bi, and H. Hao. 2018. Dynamic analyses of operating offshore wind turbines including soil-structure interaction. *Engineering Structures* 157: 42–62.

Chapter 10
Summary and Prospect

10.1 Summary

The chapters of the book can be summarized as follows:

(1) In Chap. 1, the research background and significance are introduced. It summarizes and analyzes the dynamic response and corresponding numerical analysis methods of typical offshore wind turbine foundations under environmental loads such as wind, waves, and earthquakes. This section outlines the research aims and contents of the book.

(2) In Chap. 2, a simple single bounding surface constitutive model is developed to predict the undrained behavior of saturated clays under cyclic loads. This new model avoids complex kinematic hardening rules and requires memorization of significant stress reversal events, making simplicity its primary advantage. An innovative interpolation function for the elastoplastic shear modulus is proposed based on bounding surface theories. The evolution of the hardening modulus is described in the deviatoric stress space through movement and updating of a mapping center using the new interpolation function, enabling the model to depict stress–strain hysteretic responses of clays under cyclic loading. This model represents an advancement over classical one-dimensional soil dynamic models, extending into three-dimensional stress space. Model parameters are typically determined through triaxial tests. The developed model effectively captures essential behaviors of saturated clay, including reverse plastic flow, evolution of hysteretic loops, accumulation of plastic deformations, and degradation of soil stiffness. Furthermore, the newly developed constitutive model has been successfully implemented into the ABAQUS software package through the secondary development interface of UMAT.

X. Cheng et al., *Dynamic Analysis of Offshore Wind Turbine Foundations in Soft Clays*,
https://doi.org/10.1007/978-981-97-9454-6_10

(3) In Chap. 3, a novel cyclic p-y elastoplastic model is developed within the framework of single-surface bounding surface theory. This model effectively captures soil stiffness degradation under cyclic loading by integrating cumulative plastic displacement into an interpolation function of the elastoplastic resistance coefficient. The model is relatively simple with only four parameters that can be determined from standard soil properties and stress–strain responses measured in cyclic direct simple shear tests (DSS). The model is implemented numerically using the BNWF method, and its performance is validated by accurately predicting the cyclic lateral response of piles installed in soft clay, as observed in field and centrifuge tests documented in the literature. The model reliably simulates both monotonic and cyclic responses of piles under various lateral loading patterns, accurately capturing key characteristics of the pile head load–displacement curve, including nonlinearity, hysteresis, displacement accumulation, and stiffness degradation. Additionally, it predicts the evolution of lateral deflection and sectional bending moment along the pile during cyclic loading. In terms of computational efficiency, it outperforms 3D finite element methods based on advanced dynamic constitutive models.

(4) In Chap. 4, a three-dimensional finite element method is developed to analyze the lateral cyclic responses of large-diameter monopiles, employing a simplified bounding surface constitutive model for clays. The method's validity is confirmed through simulation against an existing centrifugal model test. It accurately predicts the nonlinear hysteresis responses of monopiles in clay subjected to cyclic loading, including the evolution of bending moment and lateral deflection profiles throughout loading cycles. Various lateral cyclic loads are applied to monopiles, encompassing different loading patterns such as one-way and two-way loading, symmetric and asymmetric loading, and variable-amplitude and constant-amplitude loading. The study systematically investigates the influence of these diverse cyclic loading patterns on the lateral responses of large-diameter monopiles.

(5) In Chap. 5, a numerical method was developed to simulate the dynamic interaction of soil-pile-structure systems using an elastoplastic bounding surface constitutive model capable of accurately capturing cyclic degradation in clay stiffness. The method's effectiveness was validated through comparison with existing literature simulation results. The dynamic responses of Monopile Offshore Wind Turbines (MOWTs) to seismic and environmental loading were then simulated using this numerical approach. The study investigated how soil plasticity and stiffness degradation affect the dynamic responses of MOWTs, as well as analyzing the influence of environmental loads on their seismic responses. The findings indicated significant rotations and settlements in the monopile, with soil plasticity and stiffness degradation notably increasing peak responses in lateral deflection, rotation angle, and bending moment under combined seismic and environmental loads. The input bedrock peak accelerations were amplified to varying degrees from the monopile base to the tower top, with the most significant amplification occurring along the height when considering soil stiffness degradation. Moreover, environmental loads

such as winds and waves substantially increased the absolute values of seismic response metrics, including deflection, rotation angle, bending moment, and acceleration amplification factors.

(6) In Chap. 6, the response of tripod pile foundations for offshore wind turbines in clay under monotonic, cyclic, and seismic loads is thoroughly investigated using three-dimensional finite element analysis. The study reveals that the direction of loading significantly influences the ultimate bearing capacity of the tripod pile foundation, where the capacity depends on whether Pile A is in tension and Pile B is in compression, or vice versa. The evolution of axial force, bending moment, and lateral displacement profiles of Pile A and Pile B shows distinct characteristics with increasing cycles under both lateral one-way and two-way cyclic loading conditions. Under seismic loads, the foundation accumulates rotation toward the Pile A side due to its lower vertical bearing capacity compared to Pile B. At the tower top, maximum lateral displacement and rotation angles are observed, while the maximum bending moment occurs at the top of the tripod support. These findings provide valuable insights into the dynamic behavior of tripod pile-supported OWTs subjected to various loading conditions in offshore environments.

(7) In Chap. 7, a simplified kinematic hardening model from the ABAQUS library was utilized to effectively capture the undrained dynamic nonlinear behavior of clays. The methodology for determining model parameters was introduced, and its predictive capability was validated through simulations of cyclic simple shear tests. Subsequently, a three-dimensional finite element model was developed, incorporating appropriate boundary and contact conditions, to analyze the seismic responses of a single suction bucket foundation in clays using the aforementioned soil model. The finite element analysis uncovered the rotational bearing mechanism of the suction bucket foundation. Various factors influencing the seismic responses of the suction bucket were investigated, including soil strength, seismic intensity, seismic frequency, and bucket geometry. It was observed that significant accumulated rotation occurs in the suction bucket during seismic events. Increasing the length or diameter of the bucket was found to effectively reduce the final rotation angle, although it did not significantly impact settlement. These findings contribute valuable insights into optimizing suction bucket foundation design to enhance its performance under seismic loading conditions.

(8) In Chap. 8, a three-dimensional numerical method is developed to analyze the lateral cyclic response of tripod suction bucket foundations. The numerical model's accuracy is confirmed through validation against results from scaled model tests documented in the literature. Subsequently, the rotational mechanism of tripod suction buckets under lateral monotonic and cyclic loading is thoroughly investigated. The model employs both 'bonded contact' and 'separable contact' conditions between the soil plug and the inner wall to simulate scenarios representing the upper and lower limits of the suction bucket's tensile capacity. This approach allows for comprehensive exploration of the influences of bucket-soil contact conditions, loading direction, and cyclic loading modes

(one-way and two-way) on cyclic behavior. The analysis reveals that these factors significantly affect the cyclic responses, permanent rotation angles, and failure mechanisms of tripod bucket foundations. These findings contribute valuable insights into optimizing the design and performance of tripod suction bucket foundations subjected to lateral cyclic loading conditions.

(9) In Chap. 9, the dynamic behavior of tripod bucket foundations installed in clay is investigated using advanced three-dimensional nonlinear finite element analysis. The analysis focuses on understanding the rotational mechanisms under various loading conditions, including environmental loads, seismic loads, or their combined effects. Dynamic responses are evaluated considering ground motions of varying intensity and frequency contents, as well as environmental loads of varying magnitudes. It was observed that while the offshore wind turbine system may not experience instability failures due to combined seismic and environmental loads, it may fail to meet serviceability limit state (SLS) requirements due to significant permanent rotations induced by differential settlement between the leeward bucket and the windward bucket. The study highlights the significant influence of ground motion frequency content on the dynamic response of tripod foundations. Larger permanent rotations were noted when the seismic predominant frequencies aligned closely with frequency ranges such as 1p and 3p of the wind turbine structure. The rotation angles of the wind turbine structure were found to depend on both the predominant frequency and intensity of the ground motion. These findings underscore the critical role of dynamic analysis in optimizing tripod bucket foundation designs to ensure structural integrity and performance under varying environmental conditions.

(10) In this chapter, the entirety of the book's chapters is succinctly summarized, encapsulating their key findings and contributions. Additionally, the chapter presents an outlook on future research directions and potential avenues for further exploration in the field.

10.2 Prospect

Based on the research work presented in this book, future work can be carried out:

(1) It should be noted that in the research presented in this book, the dynamic behavior of soft clay was described using a total stress-based bounding surface constitutive model. The clay was assumed to be undrained during the numerical calculations. While this assumption is reasonable for short-term dynamic loads such as earthquakes, it may lead to conservative results for long-term cyclic loads from wind and waves. The influence of soil pore pressure on foundation-soil dynamic interaction was therefore not adequately accounted for. Future research could benefit from employing effective stress-based constitutive models to address these limitations more comprehensively.

(2) The authors have developed a cyclic elastoplastic p-y model for lateral loaded piles in soft clay. However, in the case of wind turbine foundations with multiple piles (such as tripods or jackets), the loads from the offshore wind turbine and its support structure are primarily transmitted axially through the braces to the pile foundation. In such scenarios, the t-z and Q-z curves become more significant compared to the p-y curve. Therefore, future research should focus on developing cyclic t-z and Q-z models to more accurately simulate the dynamic response of multi-pile foundations. Additionally, there is a need to develop user-defined spring elements corresponding to the p-y, t-z, and Q-z models to enhance integration with commercial software like ABAQUS. This approach would enable the finite element method to effectively solve complex boundary problems associated with multi-pile foundation systems for wind turbines.

(3) The authors primarily elucidated the dynamic response laws and disaster mechanisms of offshore wind turbine foundations through refined three-dimensional numerical simulations. As an essential complement, it is recommended to conduct relevant model tests, such as scaled and centrifugal model tests simulating offshore foundations under cyclic and seismic loads. Integrating numerical simulations with model experiments would enhance the accuracy of assessing the dynamic bearing characteristics of wind turbine foundations. This combined approach ensures a more robust validation of theoretical models and provides valuable insights for optimizing foundation designs in offshore environments.

(4) The research in this book predominantly addresses foundations on soft clay; however, wind turbine foundations often rest on sandy soil, presenting distinct challenges in dynamic response to wind, waves, and earthquakes, notably due to potential liquefaction. Compared to soft clay, the dynamic behavior of wind turbine foundations in sandy soils is more complex, especially under conditions where liquefaction may occur. Simulating the constitutive behavior of sand undergoing liquefaction poses a significant challenge in three-dimensional finite element analysis. Therefore, there remains considerable research to be conducted on the dynamic response laws and disaster mechanisms of offshore wind turbine foundations situated on liquefiable soils.

The manufacturer's authorised representative in the EU is Springer Nature Customer Service Centre GmbH, Europaplatz 3, 69115 Heidelberg, Germany. If you have any concerns regarding our products, please contact ProductSafety@springernature.com

Printed and bound by CPI Group (UK) Ltd, Croydon, CR0 4YY
01/10/2026
02225907-0002